Plant Cell Monographs
Volume 16

Series Editor: David G. Robinson
Heidelberg, Germany

For further volumes:
http://www.springer.com/series/7089

Plant Cell Monographs

Recently Published Titles

Lipid Signaling in Plants
Volume Editor: Munnik, T.
Vol. 16, 2010

Cell Biology of Plant Nematode Parasitism
Volume Editor: Berg, R. H., Taylor, C. G
Vol. 15, 2009

Functional Organization of the Plant Nucleus
Volume Editor: Meier, I.
Vol. 14, 2009

The Chloroplast
Interactions with the Environment
Volume Editors: Sandelius, A. S., Aronsson, H.
Vol. 13, 2009

Root Hairs
Volume Editors: Emons, A. M. C., Ketelaar, T.
Vol. 12, 2009

Plant Microtubules
Development and Flexibility
2nd Edition
Volume Editor: Nick, P.
Vol. 11, 2008

Plant Growth Signalling
Volume Editors: Bögre, L., Beemster, G.
Vol. 10, 2008

Cell Division Control in Plants
Volume Editors: Verma, D. P. S., Hong, Z.
Vol. 9, 2008

Endosperm
Volume Editor: Olsen, O.-A.
Vol. 8, 2007

Nitric Oxide in Plant Growth Development and Stress Physiology
Volume Editors: Lamattina, L., Polacco, J.
Vol. 6, 2007

The Expanding Cell
Volume Editors: Verbelen, J.-P., Vissenberg, K.
Vol. 5, 2007

The Plant Endoplasmic Reticulum
Volume Editor: Robinson, D. G.
Vol. 4, 2006

The Pollen Tube
A Cellular and Molecular Perspective
Volume Editor: Malhó, R.
Vol. 3, 2006
Vol. 5/L, 2006

Somatic Embryogenesis
Volume Editors: Mujib, A., Samaj, J.
Vol. 2, 2006

Plant Endocytosis
Volume Editors:
Šamaj, J., Baluška, F., Menzel, D.
Vol. 1, 2005

Teun Munnik
Editor

Lipid Signaling in Plants

Springer

Editor
Teun Munnik
University of Amsterdam
Swammerdam Institute for Life Sciences (SILS)
Section Plant Physiology
Science Park 904
1098 XH, Amsterdam, The Netherlands
t.munnik@uva.nl

Series Editor
David G. Robinson
Ruprecht-Karls-University of Heidelberg
Heidelberger Institute for Plant Sciences (HIP)
Department Cell Biology
Im Neuenheimer Feld 230
69120 Heidelberg
Germany

ISSN 1861-1370 e-ISSN 1861-1362
ISBN 978-3-642-03872-3 e-ISBN 978-3-642-03873-0
DOI 10.1007/978-3-642-03873-0
Springer Heidelberg Dordrecht London New York

Library of Congress Control Number: 2009933122

Cover design: WMXDesign GmbH, Heidelberg, Germany

Printed on acid-free paper

Springer is part of Springer Science+Business Media (www.springer.com)

To my friend and mentor, Alan Musgrave

About the Editor

Teun Munnik is an associate professor plant cell biology at the Swammerdam Instituut for Life Sciences of the University of Amsterdam, The Netherlands. He obtained his B.Sc. degree in Botany from the University of Applied Sciences and his Ph.D. in Biology from the University of Amsterdam. His research pioneered phospholipid-based signaling mechanisms in plants, involving labs of Alan Musgrave (University of Amsterdam, NL), Robin F. Irvine (Cambridge University, UK), Heribert Hirt (Vienna Biocenter, AT) and George M. Carman (Rutgers University, NJ, US). Dr. Munnik serves on the editorial board of several international journals and has published over 70 original research papers and 7 book chapters. His current research is focused on studying phospholipid signaling in plant stress and development, using the model plant, *Arabidopsis thaliana*.

Preface

Phospholipids have long been known for their key role in maintaining the bilayer structure of membranes and in physically separating the cytosol from organelles and the extracellular space. In the past decade, a completely novel and unexpected function emerged, fullfilling a crucial role in cell signaling. It was the discovery in animal cells, that agonist-activated cell surface receptors led to the activation of a phospholipase C (PLC), to hydrolyze the minor lipid, phosphatidylinositol 4,5-bisphosphate into two second messengers, inositol 1,4,5-trisphosphate ($InsP_3$) and diacylglycerol (DAG). While $InsP_3$ diffuses into the cytosol, where it releases Ca^{2+} from an intracellular store by activating a ligand-gated Ca^{2+}-channel, DAG remains in the membrane to recruit and activate members of the protein kinase C family.

Over the years, a variety of other lipid based-signaling cascades were discovered. These include, phospholipase A, generating lyso-phospholipids and free fatty acids (to be converted into prostaglandins and leukotrienes), phospholipase D, to generate the lipid second messenger, phosphatidic acid (PA), and phosphoinositide 3-kinase (PI3K), generating a distinct set of polyphosphoinositides (PPI) phosphorylated at the D3-position of the inositol ring, all with separate signaling functions. Sphingolipids, representing another important group of signaling lipids, also came across.

The majority of these lipid-based signaling pathways have been discovered in plant cells too. Moreover, they have been found to be activated in response to a wide variety of biotic and abiotic stress signals, but also to be basically involved in plant growth and development. While many of the enzymes, lipids, and their targets involved are well conserved, major differences with the mammalian paradigms have also emerged.

This book highlights the current status of plant lipid signaling. All chapters have been written by experts in the field and cover information for both beginners and advanced lipidologists. Part I includes phospholipases (Chaps. 1–3), part II, lipid kinases (Chaps. 4–7), part III, lipid phosphatases (Chaps. 8–9), part IV, inositolphosphates and PPI metabolism (Chaps. 10–13), part V, PA signaling (Chaps. 14–17), and part VI, additional lipid signals, e.g. oxylipins, NAPE and sphingolipids (Chaps 18–20). It has been a great pleasure to be the editor of this book and to be a witness of this lipid-signaling adventure.

Amsterdam, June 2009 Teun Munnik

Contents

Part III Phosphatases

Part IV PPI Metabolism

Part V PA Signaling

Part VI Additional Lipid Signals

Contributors

Steven A. Arisz Section Plant Physiology, Swammerdam Institute for Life Sciences, University of Amsterdam, Science Park 904, NL-1098 XH, Amsterdam, The Netherlands

Elison B. Blancaflor Samuel Roberts Noble Foundation, Plant Biology Division, Ardmore, OK 73401, USA, eblancaflor@noble.org

László Bögre Royal Holloway, University of London, School of Biological Sciences, Egham Hill, Egham, TW20 0EX, UK, l.bogre@rhul.ac.uk

Kent D. Chapman University of North Texas, Center for Plant Lipid Research, Department of Biological Sciences, Denton, TX 76203, USA, chapman@unt.edu

Ayelen M. Distéfano Instituto de Investigaciones Biológicas, Facultad de Ciencias Exactas y Naturales, Universidad Nacional de Mar del Plata, CC 1245, 7600 Mar del Plata, Argentina

Peter E. Dowd Department of Botany, The University of Wisconsin, 132 Birge Hall, 430 Lincoln Drive, Madison, Wisconsin 53706, USA, pdowd@wisc.edu

Ivo Feussner Department of Plant Biochemistry, Albrecht-von-Haller-Institute for Plant Sciences, Georg-August-University Göttingen, Justus-von-Liebig-Weg 11, 37077 Göttingen, Germany, ifeussn@uni-goettingen.de

Carlos García-Mata Instituto de Investigaciones Biológicas, Facultad de Ciencias Exactas y Naturales, Universidad Nacional de Mar del Plata, CC 1245, 7600 Mar del Plata, Argentina

Glenda E. Gillaspy Department of Biochemistry and Fralin Biotechnology Center, Virginia Tech, Blacksburg, VA, USA, gillaspy@vt.edu

Simon Gilroy Department of Botany, The University of Wisconsin, 132 Birge Hall, 430 Lincoln Drive, Madison, Wisconsin 53706, USA

Ingo Heilmann Department of Plant Biochemistry, Albrecht-von-Haller-Institute for Plant Sciences, Georg-August-University Göttingen, Justus-von-Liebig-Weg 11, 37077 Göttingen, Germany, iheilma@uni-goettingen.de

Yueyun Hong College of Life Science and Technology, National Key Laboratory of Crop Genetic Improvement, Huazhong Agricultural University, Wuhan 430070, China

Yang Ju Im Department of Plant Biology, North Carolina State University, Raleigh NC 27695, USA

Till Ischebeck Department of Plant Biochemistry, Albrecht-von-Haller-Institute for Plant Sciences, Georg-August-University Göttingen, Justus-von-Liebig-Weg 11, 37077 Göttingen, Germany

Emmanuelle Jeannette Physiologie Cellulaire et Moléculaire des Plantes, UR5 Université Pierre et Marie Curie, 3 rue Galilée, F- 94200 Ivry-sur-Seine, France, emmanuelle.jeannette@upmc.fr

Edgar Eduard Kooijman Department of Biological Sciences, Kent State University, Kent OH, USA, ekooijma@kent.edu

Lorenzo Lamattina Instituto de Investigaciones Biológicas, Facultad de Ciencias Exactas y Naturales, Universidad Nacional de Mar del Plata, CC 1245, 7600 Mar del Plata, Argentina

Ana M. Laxalt Instituto de Investigaciones Biológicas, Facultad de Ciencias Exactas y Naturales, Universidad Nacional de Mar del Plata, CC 1245, 7600 Mar del Plata, Argentina, amlaxalt@mdp.edu.ar

Yuree Lee POSTECH-UZH Cooperative Laboratory, Division of Molecular Life Sciences, Pohang University of Science and Technology, Pohang 790–784, Korea

Youngsook Lee POSTECH-UZH Cooperative Laboratory, Division of Molecular Life Sciences, Pohang University of Science and Technology, Pohang 790–784, Korea, ylee@postech.ac.kr

Weiqi Li Kunming Institute of Botany, Chinese Academy of Sciences, Kunming 650204 and Department of Biology, Honghe University, Mengzi, Yunnan 661100, P.R., China

M. Luciana Lanteri Instituto de Investigaciones Biológicas, Facultad de Ciencias Exactas y Naturales, Universidad Nacional de Mar del Plata, CC 1245, 7600 Mar del Plata, Argentina

Ryan L. McCarthy Department of Plant Biology, University of Georgia, Athens, GA 30602, USA

Louise V. Michaelson Biological Chemistry Department, Rothamsted Research, Harpenden, Herts AL5 2JQ, UK, louise.michaelson@bbsrc.ac.uk

Alina Mosblech Department of Plant Biochemistry, Albrecht-von-Haller-Institute for Plant Sciences, Georg-August-University Göttingen, Justus-von-Liebig-Weg 11, 37077 Göttingen, Germany

Teun Munnik Section Plant Physiology, Swammerdam Institute for Life Sciences, University of Amsterdam, Science Park 904, NL-1098 XH, Amsterdam, The Netherlands, t.munnik@uva.nl

Yuki Nakamura Temasek Life Sciences Laboratory, 1 Research Link, National University of Singapore, Singapore 117604, Singapore

Johnathan A. Napier Biological Chemistry Department, Rothamsted Research, Harpenden, Herts AL5 2JQ, UK

Erik Nielsen Department of Molecular, Cellular, and Developmental Biology, University of Michigan, 830 North University Avenue, Ann Arbor, MI 48109, USA, nielsene@umich.edu

Hiroyuki Ohta Center for Biological Resources and Informatics, Tokyo Institute of Technology, 4259-B-65 Nagatsuta-cho, Midori-ku, Yokohama 226-8501, Japan; Research Center for the Evolving Earth and Planets, Tokyo Institute of Technology, 4259-B-65 Nagatsuta-cho, Midori-ku, Yokohama 226-8501, Japan, hohta@bio.titech.ac.jp

Sophie Paradis Physiologie Cellulaire et Moléculaire des Plantes, UR5 Université Pierre et Marie Curie, 3 rue Galilée, F-94200 Ivry-sur-Seine, France

Imara Y. Perera Department of Plant Biology, North Carolina State University, Raleigh, NC 27695, USA, imara_perera@ncsu.edu

Brian Q. Phillippy Department of Plant Biology, North Carolina State University, Raleigh, NC 27695, USA, bqphilli@ncsu.edu

Günther F.E. Scherer Leibniz-Universität Hannover, Institut für Zierpflanzenbau und Gehölzwissenschaften, Abt. Molekulare Ertragsphysiologie, Herrenhäuser Str. 2, D-30419 Hannover, Germany, scherer@zier.uni-hannover.de

Jill Stevenson-Paulik BASF Plant Science, LLC. 26 Davis Drive, Research Triangle Park, NC 27709, USA, jill.stevenson@basf.com

Amy L. Szumlanski Department of Molecular, Cellular, and Developmental Biology, University of Michigan, 830 North University Avenue, Ann Arbor, MI 48109, USA

Arjen ten Have Instituto de Investigaciones Biológicas, Facultad de Ciencias Exactas y Naturales, Universidad Nacional de Mar del Plata, CC 1245, 7600 Mar del Plata, Argentina

Christa Testerink Section of Plant Physiology, Swammerdam Institute for Life Sciences, University of Amsterdam, Science Park 904, 1098 XH, Amsterdam, The Netherlands

Joop E.M. Vermeer Section Plant Physiology, Swammerdam Institute for Life Sciences, University of Amsterdam, Sciencepark 904, 1098 XH, Amsterdam, The Netherlands, J.E.M.Vermeer@uva.nl

Xiaobo Wan College of Life Science, State Key Laboratory of Crop Genetics and Germplasm Enhancement, Nanjing Agricultural University, Nanjing 210095, P.R. China

Xuemin Wang Department of Biology, University of Missouri, St. Louis, MO 63121 and Donald Danforth Plant Science Center, St. Louis, MO 63121, USA, wangxue@umsl.edu

Zheng-Hua Ye Department of Plant Biology, University of Georgia, Athens, GA 30602, USA, zhye@plantbio.uga.edu

Christine Zalejski Royal Holloway, University of London, School of Biological Sciences, Egham Hill, Egham, TW20 0EX, UK; Physiologie Cellulaire et Moléculaire des Plantes, UR5 Université Pierre et Marie Curie, 3 rue Galilée, F-94200 Ivry-sur-Seine, France

Wenhua Zhang College of Life Science, State Key Laboratory of Crop Genetics and Germplasm Enhancement, Nanjing Agricultural University, Nanjing 210095, P.R. China, whzhang@njau.edu.cn

Ruiqin Zhong Department of Plant Biology, University of Georgia, Athens, GA 30602, USA

Part I
Phospholipases

Phospholipase A in Plant Signal Transduction

Günther F.E. Scherer

Abstract Phospholipase A (PLA) is an acyl hydrolase, which hydrolyses phospholipids either at the hydroxyl group of the C_1 (phospholipase A_1; PLA_1) or the C_2 atom (PLA_2). Structurally different enzymes can have this activity. These enzymes are (1) the small (14 kDa) secreted phospholipases A_2 ($sPLA_2$) found in fungi, plants and animals; (2) the soluble or secreted patatin-related phospholipases A_2 ($pPLA_2$), including the homologous soluble calcium-independent phosholipases A_2 ($iPLA_2$) in animals; (3) the cytosolic or calcium-activated phospholipases A_2 ($cPLA_2$); (4) the lipase-like phospholipase A_1 and (5) the bacterial dimeric phospholipase A_2. Since, bacterial phospholipase A_2 is not found in plants, it is not discussed here. Both $pPLA_2$ and the homologous $iPLA_2$ hydrolyse in vitro phospholipids at the C_1- and C_2-position so that the plant enzymes are often called PLA (non-specified A), but indications for PLA_2 specificity in vivo exist and hence called $pPLA_2$ here. Although few facts are known about the functions of $sPLA_2$ (four genes in Arabidopsis), there is rapidly accumulating evidence that the $pPLA_2$ in plants (ten genes in Arabidopsis) have function in several signal transduction pathways, such as auxin, pathogen and, perhaps, light signaling. The known localisation of five different enzymes is in the cytosol. Thus, the $pPLA_2$ of plants takes over the function in plant signal transduction, which is fulfilled by the $cPLA_2$ in animal cells. Evidence that the breakdown products, free fatty acid and lysophospholipids are second messengers is fragmentary. The PLA_1 group in plants has a preference for hydrolysis of galactolipids and is localised to chloroplasts, so they could be the enzymes to release linolenic acid as a precursor for jasmonate synthesis.

G.F.E. Scherer
Institut für Zierpflanzenbau und Gehölzwissenschaften, Leibniz-Universität Hannover, Abt. Molekulare Ertragsphysiologie, Herrenhäuser Str. 2, D-30419, Hannover, Germany
e-mail: scherer@zier.uni-hannover.de

T. Munnik (ed.), *Lipid Signaling in Plants*, Plant Cell Monographs 16,
DOI 10.1007/978-3-642-03873-0_1,

1 Structure and Function of $sPLA_2$ in Plants

The secreted $sPLA_2$ are true PLA_2 in that the catalytic mechanism allows only hydrolysis at the C2 position of phospholipids (Mansfeld et al. 2006; Mansfeld and Ulbrich-Hofmann 2007) (Fig. 1). The group of PLA_2 enzymes longest known is the small secreted enzymes, which, in animals, helps in digestion through pancreatic juices or through toxins from insects, scorpions and snakes (Dennis 1994). Their molecular weight is around 14 KDa and were later also found in plants (Kim et al. 1994, 1999; Ståhl et al. 1998; Ståhl et al. 1999). In Arabidopsis, four $sPLA_2$ enzymes were identified (Ryu 2004). Plant and animal enzymes have an overall homologous structure although sequence conservation in parts other than the catalytic centres and the calcium-binding loop is not very high (Ståhl et al. 1998, 1999; Bahn et al. 2003; Mansfeld and Ulbrich-Hofmann 2007). The histidine residue of the catalytic his–asp dyad is found in plant enzymes (Ståhl et al. 1998; Bahn et al. 2003; Mansfeld and Ulbrich-Hofmann 2007) although not in all of them. The pattern of 12 cysteines, destined to form six cysteine bridges, is also present in plant sequences (Mansfeld et al. 2006). Calcium is required for the hydrolysis, and the pH optimum was determined in several cases as pH 8–10 (Ståhl et al. 1998; Bahn et al. 2003; Lee et al. 2003, 2005; Fujikawa et al. 2005; Ryu et al. 2005; Mansfeld et al. 2006). Plant enzymes are secreted to the cell wall (Bahn et al. 2003; Lee et al. 2003), although not all have been analysed yet, so that though secretion is predicted by sequence, additional compartments may be targeted. Because of the alkaline pH optimum, their activity status and function at the apoplast remain unclear. Zwitterionic phospholipids are preferred over anionic phospholipids substrates and relatively short chain (12 C atoms), and desaturated fatty acids are cleaved preferentially, excluding, however, arachidonic acid (Mansfeld and Ulbrich-Hofmann 2007; Dennis 1997; Ghosh et al. 2006; Lio and Dennis 1998; six and Dennis 2000; Srijder and Dijkstra 2000; Winstead et al. 2000).

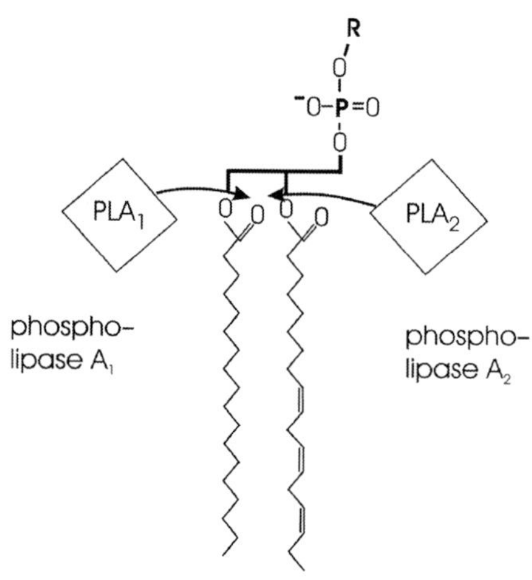

Fig. 1 Scheme of phospholipid hydrolysis by phospholipase A_1 or A_2

Possible involvement of $sPLA_2$ in light signaling of guard cells was mentioned in the literature early (Suh et al. 1998), but the function of $sPLA_2$ in plants must be regarded as still largely unknown. Only one publication reports a functional investigation by analysis of a knockout *Arabidopsis* mutant and comparison to corresponding overexpressing and promoter::GUS lines (Lee et al. 2003). Transcription of the gene was upregulated by auxin within 1 h. The authors noticed small transient changes in elongation and defects in gravitropism in both knockout and over-expressing lines. As there is a hypothesis that $pPLA_2$ mediates auxin functions (Paul et al. 1998; Scherer and André 1989; Holk et al. 2002; Scherer 2002), this could also be carried out by $sPLA_2$, perhaps in parallel. Although the majority of known plant pPLAs are cytosolic (apart from potato patatins), and post-translational activation mechanisms can be proposed (see below), rapid activation by signals via receptors is more difficult to imagine in case of $sPLA_2$ enzymes located in the cell wall, as calcium is already very abundant, and a shift towards an alkaline pH is not characteristic of auxin (Hager et al. 1971). However, during slower responses, such as with growth measured over hours, a "chronic" change in lipid breakdown products could be possible and might exert such physiological functions. The transcription of this $sPLA_2$ responds to auxin within 1 h, but the functional significance for this remains to be investigated.

2 Animal $cPLA_2$ as Role Model for Signal Transduction

Calcium-independent PLA was originally named to contrast with the group of soluble animal PLAs, the cytosolic or calcium-dependent PLA_2 (Clark et al. 1991; Sharp et al. 1991). The latter have a so-called C2 domain which binds calcium. Upon calcium binding, these enzymes bind to the ER or the nuclear envelope and become hydrolytically active (Glover et al. 1995; Schieviella et al. 1995; Evans et al. 2001; Sheridan et al. 2001). Thus, rises in cytosolic calcium will activate $cPLA_2$ so that it could be linked to many receptors in animal cells. Other activation mechanisms of this enzyme include phosphorylation (Lin et al. 1992; Qin et al. 1993; Qui and Leslie 1994) and, indirectly, G proteins (Handlogten et al. 2001; Misra and Pizzo 2002). It is the major enzyme in animal signal transduction to generate arachidonic acid (20:4) which is converted to prostaglandins and leucotrienes (Dessen 2000). To what extent $iPLA_2$ liberates arachidonic acid in animal systems is still under debate but recent reports indicate this is possible (Murakami et al. 1998; Balsinde and Balboa 2005). The obvious analogy to plant systems of animal $cPLA_2$ and $iPLA_2$ functions in signal transduction is the liberation of linolenic acid (18:3) in plants and biosynthesis of jasmonic acid (Wasternack 2007) as a part of plant signal transduction. However, the plant enzyme starting JA biosynthesis by liberating linolenic acid is, to some extent, an open question and may rather be a function of certain PLA_1 enzymes (Ishiguro et al. 2001, see below). A $cPLA_2$ gene is not found in the plant genome (Holk et al. 2002), but it served as a role model for $pPLA_2$ as the plant signal transduction by PLA_2.

3 Structure and Function of $pPLA_2$s in Plants

3.1 Structure and Enzymatic Properties of $pPLA_2$

Patatin, was identified as a phospholipase A early (Racusen 1984). It is the potato storage protein apparently having the dual function of storage protein and acyl hydrolase. The enzymes of this group with no additional domains have a molecular weight of around 50 $\pm$ 5 kDa but several have acquired additional domains. Patatins could be sequenced rather early (Rosahl et al. 1986; Andrews et al. 1988), so that later, after sequencing the, firstly, so-called calcium-independent $iPLA_2$ from animal organisms (Tang et al. 1997) had to be renamed into $pPLA_2$ although this is not strictly followed in the literature (as it is here for convenience of distinguishing animal and plant enzymes).

With the advent of whole genomes being sequenced, it can now be clearly stated that pPLA's are found in apparently all organisms, down to yeasts and bacteria. Not all animals possess the $cPLA_2$, e.g. *C. elegans* (unpublished observation), so plants are not "unusual" in this respect (Holk et al. 2002). The catalytic centres of $cPLA_2$ and $pPLA_2$ have some sequence similarity, in that they possess a lipase-type catalytic serine in the GXSXG constellation, but contain distinct phosphate-binding motifs, i.e. DGGGXR for $pPLA_2$ and SGGGXR for $cPLA_2$ (Fig. 2; Holk et al. 2002). A number of $pPLA_2$ enzymes and genes from various species have been isolated (Rosahl et al. 1986; Andrews et al. 1986; Senda et al. 1996; Kostyal et al. 1998; Sowka et al. 1998; Dhondt et al. 2000; Jung and Kim 2000; Hirschberg et al. 2001; Huang et al. 2001; Matos et al. 2001; Jekel et al. 2003; La Camera et al. 2005). The third amino acid necessary for catalysis, identified in both types of enzymes, is aspartic acid (Murakami et al. 1998; Dessen et al. 1999; Rydel et al. 2003).

Different nomenclatures exist for plant pPLA's. Our nomenclature at the advent of full knowledge of the *Arabidopsis* genome, containing the ten *Arabidopsis* sequences, was to name the sequences after the chromosome number as Roman numbers and, in addition, with letters A, B, C etc. as suffix (Holk et al. 2002). Ryu (2004) renamed these genes in a simpler fashion and La Camera et al. (2005) used the prefix PN and Arabic numbers, e.g. *AtPLAIIA* (our nomenclature) is named *PNPLA2a*. For the time being, the first nomenclature is used until a more final solution can be perhaps generated.

We and others agree that the plant pPLA gene family is divided into three subgroups (Holk et al. 2002; Ryu 2004; La Camera et al. 2005). Group 1 contains only one gene (*AtPLAI*), having an additional C-terminal LRR domain containing a G protein-binding motif and a C-terminal ankyrin motif. The second group comprises five genes (*AtPLAIIA, AtPLAIVA, AtPLAIVB, AtPLAIVC, AtPLAV*) and the third group contains four genes (*AtPLAIIB, AtPLAIIIA, AtPLAIIIB, AtPLAIVD*). The gene from group 1 is most closely related to the animal genes so that this may be the evolutionarily oldest. Even though animal *iPLA* genes also may contain a N-terminal ankyrin repeat domain, it is not homologous to the plant C-terminal domain. Group 2 and group 3 are distinguished by their exon–intron pattern, group

The Catalytic Centers of iPLAs and cPLAs

```
      S/DGGGhRGhh                                  GXSXG
PKQGLRILTMDGGGMKGLA   TVQILKEIE     KGSGK     PIHELFDLICGTSTGGMLAIALGVKLMTLEQ     ATPLAI
YGNLVTILSIDGGGIRGLI   PAVILGFLESELQKLDGEE   ARLADYFDVIAGTSTGGLVTAMLTAPNKEGRP      ATPLAIIA
CGSLVTILSLDGGGVRGII   AGVILAFLEKQLQELDGEE   ARLADYFDVIAGTSTGGLVTAMLTVPDETGRP      ATPLAIVA
YGTLVTILSLDGGGVRGII   AGVILAYLEKQLQELDGE   HVRVADYFDVIAGTSTGGLVTAMLTAPDENGRP      ATPLAIVB
YGQLVTILSIDGGGIRGII   PGTILAYLESQLQELDGEE   ARLVDYFDVISGTSTGGLIVAMLTAQDQSGGH      ATPLAIVC
QRGKVCVLSIDSGGMRGII   PGKALAYLEHALKSKSGDDPNARIADYFDVASGSGIGGIFTAMLFASSDGNRP    ATPLAIIB
QRGKICILSIDGGGMRGIL   PGKALAYLEHALKSKSGD PNARIADYFDVAAGSGIGGIYTAMLFGSRDGNRP    ATPLAIIIA
SSRKTRILSIDGGGTTGIV   AAASILHLEHQIRLQTGD PHAHISDFFDIVAGTGIGGILAALLVADDGSGRP    ATPLAIIIB
PRGRICVLSIDGGGMRGLL   AGKSLIYLEQMLKEKSGD PNARIADYFDVAAGSGVGGVFAAMIFATRDGNRP    ATPLAIVD
NTKKSSLALLEGGGVKGEI HLEKLKVIEEI       TGK PTCK    VFDFTGGTSVGGLILILLNLPDSDNPGKP   RICKETTSIA
QKAPIEYIAFSGGGAKGAI YSGVYEAAKKT                 GILDNVKAVAGSSVGAITAAVVALGTPPKRFEE  RICKETTSIA
RDVPVVAILGSGGGFRAMVGFSGVMKALYES                 GILDCATYVAGLSGSTWYMST LYSHPDFPEKGP   RAT
RDVPVVAILGSGGGFRAMVGFSGVMKALYES                 GILDCATYIAGLSGSTWYMST LYSHPDFPEKGP   MOUSE
RDVPVVAILGSGGGFRAMVGFSGVMKALYES                 GILDCATYVAGLSGSTWYMST LYSHPDFPEKGP   MAN
SSLPNIGIAVSGGGWRAMVGFAGVMKALYES                 GVLDCATYIAGLSGSTWYMST LYSHPDFPEKGP   CHICKEN
REVPTIAILGSGGGFRAMVGFSGVMKALYES                 GVFDCATYVAGLSGSTWYMSM LYSHPEFPAKGP   FISH
DTAPKVAIACSGGGYRAMLSGAGMISAM DNRTDGANDHG LGGLLQSSTYLAGLSGGNWLVGT LAYNNWTSVQAI   KLUVEROM
SQLPNVAIAISGGGYRALMNGAGFVAAA DNRIQNTTGAGGIGGLLQSSTYLAGLSGGGWLVGS LFSNNFSSIETL   NEUROSP
SEGPSIGIALSGGGLRAMILGSGALSAM DARHDNHT    VLTGLLQASDYLVGTDGSAWTVGGIALNNFSTINDF   SCHIZOS
SDIPRIAVACSGGGYRAMLSGAGMLAAM DNRTDGANEHG LGGLLQSTTYLAGLSGGNWLVGT LAWNNWTSVQDI   TORULOPS.
SDIPRIAVACSGGGYRALMNGAGAVKAF DSRTDNATATGHLGGLLQSATYISGLSGGSWLLGS IYINNWTSVQDI   PENICILL.
RDVPVIAVLGSGGGFRAMLSGAGMLAAM DNRTDGANEHG LGGLLOGATYLAGLSGGNWLTST LAWNNFTTVDKL   SACCHAROM.
```

Fig. 2 Selected amino acid sequences of catalytic centres of patatin-related phospholipases A_2 (*Arabidopsis* and *Rickettsia* as a prokaryote) and cytosolic phospholipases A_2 from higher animals and some fungi. Amino acids are *colour-coded*. Acidic amino acids are *red*, basic ones are *blue*, polar ones are *green*, and aliphatic and aromatic ones are *black*. This highlights the similarity in the polar/non-polar patterns. Note that the lower Rickettsia sequence has similarity to the $cPLA_2$ sequence although it is overall a $pPLA_2$. (Gene numbers and accession numbers: At1g61850 (AtPLAI): NP_176378. At2g26560 (AtPLAIIA): NP_180224. At2g39220 (AtPLAIIB): NP_181455. At3g54950 (AtPLAIIIA): NM_115352. At3g63200 (AtPLAIIIB): NP_567142. At4g37070 (AtPLAIVA): 4006869. At4g37060 (AtPLAIVB): 4006870. At4g37050 (AtPLAIVC): NP_195422. At4g29800 (AtPLAIVD): NP_194709. At5g43590 (AtPLAV): NP_199172. Rickettsia prowazekii 1: LOCUS NP_220970. *Rickettsia prowazekii* 2: LOCUS NP_220907. Rat: 1743846. Mouse: BAB23486. Man: 3142700. chicken: NP_990754. zebrafish: NP_571370. *Kluyveromyces lactis*: XP_452462. Neurospora crassa: AAC03053. Schizosaccharomyces pombe: NP_593196. Torulaspora delbrueckii: LOCUS Q11121. *Penicillium chrysogenum* LOCUS: P39457. *Saccharomyces cerevisiae*: LOCUS: EDN64399)

3 having only one intron, and both group 1 and group 2 have five introns in the pPLA domain. A further distinction is that group 3 has, instead of the canonical GXSXG motif, a GSXG (X = aliphatic amino acid) in the catalytic centre (Holk et al. 2002). This catalytic centre is unique for the plant group three enzymes, and its significance is presently unknown (substrate specificity?).

The comparison of the three-dimensional structures of $cPLA_2$ and $pPLA_2$ suggests a steric relationship between the catalytic centre motif GXSXG with the phosphate binding-sequence motifs DGGGXR ($pPLA_2$) or SGGGGXR ($cPLA_2$) (Dessen et al. 1999; Rydel et al. 2003). An unknown common ancestor may have existed for the active site but the other parts of the two types of enzymes are not homologous at all in terms of amino acid sequence.

Calcium dependence and substrate specificity of plant $pPLA_2$'s must be viewed together with the subcellular localisation known for a few $pPLA_2$ enzymes. The first classification of the homologous animal enzymes as $iPLA_2$ (=calcium-independent PLA_2) indicates the independence of the reaction mechanism of calcium

(Ackermann et al. 1994), yet in almost all plant enzyme assays millimolar calcium is included (Dhondt et al. 2000; Jung and Kim 2000; Hirschberg et al. 2001; Huang et al. 2001; Matos et al. 2001; La Camera et al. 2005) with few exceptions (Viehweger et al. 2006; Heinze et al. 2007). When tested with and without calcium, activity is more than 10-fold higher in the presence of millimolar calcium (Holk et al. 2002 and unpublished observations S. Rietz and G. E. F. Scherer). Whether this is a true calcium dependence remains doubtful because all five *Arabidopsis* enzymes expressed as GFP hybrid proteins localise to the cytosol or cytosolic membrane surfaces (Holk et al. 2002; La Camera et al. 2005) and none is predicted to be secreted when searched by prediction algorithms. Hence, the vacuolar localisation of patatin rather seems to be the exception, not the rule. Activation by increases of cytosolic calcium remains a possibility but the increase between 0.1 and 10 μM or even to 100 μM is still small in comparison to maximal stimulation usually obtained in vitro by 1 mM, which is a very unlikely cytosolic calcium concentration. Moreover, published pH optima are in the near neutral range also indicating cytosolic properties (Senda et al. 1996; Holk et al. 2002; Rietz et al. 2004; La Camera et al. 2005). This puzzle remains to be solved in future work, but it must be pointed out that enzymatic tests without calcium do show activity (Viehweger et al. 2006; Heinze et al. 2007). A similar puzzle, considering cytosolic localisation of pPLAs, is the preference of plant $pPLA_2$ for galactolipids in vitro (Matos et al. 2001; La Camera et al. 2005; Yang et al. 2007, S. Rietz and G. F. E. Scherer, unpublished observation) since galactolipids are not present in the plasma membrane or endomembranes under normal growth conditions, but only under severe phosphate stress (Andersson et al. 2005). The highly conserved element of catalytic centre, DGGGXR, binds to the phosphate of phospholipids (Rydel et al. 2003), which is not pointing to a preference for galactolipids as substrates either.

The precursor for jasmonic acid is linolenic acid, liberated from galactolipids in the chloroplast thylakoids (Wasternack 2007). So the hypothesis that $pPLA_2$ stimulation could increase the production of jasmonic acid (Dhondt et al. 2000, 2002; La Camera et al. 2005; Yang et al. 2007) would only hold unless chloroplast-localised $pPLA_2$s are found. Only AtPLAI is partially chloroplast-localised to the envelope (Holk et al. 2002) so that the decreased level of jasmonic acid in the *AtplaI* knockout *Arabidopsis* in response to *Botrytis* challenge might perhaps be explained by this finding (Yang et al. 2007). However, *AtPLAI* is expressed around bundles and in the root stele (our unpublished results), which are not chloroplast-rich tissues. Chloroplast localisation would remain to be shown or explained for some other plant $pPLA_2$'s if the hypothesis is true that $pPLA_2$'s liberate the precursor for JA biosynthesis.

3.2 Function of $pPLA_2$ in Auxin Signal Transduction

Two main functions of $pPLA_2$s have emerged in plants, i.e. in auxin action and in its defence against pathogens. In animal systems, $iPLA_2$ is assumed to play a role in membrane lipid remodelling (Balsinde et al. 1995). For plant $pPLA_2$, this cannot be excluded but so far there is no such data available yet.

Historically, the finding of rapid activation by auxin of $pPLA_2$ within 5 min initiated research on $pPLA_2$ in signal transduction (Scherer and André 1993). Activation of $pPLA_2$ was shown as accumulation of phospholipids' breakdown products, free fatty acids and/or lysolipids. Either, accumulation of radioactive lysolipid, labelled at the headgroup ethanolamine or choline, or the accumulation of either fluorescent LPC or free fatty acid was used as a parameter of $pPLA_2$ activation in cell cultures (Scherer and André 1989; Paul et al. 1998) or hypocotyls segments (Scherer 1992, 1995). The magnitude of activation of $pPLA_2$ by several active and inactive auxins paralleled the auxin efficiency in activating elongation so that specificity for active auxins of $pPLA_2$ activation was demonstrated (Holk et al. 2002). In work with cell cultures, two differently-labelled phosphatidylcholine (PC) substrates were used, one carrying two fluorescently labelled fatty acids (bis-BODIPY-PC), and the second carrying only a labelled fatty acid at the C_2 position (mono-BODIPY-PC). With both labelled substrates, an increase of free fatty acids after auxin application was found, indicating that release of free fatty acid from the C_2 atom of the substrate took place. Even though this is not a full proof that $pPLA_2$ is in vivo indeed a PLA_2 and not a PLA_1, it provides a first hint that this is the case. Activation of $pPLA_2$ by auxin was also measured in vitro in isolated, prelabelled membrane fractions as accumulation of radioactive lysolipid (André and Scherer 1991; Scherer and André 1993) at micromolar auxin concentrations. In contrast, in vivo lysolipid accumulation was only detectable at rather high auxin concentrations (>100 μM) but, when accumulation of fluorescent free fatty acid was quantified as $pPLA_2$ activity, micromolar auxin concentration was sufficient to observe activation. This paradox indicates that lysolipids might be rapidly reacylated in vivo but not in vitro so that only high auxin concentrations lead to detectable lysolipid accumulation in vivo and, thus, the fatty acid should be the relevant prospective second messenger in auxin signal transduction. In isolated membranes, auxin activation of $pPLA_2$ was enhanced by GTPγS but prevented by GDP, similarly as observed after elicitor treatment (Heinze et al. 2007, see below). However, this does not mean that a trimeric G protein is involved in $pPLA_2$-mediated auxin signal transduction. Such arguments, for Gα-involvement as $pPLA_2$ interactor are much stronger for the elicitor-activated enzyme(s).

Inhibitors of $pPLA_2$ inhibited the purified enzyme (Holk et al. 2002), the activation of pPLA by auxin (Paul et al. 1998), hypocotyl elongation (Yi et al. 1996; Scherer and Arnold 1997; Holk et al. 2002) and activation of early auxin-induced genes (Scherer et al. 2007). The elongation responses were auxin-specific, because $pPLA_2$ inhibitors did not inhibit elongation induced by gibberellic acid, or cotyledon expansion induced by cytokinin (Scherer and Arnold 1997). Moreover, auxins active in elongation stimulated $pPLA_2$ but inactive auxins did not (Paul et al. 1998).

The role of $pPLA_2$ in auxin signaling is summarised in Fig. 3. Still, various questions remain open including, which protein could be the receptor for this response, what downstream processes are regulated by $pPLA_2$, and which one of

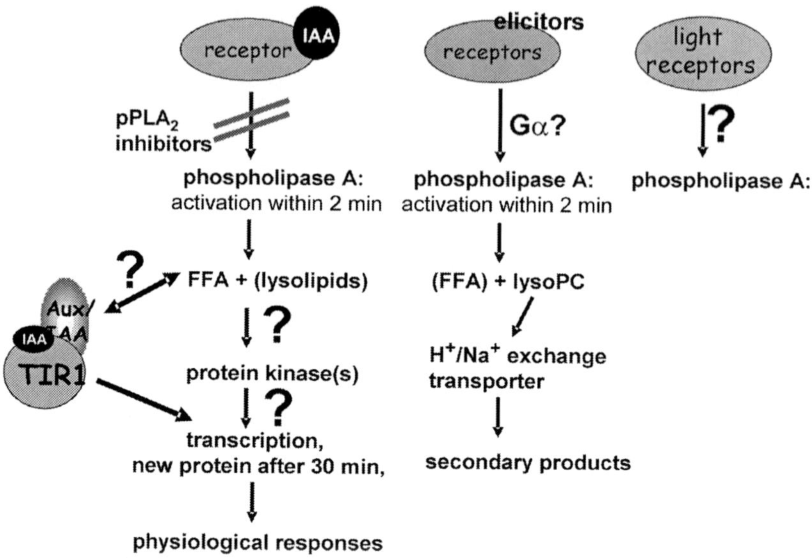

Fig. 3 Summary of models for the function of $pPLA_2$ in plant signal transduction. *Question marks* indicate unknown steps

the ten genes/proteins is responsible for what function? Two auxin receptors are known, the AUXIN-BINDING-PROTEIN1 (=ABP1) (Napier et al. 2002) and the TRANSPORT-INHIBITOR-RESISTANT1 (=TIR1) (Dharmasiri et al. 2005; Kepinski and Leyser 2005; Tan et al. 2007). For both, the three-dimensional structures and the auxin binding pockets are known (Woo et al. 2002; Tan et al. 2007). TIR1 is an F-box protein, which is actually an auxin-activated E3 ubiquitin ligase. Upon auxin binding, a sandwich complex between TIR1 and the substrate proteins, the Aux/IAA negative transcription factors, is formed and ubiquitination is stimulated by auxin binding, leading to enhanced degradation of these negative transcriptional co-regulators in the proteasome. Formation of the negatively regulating ARF-Aux/IAA dimers is decreased so that early auxin-activated genes are upregulated in transcription (Dharmasiri and Estelle 2004). So, this type of receptor is directing transcription, which qualifies it as receptor by both binding auxin and by regulating genes and synthesis of new proteins, which is a necessity to explain the multitude of auxin functions. However, such a receptor does not explain events triggered by auxin faster than 30 min because this is the minimal time span needed for cells to manufacture a new protein from a fast auxin-activated promoter (Calderon-Villalobos et al. 2006). Auxin activation of $pPLA_2$ happens within 2–5 min so that TIR1 cannot be the receptor for this fast response. TIR1-dependent degradation of IAA1-luciferase hybrid protein was measured in the presence of $pPLA_2$ inhibitors and was not affected during the first 10 min and did not affect auxin-dependent TIR1-activation directly showing that $pPLA_2$ activation is independent of the receptor TIR1 (Scherer et al. 2007). However, long-term IAA1-luciferase degradation after 2 h was inhibited by $pPLA_2$ inhibitors as was the transcriptional regulation of early auxin-regulated genes including the artificial DR5 promoter, leaving room for the speculation that some other step in the regulation of proteolytic degradation of Aux/IAA proteins might be affected by $pPLA_2$ signaling (Fig. 3). The only other identified auxin receptor, ABP1, was not yet shown to regulate genes but to regulate within minutes the activity of ion channels (Napier et al. 2002). The structure of ABP1, however, does not suggest how it could regulate the activity of any next step, especially, because it is a small glycoprotein binding to the extracytoplasmic face of the membrane, presumably, to an unknown transmembrane docking protein. Hence, it must remain unclear, presently, which auxin receptor triggers $pPLA_2$ activation.

Present research of the author's laboratory is aimed at identifying the relevant auxin receptor but also events downstream of $pPLA_2$ action(s). For this second goal, we use insertional mutants of the genes to elucidate $pPLA_2$ functions. Knock-out plants of the gene *AtPLAI* show defects in light suppression of elongation in hypocotyls, shoot phototropism, shoot gravitropism and nutation (A. Holk, S. Rietz an G.F.E. Scherer, in preparation). Taken together, auxin transport rather than elongation seems to be affected in this particular knockout mutant. Other insertional mutants of *AtPLA* genes show defects in lateral root formation, a classical auxin function, but also in light-mediated responses, pointing out the exciting possibility of $pPLA_2$ involvement in light signaling. Such mutants will also be useful of identifying down-stream elements dependent on $pPLA_2$-directed signaling. Summarising auxin and $pPLA_2$ function, it should be mentioned that not many rapid events (<30 min) in auxin signal transduction are known and that several prominent physiological auxin responses are blocked by $pPLA_2$ inhibitors as pointed out above, indicating a potentially high importance of $pPLA_2$ in auxin signal transduction (Scherer 2002; Badescu and Napier 2006).

3.3 *Function of $pPLA_2$ in Pathogen Defence Signaling*

The second known function of $pPLA_2$s is in plant defence signal transduction (Lee et al. 1992; Roy et al. 1995). Chandra et al. (1996) observed very rapid (1 min) changes in fluorescence indicators upon application of several different elicitors to soybean cell cultures and interpreted them as PLA activation. Lee et al. (1997) showed that radioactive lysophospholipids accumulated as a systemic response, minutes after wounding. Phosphatidic acid also accumulated, indicating that $pPLA_2$ was activated together with PLD. Similarly, Narvaez-Vasquez et al. (1999) showed that lysophospholipid accumulation upon wounding is a biphasic process. Using both PC labelled fatty acids with the fluorophor BODIPY, rapid accumulation of both fluorescent free fatty acid and lysophosphatidycholine upon elicitor treatment was shown by several groups (Senda et al. 1998; Kasparovsky et al. 2004; Viehweger et al. 2006). Viehweger et al. (2006) showed the transient accumulation of the natural LPC peaking at 2–5 min. Transient accumulation of a second messenger is a hallmark of signal transduction because signals must also be downregulated again and this group showed this for a $pPLA_2$-derived messenger. Plasma membrane generated LPC is proposed to migrate to the vacuolar membrane where it opens a Na^+/H^+ exchange transporter, which acidifies the cytoplasm. Transient accumulation of fluorescent-labelled lipid analogues was found neither with auxin as a stimulus nor with elicitor (Holk et al. 2002; Viehweger et al. 2002, 2006), indicating that downregulating mechanisms/enzymes do not metabolise these fluorescent derivatives in the same way or velocity as natural metabolites. Additionally, when LPC was added to permeabilised cells, the biosynthesis of enzymes necessary for phenthridines was induced, proving biological activity of the second messenger. The second system where LPC was shown to have second-messenger activity, was in the process of mycorrhiza establishment in potatoes (Drissner et al. 2007). LPC accumulated during mycorrhiza induction and exogenous addition of LPC lead to medium alkalinisation and expression of phosphate transporters which are required for the establishment of arbuscular mycorrhiza.

There could be several activation mechanisms operating on the ten different $pPLA_2$ enzymes in *Arabidopsis*. Roos' group collected substantial evidence that the elicitor-activated $pPLA_2$ interacts with the $G\alpha$ subunit of the single trimeric $G\alpha$ protein in plants (Viehweger et al. 2006; Heinze et al. 2007). They used antisense suppression of $G\alpha$ and expression of antibodies against $G\alpha$ in the cytosol to show that cells with decreased $G\alpha$ levels are less sensitive for elicitor stimulation. With a set of several antibodies against $G\alpha$, they further showed direct interaction of an as yet non-identified $pPLA_2$ with $G\alpha$. The positive effect of GTPγS on auxin-activated $pPLA_2$ in isolated vesicles (Scherer and André 1993) is not sufficient to suggest a $G\alpha$-based activation mechanism. Phosphorylation is another proposed mechanism for regulating $pPLA_2$ activity.

It should be mentioned here, that all phospholipase types, A, C, and D, are involved in plant–pathogen signaling (Laxalt and Munnik 2002; Wang 2005; Bargmann and Munnik 2006). In addition, cytosolic calcium and nitric oxide have been recognised as important second messengers (Blume et al. 2000; Wendehenne et al. 2001), and many more signal substances are known to play a role. $pPLA_2$'s functional position, linkage, and importance in this type, in general, remains to be worked out.

Investigation of individual $pPLA_2$s in wild type, transgenic or knockout plants identified several *pPLA*$_2$ genes/enzymes to have a function in plant defence. Dhondt et al. (2000, 2002) showed in wild type plants that two tobacco $pPLA_2$ genes were upregulated upon pathogen infection (tobacoo mosaic virus, *Botrytis cinerea*, *Erwinia corotovora*). When the elicitor β-megaspermin was applied, both $pPLA_2$ genes were upregulated after 6 h, while enzyme activity increased after 12–18 h. A later increase of OPDA and JA, peaking at 24 h, was observed. Though these data suggest the involvement of a $pPLA_2$ in the liberation of linolenic acid and subsequent formation of JA, it was not shown that these pPLA2s were actually located in the chloroplasts where the linolenic acid is liberated (Wasternack 2007), so the final proof for the authors' hypothesis is still lacking.

In *Arabidopsis*, *AtPLAIIA* (=*PNPLA2*) was identified to respond transcriptionally to application of *Botrytis cinerea* and *Pseudomonas syringae pv. tomato* (avrRpt2) (La Camera et al. 2005). When *AtPLAIIA* was overexpressed in *Arabidopsis*, the plants were, unexpectedly, more sensitive to the pathogens and knockout plants were less sensitive. The transcriptional upregulation of this gene not only in response to SA, wounding and JA, but also to phosphate and iron deficiency was also observed by other laboratories (Narusaka et al. 2003; Rietz et al. 2004). This protein was shown to be localised in the cytosol and not in chloroplasts (Holk et al. 2002; La Camera et al. 2005). Considering another *Arabidopsis* gene in *AtplaI* knockout plants, the basal rate of JA synthesis was not distinguishable from wild type rate but upon challenge with *Botrytis cinerea* or *Pseudomonas syringae*, it was lower than in wild type (Yang et al. 2007). AtPLAI–GFP hybrid protein is localised to the cytosol and, partially, to the chloroplast envelope (Holk et al. 2002).

When the *Arabidopsis* genes *AtPLAI*, *AtPLAIIA*, *AtPLAIIB* and *AtPLAIVC* were expressed in Petunia, no effect upon application of *Botrytis cinerea* of *Pseudomonas syringae* pv. tomato was observed in antisense plants, however, in overexpressing Petunias, pathogen damage symptoms occurred more rapid when genes *AtPLAIIA*, *AtPLAIIB* (*Botrytis*, *Pseudomonas syringae*) or *AtPLAIVC* (*Pseudomonas syringae* only) were used. *AtPLAI* had no effect when either overexpressed or anti-sensed (Zahn et al. 2005). Like in *Arabidopsis* (La Camera et al. 2005), overexpression of the other genes again increased sensitivity to pathogens while no significant increase in JA or OPDA biosynthesis occurred, and SA levels only doubled. This indicates a signal transduction function of pPLA's in pathogen defence rather then liberating linolenic acid for JA production. In work with *Eschscholzia* cell cultures, Roos' group showed that at low yeast elicitor concentration a Gα protein-coupled $pPLA_2$ was activated but no JA biosynthesis occurred, whereas at high elicitor concentration both $pPLA_2$ and JA biosynthesis were activated (Färber et al. 2003). Taken together, this clearly identifies *pPLA*$_2$ genes having a function in plant defence but does not suggest that $pPLA_2$ enzymes liberate linolenic acid as the precursor to JA biosynthesis. Moreover, application of pathogens to different systems (Zahn et al. 2005; La Camera et al. 2005) does not allow a dosage comparison, so the observation by Färber et al. (2003), suggesting two pathways depending on elicitor dosage, may actually reconcile the seemingly different results. In general, one must discriminate between signal transduction which happens within minutes and late events like transcription/translation regulation. As mentioned, new protein is synthesised no earlier than 30–45 min (Calderon-Villalobos et al. 2006) so that events measured on a timescale of hours are never easy to interpret as signal transduction. Chloroplast localisation of the (galacto)lipase liberating linolenic acid is a precondition for a function in JA biosynthesis. It seems that PLA_1 enzymes are better candidates (see below).

Pathogen application to *Sorghum* increased transcription of three *pPLA*$_2$ genes (Salzman et al. 2005). Yet another example of long-term increase of pPLA enzymatic activity was found in orange abscission zones during the fruit ripening (Alferez et al. 2005). Clearly such transcriptional/ translational regulation of *pPLA*$_2$ genes indicates that additional functions are to be expected in genes of this gene family.

4 Plant PLA_1

Certain enzymes hydrolyse phospholipids (and galactolipids) at the C_1 -position of the glycerol backbone, hence they are called PLA_1. Structurally, they are a subgroup of the lipases and possess a canonical GXSXG as catalytic centre (Aoki et al. 2007). There is only a handful of papers on plant PLA_1. The first publication describes mainly the catalytic properties of PLA_1 (Tavernier and Pugin 1995). Another, describes a lecithin:cholesterol acyltransferase-like gene, coding for a protein with clear PLA_1 activity. When over-expressed in yeast, it decreased the phospholipids content and increased the levels of free fatty acids and lysophospholipids (Noiriel et al. 2004). Three *Arabidopsis* mutants were sequenced and the genes found to code for lipases having PLA_1 enzymatic activity, i.e. *DAD1*, (Ishiguro et al. 2001), *SRG2* (Kato et al. 2002) and *DONGLE* (*DGL*) (Hyun et al. 2008). *dad1* plants have diminished JA content in their flowers and show defects in theka dehiscence so that pollination is often prevented. Because it can be rescued by application of linolenic acid or JA, JA-biosynthetic enzymes are not defective but the enzyme liberating linolenic acid from galactolipids in chloroplasts, as that is where the DAD1–GFP hybrid protein was found to be localised. *SGR2* codes for a PA-specific PLA_1 and the mutant is defective in shoot gravitropism. Another functionally interesting PLA_1 cDNA was isolated from bell pepper where it is transiently expressed in a few-days-old seedlings (Seo et al. 2008). When over-expressed in *Arabidopsis* roots, leaves and petioles grew more rapidly than in wild type and the inflorescence was elongated and early bolting. The expression pattern of the homologous gene in *Arabidopsis* is unknown. Interestingly, the cells in the over-expressing plants were smaller and a higher cell division rate was suggested. While the better interpretation needs the knowledge about the homologous *Arabidopsis* gene, all the findings on plant PLA_1 taken together, clearly suggest interesting functions of lipid breakdown products as biologically active molecules or second messengers in plant growth.

Evolutionarily, animal PLA_1 enzymes are derived from lipases. The shift in substrate specificity seems to be mainly achieved by acquiring a smaller "lid" that covers the substrate-binding site when no substrate is bound. The smaller lid enables these enzymes not only to accept triglycerides but also phospholipids and galactolipids (Carrière et al. 1998), so that such lipases could act as galactolipases. When residing in the chloroplast, it makes sense to expect them to liberate linolenic acid from galactolipids to start JA biosynthesis (Ishiguro et al. 2001; Wasternack 2007).

5 Biological Activities of Free Fatty Acids and Lysolipids, the Hydrolysis Products of $pPLA_2$

If lipid breakdown products are to be second messengers, further steps or enzymes need to be activated in order to propagate the signal. Very often, protein kinases are regulated by second messengers. In animal signal transduction, the major lipid-activated protein kinase is protein kinase C (Nishizuka 1992, 1995). Protein kinase C is not only activated by diglycerides but also by arachidonic acid and lysolipids (Oishi et al. 1988; Khan et al. 1993; Nakanishi et al. 1993). Plant lipid-activated protein kinase(s) are still badly defined and there is no protein kinase C in the plant genome (Scherer 1996; van Leeuwen et al. 2004).

Earlier publications describe the activation of plant protein kinases by fatty acids (Lucantoni and Polya 1987; Klucis and Polya 1987; Polya et al. 1990) or lysophospholipids (Martiny-Baron and Scherer 1988, 1989; Nickel et al. 1991; Martiny-Baron et al. 1991; Scherer et al. 1993a). However, neither kinases nor substrate proteins have ever been molecularly identified. Lysolipid activation of purified calcium-dependent protein kinase was regarded too low to be biologically significant (Harper et al. 1993). Protein kinase was also stimulated by oxidised fatty acids (Schweizer et al. 1996). This suggests the interesting idea that not only a typical free fatty acid could have the second messenger function, but also modified fatty acids, so that biological specificity or activity would be more stringent.

Considering regulation and specificity, it seems difficult that the total pool of a given fatty acid – influenced by all lipolytic and reacylating enzymes – would have second messenger function, rather derivatisation and/or compartmentation could be necessary to achieve biological specificity. Similarly, although not all lysophospholipids activate protein kinase activity (Scherer et al. 1993b), this seems not specific enough to clearly point out a lysolipid-activated plant protein kinase. Work on elicitors and fungus–plant interaction showed a second messenger function for LPC, but a lysolipid-activated protein kinase was not investigated (Viehweger et al. 2006; Drissner et al. 2007).

Unsaturated fatty acids and lysolipids were also reported to have a number of other biological activities. Polyunsaturated fatty acids modulate stomatal aperture and K^+ channels (Lee et al. 1994) and a protein kinase associated with the fusicoccin receptor (van der Hoeven et al. 1996). A protein phosphatase MP2C, which is involved in the wound-induced MAP kinase pathway, was found to be inhibited by unsaturated fatty acids (Baudouin et al. 1999). Interestingly, phospholipase Dδ is activated by oleic acid (Wang and Wang 2001; Zhang et al. 2003), again pointing out potential cross-talk in lipid signaling. Moreover, LPE inhibits PLD activity (Farag and Palta 1993a, 1993b; Kaur and Palta 1997; Ryu et al. 1997), and retards fruit senescence. Earlier findings that LPC activates the plasma membrane H^+-ATPase (Martiny-Baron and Scherer 1988; Scherer et al. 1988; Palmgren et al. 1988; Palmgren and Sommarin 1989) are no longer interpreted as being in vivo relevant, since it is now generally accepted that the H^+-ATPase is regulated by phosphorylation (Kinoshita and Shimazaki 1999; Ueno et al. 2005; Fuglsang et al. 2007).

In summary, the signal transduction function of $pPLA_2$ supports the concept of free fatty acids and lysolipids having second-messenger functions, but the downstream follow-up steps are still poorly understood.

Acknowledgements Work of the author was supported by the Deutsche Forschungsgemeinschaft, Bundesministerium für Forschung und Entwicklung, and VW Vorab.

References

Ackermann EK, Kempner ES, Dennis EA (1994) Ca^{2+}-independent cytosolic phospholipase A_2 from macrophage-like P388D_1 cells. Isolation and characterization. J Biol Chem 269:9227–9233

Alferez F, Singh S, Umbach AL, Hockema B, Burns JK (2005) Citrus abscission and Arabidopsis plant decline in response to 5-chloro-3-methly-4-nitro-1H-pyrazole are mediated by lipid signaling. Plant Cell Environ 28:1436–1449

Andersson MX, Larsson KE, Tjellström H, Liljenberg C, Sandelius AS (2005) Phosphate-limited Oat. The plasma membrane and the tonoplast as major targets for phospholipids to glycolipid replacement and stimulation of phospholipases in the plasma membrane. J Biol Chem 280:27578–27586

André B, Scherer GFE (1991) Stimulation by auxin of phospholipase A in membrane vesicles from an auxin-sensitive tissue is mediated by an auxin receptor. Planta 185:209–214

Andrews DL, Beames B, Summers MD, Park WD (1988) Characterization of the lipid acyl hydrolase activity of the major potato (*Solanum tuberosum*) tuber protein, patatin, by cloning an abundant expression in a baculovirus vector. Biochem J 252:199–206

Aoki J, Inoue A, Makide K, Saiki N, Arai H (2007) Structure and function of extracellular phospholipase A1 belonging to the pancreatic lipase gene family. Biochimie 89:197–204

Badescu GO, Napier RM (2006) Receptors for auxin: will it all end in TIRs? Trends Plant Sci. 11:217–223

Bahn SC, Lee HY, Kim HJ, Ryu SB, Shin JS (2003) Characterization of Arabidopsis secretory phospholipase A_2-gamma cDNA and its enzymatic properties. FEBS Lett 553:113–118

Balsinde J, Balboa MA (2005) Cellular regulation and proposed biological functions of group VIA calcium-independent phospholipase A_2 in activated cells. Cell Signal 17:1052–1062

Balsinde J, Bianco ID, Ackermann EJ, Conde-Frieboes K, Dennis EA (1995) Inhibititon of calcium-independent phospholipase A_2 prevents arachidonic acid incorporation and phospholipid remodeling in P388D_1 macrophages. Proc Natl Acad Sci USA 92:8527–8531

Bargmann BOR, Munnik T (2006) The role of phospholipase D in plant stress responses. Curr OpinPlant Biol 9:515–522

Baudouin E, Meskiene I, Hirt H (1999) Unsaturated fatty acids inhibit MP2C, a protein phosphatase 2C involved in the wound-induced MAP kinase pathway regulation. Plant J 20:343–348

Blume B, Nürnberger T, Nass N, Scheel D (2000) Receptor-mediated increase in cytoplasmic free calcium required for activation of pathogen defense in parsley. Plant Cell 12:1425–1440

Calderon-Villalobos LI, Kuhnle C, Li H, Rosso M, Weisshaar B, Schwechheimer C (2006) LucTrap vectors are tools to generate luciferase fusions for the quantification of transcript and protein abundance in vivo. Plant Physiol 141:3–14

Carrière F, Withers-Martinez C, van Tilbeurgh H, Roussel A, Cambillau C, Verger R (1998) Structural basis for the substrate selectivity of pancreatic lipases and some related proteins. Biochim Biophys Acta 1376:417–432

Chandra S, Heinstein PF, Low PS (1996) Activation of phospholipase A by plant defense elicitors. Plant Physiol 110:979–986

Clark JD, Lin LL, Kriz RW, Ramesha CS, Sultzman LA, Lin AY, Milona N, Knopf JL (1991) A novel arachidonic acid-selective cytosolic PLA_2 contains a Ca^{2+}-dependent translocation domain with homology to PKC and GAP. Cell 65:1043–1051

Dennis EA (1994) Diversity of group types, regulation, and function of phospholipase A_2. J Biol Chem 269:13057–13060

Dennis EA (1997) The growing phospholipase A_2 superfamily of signal transduction enzymes. Trends Biochem Sci 22:1–2

Dessen A (2000) Structure and mechanism of human cytosolic phospholipase A_2. Biochim Biophys Acta 1488:40–47

Dessen A, Tang J, Schmidt H, Stahl M, Clark JD, Seehra J, Somers WS (1999) Crystal structure of human cytosolic phospholipase A_2 reveals a novel topology and catalytic mechanism. Cell 97:349–360

Dharmasiri N, Estelle M (2004) Auxin signaling and regulated protein degradation. Trends Plant Sci 9:302–308

Dharmasiri N, Dharmasiri S, Estelle M (2005) The F-box protein TIR1 is an auxin receptor. Nature 435:441–445

Dhondt S, Geoffroy P, Stelmach BA, Legrand M, Heitz T (2000) Soluble phospholipase A_2 activity is induced before oxylipin accumulation in tobacco mosaic virus-infected tobacco leaves and is contributed by patatin-like enzymes. Plant J 23:431–440

Dhondt S, Gouzerh G, Muller A, Legrand M, Heitz T (2002) Spatio-temporal expression of patatin-like lipid acyl hydrolases and accumulation of jasmonates in elicitor-treated tobacco leaves are not affected by endogenous levels of salicylic acid. Plant J 32:749–762

Drissner D, Kunze G, Callewaert N, Gehrig P, Tamasloukht M, Boller T, Felix G, Amrhein N, Bucher M (2007) Lyso-phosphatidylcholine is a signal in the arbuscular mycorrhizal symbiosis. Science 318:265–268

Evans JH, Spencer DM, Zweifach A, Leslie CC (2001) Intracellular calcium signals regulating cytosolic phospholipase A_2 translocation to internal membranes. J Biol Chem 276:30150–30160

Farag KM, Palta JP (1993a) Use of lysophosphatidylethanolamine, a natural lipid, to retard tomato leaf and fruit senescence. Physiol Plant 87:515–524

Farag KM, Palta JP (1993b) Use of natural lipids to accelerate ripening and enhance storage life of tomato fruit with and without etephon. Hortic Tech 3:62–65

Färber K, Schumann B, Miersch O, Roos W (2003) Selective desensitization of jasmonate- and pH-dependent signaling in the induction of benzophenanthridine biosynthesis in cells of Eschscholzia californica. Phytochemistry 62:491–500

Fuglsang AT, Guo Y, Cuin TA, Qiu Q, Song C, Kristiansen KA, Bych K, Schulz A, Shabala S, Schumaker KS, Palmgren MG, Zhu JK (2007) Arabidopsis protein kinase PKS5 inhibits the plasma membrane H + -ATPase by preventing interaction with 14–3-3 protein. Plant Cell 19:1617–1634

Fujikawa R, Fujikawa Y, Iijima N, Esaka M (2005) Molecular cloning, expression, and characterization of secretory phospholipase A_2 in tobacco. Lipids 40:901–908

Ghosh M, Tucker DE, Burchett SA, Leslie CC (2006) Properties of the Group IV phospholipase A_2 family. Progr Lip Res 45:487–510

Glover S, Bayburt T, Jonas M, Chi E, Gelb MH (1995) Translocation of the 85-kDa phospholipase A_2 from the cytosol to the nuclear envelope in rat basophilic leukemia cells stimulated with calcium ionophore or IgE/antigen. J Biol Chem 270:15359–15367

Hager A, Menzel H, Kraus A (1971) Versuche und Hypothese zur Primärwirkung des Auxins beim Streckungswachstum. Planta 100:47–75

Handlogten ME, Huang C, Shiraishi N, Awata H, Miller RT (2001) The Ca^{2+}-sensing receptor activates cytosolic phospholipase A_2 via a Gqalpha -dependent ERK-independent pathway. J Biol Chem 276:13941–13948

Harper JF, Binder BM, Sussman MR (1993) Calcium and lipid regulation of an Arabidopsis protein kinase expressed in Escherichia coli. Biochem 32:3282–3290

Heinze M, Steighardt J, Gesell A, Schwartze W, Roos W (2007) Regulatory interaction of the Galpha protein with phospholipase A_2 in the plasma membrane of Eschscholzia californica. Plant J 52:1041–1051

Hirschberg HJ, Simons JW, Dekker N, Egmond MR (2001) Cloning, expression, purification and characterization of patatin, a novel phospholipase A. Eur J Biochem 268:5037–5044

Holk A, Rietz S, Zahn M, Paul RU, Quader H, Scherer GFE (2002) Molecular identification of cytosolic, patatin-related phospholipases A from *Arabidopsis* with potential functions in plant signal transduction. Plant Physiol 130:90–101

Huang S, Cerny RE, Bhat DS, Brown SM (2001) Cloning of an *Arabidopsis* patatin-like gene, *STURDY*, by activation tagging. Plant Physiol 125:573–584

Hyun Y, Choi S, Hwang HJ, Yu J, Nam SJ, Ko J, Park JY, Seo YS, Kim EY, Ryu SB, Kim WT, Lee YH, Kang H, Lee I (2008) Cooperation and functional diversification of two closely related galactolipase genes for jasmonate biosynthesis. Dev Cell 14:183–192

Ishiguro S, Kawai-Oda A, Ueda J, Nishida I, Okada K (2001) The DEFECTIVE IN ANTHER DEHISCIENCE gene encodes a novel phospholipase A_1 catalyzing the initial step of jasmonic acid biosynthesis, which synchronizes pollen maturation, anther dehiscence, and flower opening in Arabidopsis. Plant Cell 13:2191–2209

Jekel PA, Hofsteenge J, Beintema JJ (2003) The patatin-like protein from the latex of Hevea brasiliensis (Hev b 7) is not a vacuolar protein. Phytochem 63:517–522

Jung KM, Kim DK (2000) Purification and characterization of a membrane-associated 48-kilodalton phospholipase A_2 in leaves of broad bean. Plant Physiol 123:1057–1067

Kasparovsky T, Blein J-P, Mikes V (2004) Ergosterol elicits oxidative burst in tobacco cells via phospholipase A_2 and protein kinase C signal pathway. Plant Physiol Biochem 42:429–435

Kato T, Morita MT, Fukaki H, Yamauchi Y, Uehara M, Niihama M, Tasaka M (2002) SGR2, a phospholipase-like protein, and ZIG/SGR4, a SNARE, are involved in the shoot gravitropism of Arabidopsis. Plant Cell 14:33–46

Kaur N, Palta JP (1997) Postharvest dip in lysophosphatidylethanolamine, a natural phospholipid, may prolong vase-life of snapdragon flowers. HortScience 32:888–890

Kepinski S, Leyser O (2005) The Arabidopsis F-box protein TIR1 is an auxin receptor. Nature 435:446–451

Khan WA, Blobe C, Halpern A, Taylor W, Wetsel WC, Burns D, Loomis C, Hannun YA (1993) Selective regulation by protein kinase C isoenzymes by oleic acid in human platelets. J Biol Chem 268:5063–5068

Kim DK, Lee HJ, Lee Y (1994) Detection of two phospholipase A_2 (PLA_2) activities in leaves of higher plants Vicia faba and comparison with mammalian PLA_2's. FEBS Lett 343:213–218

Kim JY, Chung YS, Ok SH, Lee SG, Chung WI, Kim IY, Shin JS (1999) Characterization of the full-length sequences of phospholipase A2 induced during flower development. Biochim Biophys Acta 1489:389–392

Kinoshita T, Shimazaki K (1999) Blue light activates the plasma membrane H^+-ATPase by phosphorylation of the C-terminus in stomatal guard cells. EMBO J 18:5548–5558

Klucis E, Polya GM (1987) Calcium-independent activation of two plant leaf calcium-regulated protein kinases by fatty acids. Biochem Biophys Res Commun 147:1041–1047

Kostyal DA, Hickey VL, Noti JD, Sussman GL, Beezhold DH (1998) Cloning and characterization of a latex allergen (Hev b 7): homology to patatin, a plant PLA_2. Clin Exp Immunol 112:355–362

La Camera S, Geoffroy P, Samaha H, Ndiaye A, Rahim G, Legrand M, Heitz S (2005) A pathogen-inducible patatin-like lipid acyl hydrolase facilitates fungal and bacterial host colonization in Arabidopsis. Plant J 44:810–825

Laxalt AM, Munnik T (2002) Phospholipid signaling in plant defence. Curr Opin Plant Biol 5:332–338

Lee S-S, Kawakita K, Tsuge T, Doke N (1992) Stimulation of phospholipase A_2 in strawberry cells treated with AF-toxin 1 produced by *Alternaria alternata* strawberry phenotype. Physiol Mol Plant Pathol 41:283–294

Lee Y, Lee HJ, Crain RC, Lee A, Korn SJ (1994) Polyansaturated fatty acids modulate stomatal aperture and two distinct K^+ channel currents in guard cells. Cell Signal 6:181–186

Lee S, Suh S, Kim S, Crain RC, Kwak JM, Nam H-G, Lee Y (1997) Systemic elevation of phosphatidic acid and lysophospholipid levels in wounded plants. Plant J 12:547–556

Lee HY, Bahn SC, Kang Y-M, Lee KH, Kim HJ, Noh EK, Palta JP, Shin JS, Ryu SB (2003) Secretory low molecular weight phospholipase A2 plays important roles in cell elongation and shoot gravitropism in *Arabidopsis*. Plant Cell 15:1990–2002

Lee HY, Bahn SC, Shin JS, Hwang I, Back K, Doelling JH, Ryu SB (2005) Multiple forms of secretory phospholipase A_2 in plants. Progr Lip Res 44:52–67

Lin L-L, Wartmann M, Lin AY, Knopf JL, Seth A, Davis RJ (1992) $cPLA_2$ is phosphorylated and activated by MAP kinase. Cell 72:269–278

Lio YC, Dennis EA (1998) Interfacial activation, lysophospholipase and transacylase activity of groupVI Ca^{2+}-independent phospholipase A_2. Biochim Biophys Acta 1392:320–332

Lucantoni A, Polya GM (1987) Activation of wheat embryo Ca^{2+}-regulated protein kinase by unsaturated fatty acids in the presence and absence of calcium. FEBS Lett 221:33–36

Mansfeld J, Ulbrich-Hofmann R (2007) Secretory phospholipase A_2-alpha from Arabidopsis thaliana: functional parameters and substrate preference. Chem Phys Lipids 150:156–166

Mansfeld J, Gebauer S, Dathe K, Ulbrich-Hofmann R (2006) Secretory phospholipase A_2 from Arabidopsis thaliana: insights into the three-dimensional structure and the amino acids involved in catalysis. Biochemistry 45:5687–5694

Martiny-Baron G, Scherer GFE (1988) A plant protein kinase and plant microsomal H^+ transport are stimulated by the ether lipid platelet-activating factor. Plant Cell Rep 7:579–582

Martiny-Baron G, Scherer GFE (1989) Phospholipid-stimulated protein kinase in plants. J Biol Chem 264:18052–18059

Martiny-Baron G, Hecker D, Manolson MF, Poole RJ, Scherer GFE (1992) Proton transport and phosphorylation of tonoplast polypeptides from zucchini are stimulated by the phospholipid platelet-activating factor. Plant Physiol 99:1635–1641

Matos AR, d'Arcy-Lameta A, França M, Pêtres S, Edelman L, Kader JC, Zuily-Fodil Y, Pham-Ti AT (2001) A novel patatin-like gene stimulated by drought stress encodes a galactolipid hydrolase. FEBS Lett 491:188–192

Misra UK, Pizzo SV (2002) Regulation of cytosolic phospholipase A_2 activity in macrophages stimulated with receptor-recognized forms of alpha 2-macroglobulin: role in mitogenesis and cell proliferation. J Biol Chem 277:4069–4078

Murakami M, Shimbara S, Kambe T, Kuwata H, Winstead MV, Tischfield JA, Kudo I (1998) The functions of five distinct mammalian phospholipase A_2s in regulating arachidonic acid release. Type IIA and Type V secretory phospholipase A2s are functionally redundant and act in concert with cytosolic phospholipase A2. J Biol Chem 273:14411–14423

Nakanishi H, Brewer KA, Exton JH (1993) Activation of the zeta-isoform of protein kinase C by phosphatidylinositol 3, 4, 5-trisphosphate. J Biol Chem 268:13–16

Napier RM, David KM, Perrot-Rechenmann C (2002) A short history of auxin-binding proteins. Plant Mol Biol 49:339–348

Narusaka Y, Narusaka M, Seki M, Ishida J, Nakashima M, Kamiya A, Enju A, Sakurai T, Satoh M, Kobayashi M, Tosa Y, Park P, Shinozaki K (2003) The cDNA microarray analysis using an *Arabidopsis pad3* mutant reveals the expression profiles and classification of genes induced by *Alternaria brassicicola* attack. Plant Cell Physiol 44:377–387

Narvaez-Vasquez J, Florin-Christensen J, Ryan CA (1999) Positional specificity of a phospholipase A activity induced by wounding, systemin, and oligosaccharide elicitors in tomato leaves. Plant Cell 11:2249–2260

Nickel R, Schütte M, Hecker D, Scherer GFE (1991) The phospholipid platelet-activating factor stimulates proton extrusion in cultured soybean cells and protein phosphorylation and ATPase activity in plasma membranes. J Plant Physiol 139:205–211

Nishizuka Y (1992) Intracellular signaling by hydrolysis of phospholipids and activation of protein kinase C. Science 258:607–614

Nishizuka Y (1995) Protein kinase C and lipid signaling for sustained cellular responses. FASEB J 9:484–496

Noiriel A, Benveniste P, Banas A, Stymne S, Bouvier-Navé P (2004) Expression in yeast of a novel phospholipase A1 cDNA from Arabidopsis thaliana. Eur J Biochem 271:3752–3764

Oishi K, Raynor RL, Charp PA, Kou JF (1988) Regulation of protein kinase C by lysophospholipids. Potential role in signal transduction. J Biol Chem 263:6865–6871

Palmgren MG, Sommarin M (1989) Lysophosphatidylcholine stimulates ATP-dependent proton accumulation in isolated oat root plasma membrane vesicles. Plant Physiol 90:1009–1014

Palmgren MG, Sommarin M, Ulskov P, Joergensen PL (1988) Modulation of plant plasma membrane H^+-ATPase from oat roots by lysophosphatidylcholine, free fatty acids and phospholipase A_2. Physiol Plant 74:11–19

Paul R, Holk A, Scherer GFE (1998) Fatty acids and lysophospholipids as potential second messengers in auxin action. Rapid activation of phospholipase A_2 activity by auxin in suspension-cultured parsley and soybean cells. Plant J 16:601–611
Polya GM, Minichiello J, Nott R, Klucis E, Keane PJ (1990) Differential inhibition of plant calcium-dependent protein kinases by long-chain fatty acids and other amphiphiles. Plant Sci 71:45–54
Qin Z-H, de Carvalho MS, Leslie CC (1993) Regulation of phospholipase A_2 activation by phosphorylation in mouse peritoneal macrophages. J Biol Chem 268:24506–24513
Qui Z-H, Leslie CC (1994) Protein kinase C-dependent and -independent pathways of mitogen-activated protein kinase activation in macrophages by stimuli that activate phospholipase A_2. J Biol Chem 269:19480–19487
Racusen D (1984) Lipid acyl hydrolase of patatin. Can J Bot 62:1640–1644
Rietz S, Holk A, Scherer GFE (2004) Expression of the patatin-related phospholipase A-gene *At*PLA IIA in Arabidopsis *thaliana* is up-regulated by salicylic acid, wounding, ethylene, and by deficiency of iron and phosphate. Planta 219:743–753
Rosahl S, Schmidt R, Schell J, Willmitzer L (1986) Isolation and characterization of a gene from *Solanum tuberosum* encoding patatin, the major storage protein of potato tubers. Mol Gen Genet 203:214–220
Roy S, Pouénat M-L, Caumont C, Cariven C, Prévost M-C, Esquerré-Tugayé M-T (1995) Phospholipase activity and phospholipid patterns in tobacco cells treated with fungal elicitor. Plant Sci 107:17–25
Rydel TJ, Williams JM, Krieger E, Moshiri F, Stallings WC, Brown SM, Pershing JC, Purcell JP, Alibhai MF (2003) The crystal structure, mutagenesis, and activity studies reveal that patatin is a lipid acyl hydrolase with a ser-asp catalytic dyad. Biochemistry 42:6696–6708
Ryu SB (2004) Phospholipid-derived signaling mediated by phospholipase A in plants. Trends Plant Sci 9:229–235
Ryu SB, Karlsson BH, Özgen M, Palta JP (1997) Inhibition of phospholipase D by lysophosphatidylethanolamine, a lipid-derived senescence retardant. Proc Natl Acad Sci USA 94:12717–12721
Ryu SB, Lee HY, Doelling JH, Palta JP (2005) Characterization of a cDNA encoding Arabidopsis secretory phospholipase A_2-alpha, an enzyme that generates bioactive lysophospholipids and free fatty acids. Biochim Biophys Acta 1736:144–151
Salzman RA, Brady JA, Finlayson SA, Buchanan CD, Summer EJ, Sun F, Klein PE, Klein RR, Pratt LH, Cordonnier-Pratt MM, Mullet JE (2005) Transcriptional profiling of sorghum induced by methyl jasmonate, salicylic acid, and aminocyclopropane carboxylic acid reveals cooperative regulation and novel gene responses. Plant Physiol 138:352–368
Scherer GFE (1992) Stimulation of growth and phospholipase A2 by the peptides mastoparan and melittin and by the auxin 2, 4-dichlorophenoxyacetic acid. Plant Growth Regul 11:153–157
Scherer GFE (1995) Activation of phospholipase A by auxin and mastoparan in hypocotyl segments from zucchini and sunflower. J Plant Physiol 145:483–490
Scherer GFE (1996) Phospholipid signaling and lipid-derived second messengers in plants. Plant Growth Regul 18:125–133
Scherer GFE (2002) Secondary messengers and phospholipase A_2 in auxin signal transduction. Plant Mol Biol 49:357–372
Scherer GFE, André B (1989) A rapid response to a plant hormone: auxin stimulates phospholipase A_2 in vivo and in vitro. Biochem Biophys Res Commun 163:111–117
Scherer GFE, André B (1993) Stimulation of phospholipase A_2 by auxin in microsomes from suspension-cultured soybean cells is receptor-mediated and influenced by nucleotides. Planta 191:515–523
Scherer GFE, Arnold B (1997) Auxin-induced growth is inhibited by phospholipase A_2 inhibitors. Implications for auxin-induced signal transduction. Planta 202:462–469
Scherer GFE, Martiny-Baron G, Stoffel B (1988) A new set of regulatory molecules in plants: a plant phospholipid similar to platelet-activating factor stimulates protein kinase and proton-translocating ATPase in membrane vesicles. Planta 175:241–253

Scherer GFE, Führ A, Schütte M (1993a) Activation of membrane-associated protein kinase by lipids, its substrates, and its function in signal transduction. In Battey NH, Dickinson HG, Hetherington AM (eds) Society for Experimental Biology Seminar Series 53: Post-translational modifications in plants, vol 53. Cambridge University Press, Cambridge, UK, pp. 109–121

Scherer GFE, Hecker D, Müller J (1993b) Ca^{2+} ions and lysophospholipids activate phosphorylation of different proteins in plasma membranes and tonoplast purified by free-flow electrophoresis. J Plant Physiol 142:432–437

Scherer GF, Zahn M, Callis J, Jones AM (2007) A role for phospholipase A in auxin-regulated gene expression. FEBS Lett 581:4205–4211

Schieviella AR, Regier MK, Smith WL, Lin L-L (1995) Calcium-mediated translocation of cytosolic phospholipase A_2 to the nuclear envelope and endoplasmic reticulum. J Biol Chem 270:30747–30754

Schweizer P, Felix G, Buchala A, Müller C, Mètraux J-P (1996) Perception of free cutin monomers by plant cells. Plant J 10:331–341

Senda K, Yoshioka H, Doke N, Kawakita K (1996) A cytosolic phospholipase A_2 from potato tissues appears to be patatin. Plant Cell Physiol 37:347–353

Senda K, Doke N, Kawakita K (1998) Effect of mastoparan on phospholipase A_2 activity in potato tubers treated with fungal elicitor. Plant Cell Physiol 39:1080–1086

Seo YS, Kim EY, Mang HG, Kim WT (2008) Heterologous expression and biochemical and cellular characterization of CaPLA1 encoding a hot pepper phospholipase A_1 homolog. Plant J 53:895–908

Sharp JD, White DL, Chiou XG, Goodson T, Gamboa GC, McClure D, Burgett S, Hoskins J, Skatrud PL, Sportsman JR, Becker GW, Kang LH, Roberts EF, Kramer RM (1991) Molecular cloning and expression of human Ca^{2+}-sensitive cytosolic phospholipase A_2. J Biol Chem 266:14850–14853

Sheridan AM, Sapirstein A, Lemieux N, Martin BD, Kim DK, Bonventre JV (2001) Nuclear translocation of cytosolic phospholipase A_2 is induced by ATP depletion. J Biol Chem 276:29899–29905

Six DA, Dennis EA (2000) The expanding superfamily of phospholipase A_2 enzymes: classification and characterization. Biochim Biophys Acta 1488:1–19

Snijder HJ, Dijkstra BW (2000) Bacterial phospholipase A: Structure and function of an integral membrane phospholipase. Biochim Biophys Acta 1488:91–101

Sowka S, Wagner S, Krebitz M, Arija-Mad-Arif S, Yusof F, Kinaciyan T, Brehler R, Scheiner O, Breitenender H (1998) cDNA cloning of the 43-kDa latex allergen Hev b7 with sequence similarity to patatins and its expression in the yeast *Pichia pastoris*. Eur J Biochem 255:213–219

Ståhl U, Ek B, Stymme S (1998) Purification and characterization of low-molecular-weight phospholipase A_2 from developing seeds of elm. Plant Physiol 117:197–205

Ståhl U, Lee M, Sjödahl S, Archer D, Cellini F, Ek B, Iannacone R, MacKenzie D, Semeraro L, Tramontano E, Stymme S (1999) Plant low-molecular-weight phospholipase A_2S (PLA_2s) are structurally related to the animal secretory PLA2s and are present as a family of isoforms in rice (Oryza sativa). Plant Mol Biol 41:481–490

Suh S, Park J, Lee Y (1998) Possible involvement of phospholipase A_2 in light signal transduction of guard cells of *Commelina communis*. Physiol Plant 104:306–310

Tan X, Calderon-Villalobos LI, Sharon M, Zheng C, Robinson CV, Estelle M, Zheng N (2007) Mechanism of auxin perception by the TIR1 ubiquitin ligase. Nature 446:640–645

Tang J, Kriz RW, Wolfman N, Shaffer M, Seehra J, Jones SS (1997) A novel cytosolic calcium-independent phospholipase A_2 contains eight ankyrin motifs. J Biol Chem 272:8567–8575

Tavernier E, Pugin A (1995) Phospholipase activities associated with the tonoplast from Acer pseudoplatanus cells: identification of a phospholipase A1 activity. Biochim Biophys Acta 1233:118–122

Ueno K, Kinoshita T, Inoue S, Emi T, Shimazaki K (2005) Biochemical characterization of plasma membrane H + -ATPase activation in guard cell protoplasts of Arabidopsis thaliana in response to blue light. Plant Cell Physiol 46:955–963

van der Hoeven PC, Siderius M, Korthout HA, Drabkin AV, de Boer AH (1996) A calcium and free fatty acid-modulated protein kinase as putative effector of the fusicoccin 14–3-3 receptor. Plant Physiol 111:857–865

van Leeuwen W, Okrész L, Bögre L, Munnik T (2004) Learning the lipid language of plant signaling. Trends Plant Sci 9:378–384

Viehweger K, Dordschbal B, Roos W (2002) Elicitor-activated phospholipase A_2 generates lysophosphatidylcholines that mobilize the vacuolar H^+ pool for pH signaling via the activation of Na^+-dependent proton fluxes. Plant Cell 14:1509–1525

Viehweger K, Schwartze W, Schumann B, Lein W, Roos W (2006) The Galpha protein controls a pH-dependent signal path to the induction of phytoalexin biosynthesis in Eschscholzia californica. Plant Cell 18:1510–1523

Wang X (2005) Regulatory functions of phospholipase D and phosphatidic acid in plant growth, development, and stress responses. Plant Physiol 139:566–573

Wang C, Wang X (2001) A novel phospholipase D of Arabidopsis that is activated by oleic acid and associated with the plasma membrane. Plant Physiol 127:1102–1112

Wasternack C (2007) Jasmonates: an update on biosynthesis, signal transduction and action in plant stress response, growth and development. Ann Bot 100:681–697

Wendehenne D, Pugin A, Klessig DF, Durne J (2001) Nitric oxide: comparative synthesis and signaling in animal and plant cells. Trends Plant Sci 4:177–183

Winstead MV, Balsinde J, Dennis EA (2000) Calcium-independent phospholipase A_2: structure and function. Biochim Biophys Acta 1488:28–39

Woo EJ, Marshall J, Bauly J, Chen JG, Venis M, Napier RM, Pickersgill RW (2002) Crystal structure of auxin-binding protein 1 in complex with auxin. EMBO J 21:2877–2885

Yang W, Devaiah SP, Pan X, Isaac G, Welti R, Wang X (2007) *AtPLAI* is an acyl hydrolase involved in basal jasmonic acid production and Arabidopsis resistance to Botrytis cinerea. J Biol Chem 282:18116–18128

Yi H, Park D, Lee Y (1996) *In vivo* evidence for the involvement of phospholipase A and protein kinase in the signal transduction pathway for auxin-induced corn coleoptile elongation. Physiol Plant 96:359–368

Zahn M, Wymalasekara R, Göbel C, Feußner I, Holk A, Scherer GFE (2005) Expression of *Arabidopis* phospholipase A genes in *Petunia* x *hybrida*. Increased hypersensitive-like response after infection with *Botrytis cinerea* and *Pseudomonas syringae* pv. *tomato* DC3000 demonstrates a function for phospholipase A in pathogen defence. Physiol Mol Plant Pathol 67:2–14

Zhang W, Eang C, Qin C, Wood T, Olafsdottir G, Welti R, Wang X (2003) The oleate-stimulated phospholipase D, PLDγ and phosphatidic acid decrease H2O2-induced cell death in Arabidopsis. Plant Cell 15:2285–2295

The Emerging Roles of Phospholipase C in Plant Growth and Development

Peter E. Dowd and Simon Gilroy

Abstract In animals, the phospholipase Cs (PLCs) are recognized as key components of signaling, being involved in transducing messages delivered by hormones, neurotransmitters, and growth factors. Owing to their central role in animal biology, plant scientists have assumed an important role for these enzymes in plants. However, only recently we have begun to reveal the complexity with which PLCs can act to modulate plant behavior. This chapter focuses on describing the kinds of PLCs so far identified in plants at the molecular level and on discussing how these enzymes regulate cellular activity. The traditional idea from mammalian research is that PLCs cleave membrane phospholipids to generate signaling-related products that then go on to regulate cellular functions through specific targets, such as protein kinase C or Ca^{2+}-dependent signaling networks. However, while the plant enzymes also clearly act to generate signaling products, their activity towards modulating the levels of their substrates is an important emerging theme of regulation.

1 Introduction

This chapter highlights the recent understanding into the roles of the two major types of phospholipase C (PLCs) found in plants: the phosphoinositide-specific PLCs (PI-PLCs) and phosphatidylcholine-cleaving PLCs (PC-PLCs). Although plant PC-PLCs have been implicated in mobilizing phosphate during phosphate starvation, the precise role(s) of these enzymes remains enigmatic. However, PI-PLCs are known to regulate diverse processes such as tip growth, stomatal function, CO_2 fixation, and gravitropism. By analogy to their role in animals, the plant

P.E. Dowd (✉)
Department of Botany, The University of Wisconsin, 132 Birge Hall, 430 Lincoln Drive, Madison, WI, 53706, USA
e-mail: pdowd@wisc.edu

T. Munnik (ed.), *Lipid Signaling in Plants*, Plant Cell Monographs 16,
DOI 10.1007/978-3-642-03873-0_2, © Springer-Verlag Berlin Heidelberg 2010

PI-PLCs are thought to cleave the membrane phospholipid phosphatidylinositol-4,5-bisphosphate ($PtdInsP_2$) to release the second-messengers inositol-1,4,5-trisphophate ($InsP_3$) and 1,2-diacylglycerol (DAG). Although the evidence for the role of DAG in signaling in plants is limited, there is much more evidence for $InsP_3$ as a second-messenger (see chapter, "InsP3 in Plant Cells"), yet a plant homolog of the Ca^{2+} releasing $InsP_3$ receptor has not been identified. Observations of their functions in pollen tubes has led to an alternate model for PI-PLC action where, in addition to releasing second-messengers, these enzymes may also restrict the localization of their substrate, $PtdInsP_2$, and so also restrict a host of $PtdinsP_2$-dependent events to microdomains within the plasma membrane.

2 Phospholipase C in Plants

Two classes of PLCs have been identified in plants and categorized by the substrate on which they act. PC-PLCs can use phosphatidylcholine (PC) as their substrate; however, they are largely nonspecific in substrate preference. In contrast, the second class of PLCs, the PI-PLCs, preferentially cleave $PtdInsP_2$. Although both groups are categorized as PLCs because of where they cleave their phospholipid substrates (Fig. 1), the genes identified as members of these two groups have no common structural domains and are thought to have evolved independently.

2.1 Phosphatidylcholine-Cleaving Phospholipase C

PC-PLCs can use the structural phospholipid PC as a primary substrate and cleave it into DAG and PC (Fig. 1). PC-PLC activities are well-known in mammals (e.g., Kates 1955; Exton 1994) where they have been proposed to play crucial roles in receptor-stimulated processes that trigger responses such as cell division and differentiation (Exton 1994). Similar activities contribute to the nuclear DAG pools during chromatin restructuring (Albi et al. 2008) and to the pathogenesis of

Fig. 1 Substrates and cleavage products of PI-PLC and PC-PLC. Head groups of $PtdInsP_2$, PC, Phosphatidylethanolamine and phosphatidylserine are depicted next to a schematic of a phospholipid backbone consisting of a head group, a glycerol molecule, and two fatty acid chains of varying lengths, where n and n_1 represent different numbers of carbons on each of the fatty acid tails. Note: **X** marks the position used in the phosphodiester-bond formation between the head group and the "phosphorylated diacylgycerol," phosphatidic acid (PA). The chemical structures for the products 1,2 diacylgycerol (DAG) and 1,4,5, inositol triphosphate ($InsP_3$) from the cleavage of phospholipid phosphatidylinositol-4,5-bisphosphate ($PtdInsP_2$) by PI-PLC are shown. $PtdInsP_2$ ($PtdInsP_2$) is a phospholipid with an inositol-4,5-bisphosphate ($InsP_2$) as the head group. The chemical structures for the products 1,2 diacylgycerol (DAG) and phosphocholine from the cleavage of phosphatidylcholine (PC) by PC-PLC are shown

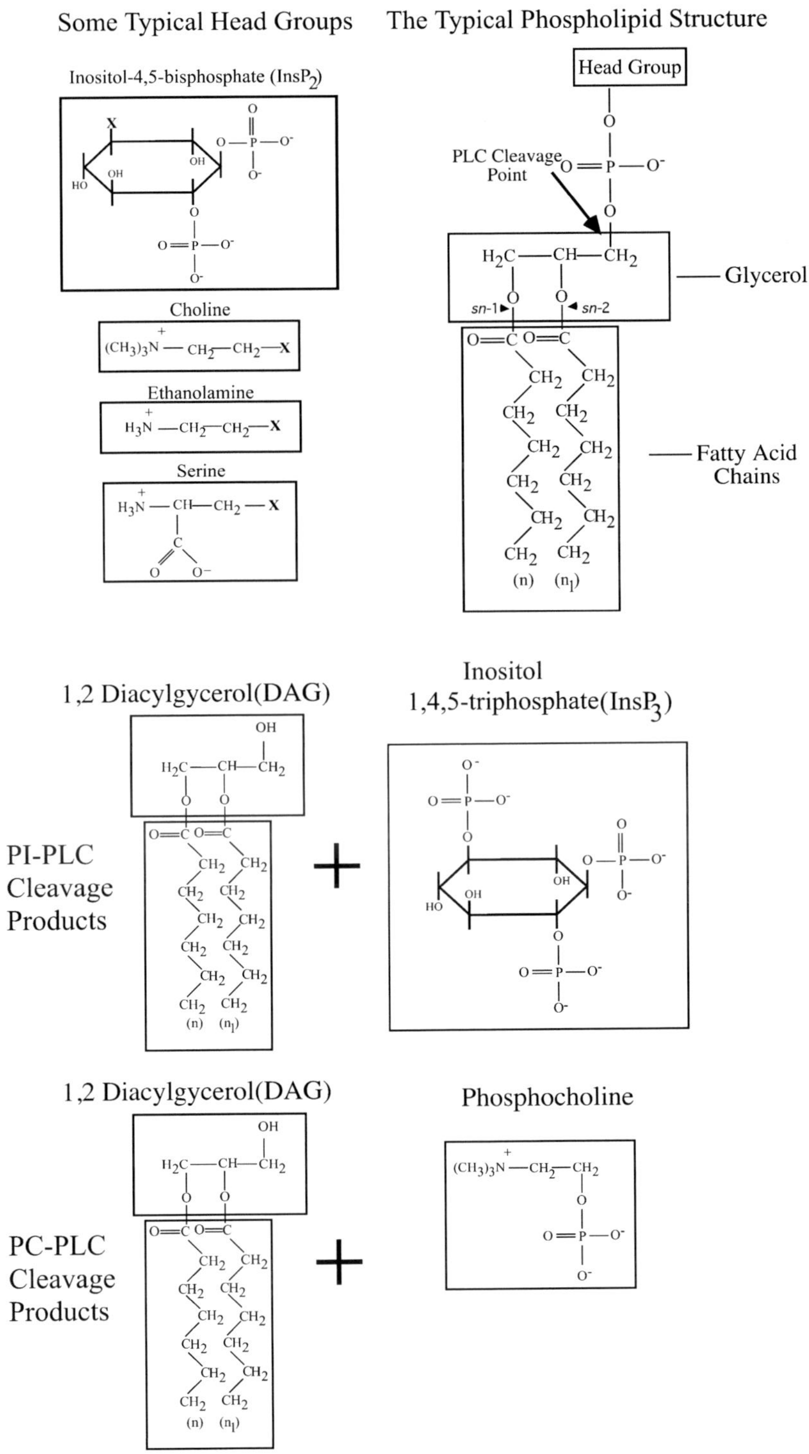

Some Typical Head Groups
The Typical Phospholipid Structure
Inositol-4,5-bisphosphate (InsP2)
Head Group
PLC Cleavage Point
Glycerol
sn-1
sn-2
Choline
Ethanolamine
Serine
Fatty Acid Chains
1,2 Diacylgycerol(DAG)
Inositol 1,4,5-triphosphate(InsP3)
PI-PLC Cleavage Products
1,2 Diacylgycerol(DAG)
Phosphocholine
PC-PLC Cleavage Products

HIV (Fantuzzi et al. 2008). In these reports, experimental evidence for PC-PLC involvement is based on biochemical PC-PLC activity assays and/or use of antibodies raised against a bacterial PC-PLC. However, despite the availability of this antibody, the molecular identity of the mammalian PC-PLC(s) remains unknown (Spadaro et al. 2006), so there is, at present, no molecular or genetic characterization of these enzymes.

In contrast to the many roles for PC-PLC in mammalian cells, the literature detailing the involvement of PC-PLCs in plant growth is sparse. For example, when plant cell cultures were fed a PC where both fatty acyl chains were fluorescently labeled, fluorescent DAG was produced, suggesting PC-PLC action (Scherer et al. 2002). This PC-PLC activity was downregulated by two fungal elicitors but its precise role in plant growth or development remains to be determined.

Despite this paucity of evidence for physiological roles for plant PC-PLCs, the genes encoding these enzymes have been cloned (Nakamura et al. 2005). Thus, when the amino acid sequence of a bacterial PC-PLC was used to find similar sequences in the *Arabidopsis* genome, six were identified. They were named nonspecific phospholipase C 1 through 6 (NPC 1–6). Interestingly, BLAST analysis of the *Arabidopsis* NPC protein sequences against the human genome reveals no significant homology. NPC 1, 2 and 6 have signal peptides and are therefore thought to be secreted. However, analysis of their protein domain structures provides few clues about their function. Thus, when the NPC sequences are analyzed by programs such as Interpro scan, the only domain found is the phosphoesterase domain required for phospholipase activity (Nakamura et al. 2005).

During phosphate starvation, one of these *PC-PLCs*, *NPC4*, was reported to be transcriptionally upregulated (Nakamura et al. 2005). The recombinant NPC4 enzyme was shown to have PC-PLC activity but also could cleave phosphatidylethanolamine (recently NPC5 was shown to cleave galactolipids as well; Gaude et al. 2008). In oat, the membrane fraction cross-reacting with an antiNPC4 antibody actually had Ca^{2+}-dependent phospholipase D (PLD)-like activity (Andersson et al. 2005). However, T-DNA insertional knockout lines for *NPC4* had a drastic reduction in PC hydrolyzing activity under phosphate starved conditions, linking NPC4 to the phosphate-dependent PC-PLC activity. Despite this reduction in PC-PLC activity, the overall lipid composition in the mutant was the same as in the wild-type (Nakamura et al. 2005). These observations suggest that either compensation for the lesion in NPC4 by other phospholipid-modulating enzymes was masking the effect of this mutant, or that the pool of PC-PLC upon which NPC4 acts is a small component of the total PC-PLC in the plant.

Taken together, these reports suggest that there is at least one family of PC-PLCs in plants and that they may play an important role during phosphate starvation. These PC-PLCs are very likely to also modulate phospholipid turnover and general cell membrane maintenance, but placing these activities in the context of plant growth and development must await further in-depth analysis of these enzymes.

2.2 Phosphoinositide-Specific Phospholipase C (PI-PLC)

The enzymes of the PI-PLC class can cleave $PtdInsP_2$, to produce two cellular regulators, DAG and the Ca^{2+}-mobilizing messenger, $InsP_3$. Unlike mammals that have six different PI-PLC isoforms, classified as β, γ, δ, ε, ζ, and η (Katan 2005), plants have only one, which from their domain structure, most closely resemble the recently discovered mammalian ζ-isoform (Wang 2004). Unfortunately, the realization of this structural similarity has not given much insight into plant PI-PLCs as PLCζ is a gamete-specific protein that likely plays a role in sperm function and Ca^{2+} signaling in the animal egg (Swann et al. 2006).

All known plant PI-PLCs contain a C2 Ca^{2+}/phospholipid-binding domain (Kopka et al. 1998), the conserved X and Y catalytic domains, and a region homologous to the second loop of the Ca^{2+}-binding EF hand of PLCδ, the so-called "EF loop" (Otterhag et al. 2001). There are nine known PI-PLC sequences in the *Arabidopsis thalania* genome (Mueller-Roeber and Pical 2002) and multiple isoforms are found in almost every plant species in which molecular characterization of *PI-PLC* genes has been performed (e.g., Kopka et al. 1998; Shi et al. 1995; Pical et al. 1997; Pan et al. 2005).

From biochemical characterizations of extracts, there appear to be two distinct classes of plant PI-PLCs (Helsper et al. 1987; Drøbak 1992; Munnik et al. 1998). One is found predominately in the cytosol, which requires millimolar Ca^{2+} for activity and uses phosphatidylinositol (PtdIns) as a substrate. The second class is membrane bound, requires micromolar Ca^{2+} and uses phosphatidylinositol 4-phosphate (PtdIns4P) and $PtdInsP_2$ as its substrates. All of the PI-PLCs cloned to date belong to the second class. The molecular identity of the cytosolic PI-PLC activity remains elusive and may represent either the action of an enzyme that shares little homology to the known plant PI-PLCs or an artifact of the break-down of one of the known membrane-associated PI-PLCs during cell fractionation and biochemical assay.

In mammalian cells, PI-PLCs interact with a host of regulatory elements, such as tyrosine kinases, membrane adaptor proteins, and guanine exchange factors (Braiman et al. 2006; Jones and Katan 2007). Considering the critical role of the PI-PLCs in plant growth and development, it is likely that regulatory interactions will emerge as more studies are oriented towards finding the PI-PLC partners in plants. However, at present, there is little evidence indicating a physical interaction of plant PI-PLCs with proteins other than those indirectly from pharmacological treatments. For example, two PI-PLCs, have been identified in *Lilium davidii* pollen protoplasts (Pan et al. 2005). Their PLC activity was increased by addition of cholera toxin, a G-protein activator, and decreased by pertussis toxin, an antagonist of G-protein activity (Pan et al. 2005). Similarly, Apone et al. (2003) manipulated the levels of G-protein-coupled receptor (GPCR) and Gα subunit (GPA1) of *Arabidopsis* in BY2 cells, and then examined the effects on cell growth and PLC activity. In cultures over-expressing either GPCR or GPA1, DNA synthesis was increased and this increase was dependent on an increase in $InsP_3$ and PI-PLC activity. These results hint at an interaction between PI-PLC and a G-protein coupled receptor system. However, to date, there is no direct evidence of such an interaction. In tobacco, the PI-PLC, NtPLC3, has been noted to possibly directly interact with the small monomeric G-protein NtRac5 (Helling et al. 2006), though. Characterizing interacting partners of the plant PI-PLCs will certainly help to define the regulatory pathways modulating/responding to PI-PLC activity. However, at present, the clearest data on the regulation of plant PI-PLC activity from across several genera remain their well-characterized sensitivity to cytosolic Ca^{2+} level (e.g., Drøbak 1992; Coursol et al. 2000; Dowd et al. 2006; Helling et al. 2006).

3 Roles of PI-PLCs in Plants

PI-PLC activities have been implicated in processes as diverse as signal transduction in guard cells (Hunt et al. 2003; Mills et al. 2004), pathogen response (de Jong et al. 2004), gravitropism (Perera et al. 2001), Nod factor signaling (Charron et al. 2004; den Hartog et al. 2003; Engstrom et al. 2002), and carbon fixation in C4 plants (Coursol et al. 2000). The control of stomatal opening/closing by the plant hormone abscisic acid (ABA) represents one of the most thorough evaluations of PI-PLC action in plants. Thus, an increase in the concentration of cytosolic Ca^{2+} has been shown to be an early event in ABA signaling that stimulates stomatal closure. The PI-PLC inhibitor, U-73122, inhibited both ABA-induced increases in cytosolic Ca^{2+} concentration and stomatal closure (Hunt et al. 2003; Mills et al. 2004). Conversely, light is known to induce stomatal opening and this opening is in turn inhibited by ABA. In antisense transgenic lines producing lower levels of PI-PLC in guard cells, ABA-induced inhibition of light-dependent stomatal opening was reduced relative to wild-type (Hunt et al. 2003). Interestingly, these plants seem not to be affected in the initial stomatal closure but rather in their ability to keep the stomata closed (Mills et al. 2004), suggesting a role for PLC signaling in the extended response of the guard cell. Consistent with this idea, both of the products of $PtdInsP_2$ turnover by PI-PLCs have been reported to modulate guard cell activities. Thus, DAG has been reported to activate the proton pump related to light-induced stomatal opening in guard cells of *Commelina communis* and *Vicia faba* (Lee and Assmann 1991), while increasing $InsP_3$ levels are thought to induce cytoplasmic Ca^{2+} increases related to triggering stomatal closure (e.g., Gilroy et al. 1990; Blatt et al. 1990). How two messengers produced by the same PI-PLC act to coordinate opposite stomatal responses is unclear, but it is important to note that DAG may well be rapidly converted to phosphatidic acid (PA) and $InsP_3$ to $InsP_6$ and vice-versa. Both PA and $InsP_6$ are thought to play signaling roles in the guard cell (Jacob et al. 1999; Lemtiri-Chlieh et al. 2003), greatly complicating the interpretation of experiments where the levels of any of these molecules are manipulated (e.g., see chapters, "Plant Phospholipase D," "Diacylglycerol kinase," "InsP3 in Plant Cells," "Inositol Polyphosphates and Kinases," "Phosphatidic acid – an Electrostatic/Hydrogen-Bond Switch?").

Another plant cell system where PI-PLCs have been studied in some detail is in pollen and pollen tube growth. Release of $InsP_3$ in pollen tubes resulted in either a Ca^{2+} wave in the cytoplasm, which moved back from the tip through the shank (Franklin-Tong et al. 1996), or an increase in $[Ca^{2+}]$ behind the tip near the nucleus (Malho 1998). In both cases, it was hypothesized that the Ca^{2+} increase was due to release of this ion from internal stores. Considering the link between phosphoinositides, PI-PLC, $InsP_3$ production, and Ca^{2+} release seen in mammalian cells, it has been inferred that a PI-PLC should be active in regulating the Ca^{2+} dynamics that are known to be required for tip growth of pollen tubes through a similar $InsP_3$-based system (Holdaway-Clarke and Hepler 2003). However, as described in the next section, analysis of how PI-PLCs regulate tip growth in pollen tubes has led to a model of their function in plants that is quite different from the paradigm of generating DAG and $InsP_3$ as signaling molecules described for the guard cell above.

3.1 The Molecular Role of PI-PLC: Insights from Pollen Tube Growth

Figure 2 presents a classic model of PI-PLC action based on signaling in animal cells. Thus, the PI-PLC acts in concert with a heterotrimeric G-protein-coupled receptor that receives an extracellular signal. PI-PLC is then activated to cleave $PtdInsP_2$ into $InsP_3$ and DAG. The $InsP_3$ then binds to a channel-like receptor in the ER, the channel opens, and Ca^{2+} is released. The resulting cytosolic Ca^{2+} increase leads to activation/deactivation of Ca^{2+}-responsive proteins such as calmodulin and hence, signal transduction. The DAG produced by PI-PLC action also acts as a regulator, activating protein kinase C (PKC) that in turn modulates a host of targets. In support of a similar signaling cascade in plants, there is evidence for regulatory roles of both DAG and $InsP_3$ in plant cell responses, for example, in the guard cell system described above. Although, there are plant protein kinase activities that are sensitive to DAG (Deswal et al. 2004), there is no clear plant PKC homolog. However, there is some speculation that calmodulin-like domain containing protein kinases (CDPKs) may function in similar ways as PKCs in plant cells (Nanmori et al. 1994; Abo-El-Saad and Wu 1995; Yoon et al. 2006). In addition, DAG targets other than PKC are likely to exist, and the conversion of DAG to PA by DAG kinase could generate a signaling molecule known to be involved in many plant processes (Testerink and Munnik 2005; see chapters, "Diacylglycerol Kinase," "Phosphatidic acid – an Electrostatic/Hydrogen- Bond Switch?", "3-Phosphoinositide-Dependent Protein Kinase is a Switchboard from Signaling Lipids to Protein Phosphorylation Cascades").

There is also much evidence for a signaling role for $InsP_3$ in plants, with $InsP_3$ levels being reported to change in response to an enormous range of stimuli (see chapter, "InsP3 in Plant Cells"). For example, gravity stimulation causes rapid and

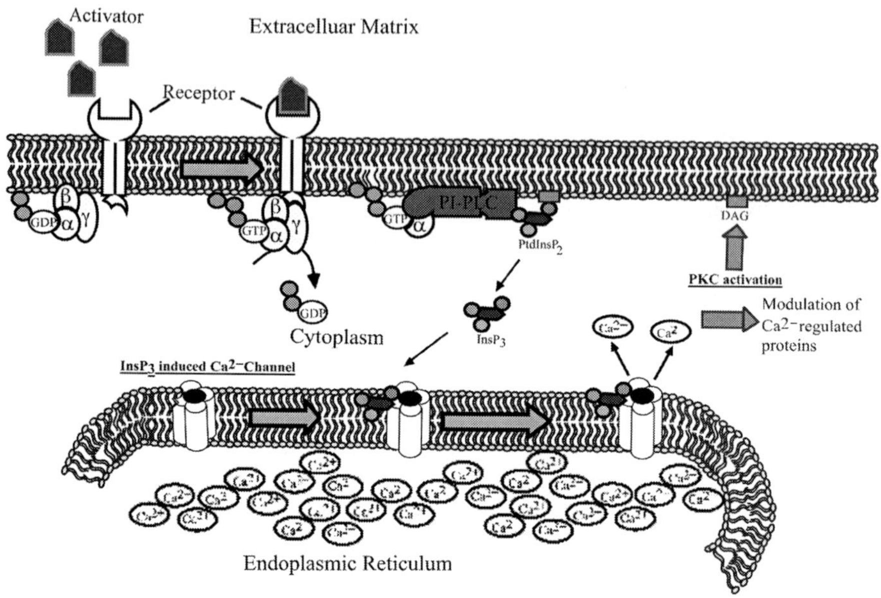

Fig. 2 Model I of the role of PI-PLC(s) in plant cell growth and regulation. This model is drawn from the known pathway of PI-PLC action in mammals. Names which are underlined indicate proteins that are integral to the mammalian PLC pathway but that for which no plant homologues have been found

complex patterns of increasing $InsP_3$ levels in the pulvinus (gravity sensing/responsive organ) of oat (Perera et al. 2001). Similarly light, cold, hormones, oxidative stress, and pathogen elicitors trigger $InsP_3$ levels to increase (reviewed in Krinke et al. 2006). $InsP_3$ also appears to play a role in the action of the thylakoid localized CAS receptor (Nomura et al. 2008; Vainonen et al. 2008), which modulates stomatal Ca^{2+} signaling in response to altered extracellular Ca^{2+} levels. CAS appears to link changes in extracellular Ca^{2+} levels to modulation of $InsP_3$ production within the cell (Tang et al. 2007), placing a potential regulator of PI-PLC activity at the stromal face of the thylakoid membrane. The precise link between this receptor and a specific PI-PLC activity/subcellular pool of $PtdInsP_2$ remains to be defined.

Possible $InsP_3$-gated Ca^{2+} conductances have been defined through electrophysiological and biochemical analysis (reviewed in Krinke et al. 2006) with these activities being predominately localized to the tonoplast, but with some evidence for $InsP_3$-gated channels on the ER and the plasma membrane. In mammalian cells, $InsP_3$ elicits its responses by binding to member of a well-defined receptor family; however, despite extensive analysis using BLAST and Hidden-Markov techniques to search for divergent sequence, the Arabidopsis genome does not appear to contain a canonical $InsP_3$ receptor (Krinke et al. 2006).

The plant plasma membrane also contains much lower levels of $PtdInsP_2$ in relation to other phospholipids when compared to animal cells (Munnik et al. 1994;

van Leeuwen et al. 2007), raising the possibility that substrate availability for the plant PI-PLCs may actually limit their ability to generate signaling active amounts of $InsP_3$ and DAG. Indeed, analysis of the role of the PI-PLCs in pollen tube growth suggests that regulating the level and distribution of $PtdInsP_2$ may be as important as producing DAG and $InsP_3$ with regards to the regulation of plant cell function.

Pollen tubes elongate by localization of growth to the extreme apex of the tube and this localization is determined by restriction of the delivery of new membrane material and the secretion of new wall components to the very tip of the elongating cell. A tip-focused gradient in cytosolic Ca^{2+} levels coupled to the localized activity of a host of regulatory and structural proteins, such as monomeric G-proteins and elements of the actin cytoskeleton, are required to maintain this spatial patterning of cell expansion. Pollen likely expresses multiple isoforms of PI-PLCs, with micro-array and promoter expression analysis suggesting *Arabidopsis* pollen produces at least three isoforms. These, PI-PLCs appear critical to supporting pollen tube growth with pharmacological inhibition of PLC activity, expression of a catalytically inactivated version ($PetPLC1^{H126A}$), and over-expression of the wild-type enzyme (NtPLC3) all disrupting growth (Dowd et al. 2006; Helling et al. 2006). Interestingly, over-expression phenotypes were observed to be mild (a 20% reduction in tube length over 6 h), consistent with the idea that due to their key regulatory role, even when over-expressed, these enzymes are tightly controlled. During tip growth, the PI-PLCs were localized to the subapical flanks of the tube leading to the proposal that their activity was to restrict $PtdInsP_2$ accumulation to the apical plasma membrane where it is required to support growth (Fig. 3). Indeed, imaging the $PtdInsP_2$ distribution by expressing the $PtdInsP_2$-binding PH-domain from mammalian PLCδ1 fused to GFP, revealed $PtdInsP_2$ to localize in the inverse pattern to the PI-PLCs in the plasma membrane (e.g., Kost et al. 1999; Dowd et al. 2006; Helling et al. 2006; see also chapter, "Imaging Lipids in Living Plants"). In mammalian cells, the signal for PI-PLC localization to the plasma membrane is thought to reside in its PH domain. However, plant PI-PLCs lack an equivalent PH domain and it appears that targeting lies in both the EF loop and C2 domains.

Domain deletion experiments have been extremely useful in understanding the role of the EF-loop and C2 domain in localizing the plant PI-PLCs to the membrane. Otterhag et al. (2001) deleted the first 100 amino acids of AtPLC2, which included the EF loop. This deletion destroyed catalytic activity suggesting a role for the EF loop in either catalysis or folding/stability of the enzyme. However, a similar deletion in a PI-PLC from petunia retained activity (Dowd et al. 2006), indicating that a functional EF loop is not a requirement for catalytic activity in all plant PLCs. Interestingly, even though inactivated as an enzyme, the truncated AtPLC2, like the full-length enzyme, still targeted to the membrane fraction (Otterhag et al. 2001). This observation led to the hypothesis that only the C2 domain may be needed for correct membrane association. Consistent with this idea, the plasma-membrane localization observed, when GFP-labeled VrPLC3 from Mung bean was heterologously expressed in *Arabidopsis* cell culture, was lost when the C2 domain was deleted (Kim et al. 2004). However, work with pollen expressed PI-PLCs suggests a role for both the EF loop and the C2 domain for targeting (Dowd et al. 2006;

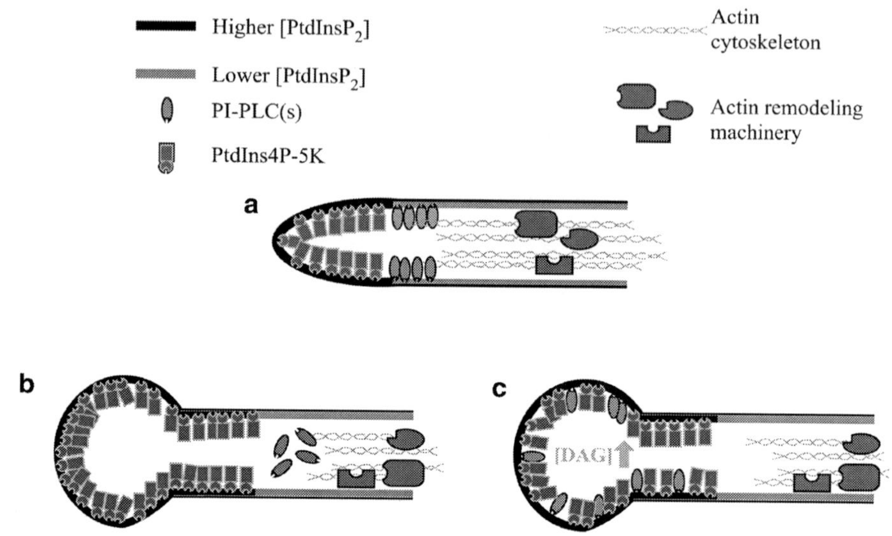

Fig. 3 Model II of the role of PI-PLC(s) and $PtdInsP_2$ in plant cell growth and regulation. This model is drawn from analysis of PI-PLC action during tip growth (i.e., elongation growth of root hairs and pollen tubes). (**a**) In a normally growing cell $PtdInsP_2$ levels are higher at the tip and lower in the shank. PI-PLC is localized in the region just behind the area of higher [$PtdInsP_2$]. The apical region elevated in $PtdInsP_2$ also represents where 4-phosphate 5-kinase (PtdIns4P-5K) is localized (Stenzel et al. 2008; Kusano et al. 2008). (**b**) and (**c**) represent two models of the effect of disrupting PI-PLC function during tip growth. (**b**) When growth is perturbed through expressing high levels of an inactive PI-PLC, endogenous PI-PLC is displaced from the membrane, leading to elevated $PtdInsP_2$ throughout the tip. This expansion of the $PtdInsP_2$ domain leads to expansion of the region of the tip capable of growth and so causes unrestricted, depolarized growth (Dowd et al. 2006). (**c**) Over expressing wild type PI-PLC does not displace the endogenous PI-PLC but causes an increase level of DAG and it is an increased level of DAG which depolarizes growth (Helling et al. 2006)

Helling et al. 2006). Thus, only when a construct containing both the EF-loop and C2 domain (EF loop:YFP:C2) was expressed in pollen tubes was full membrane association reconstituted (Helling et al. 2006). In contrast, the EF loop or C2 domain alone were ineffective at targeting to the apical membrane, suggesting that both the EF-loop and C2 domains are needed for distinctive localization.

The C2 domain is known to bind phospholipids and proteins, suggesting a possible role in membrane and/or membrane protein interaction. Both the EF loop and C2 domain also contain Ca^{2+}-binding sites suggesting the possibility of interaction with, or modulation by, membrane associated Ca^{2+}. The observation from Otterhag et al. (2001) that AtPLC2 can target to the membrane without its EF loop suggests that while the C2 domain, and possibly other regions of the protein, facilitate membrane association, the EF loop may be required to impose more subtle aspects of localization, such as to phospholipid microdomains within the membrane. Defining the precise mechanism for localizing PI-PLCs to such specific domains within the plasma membrane and how this system operates in cells other than pollen tubes remain key unanswered questions about how these enzymes function within plant cells.

All of the observations described above from the pollen tube about targeting, $PtdinsP_2$ distribution and their relationship to growth are consistent with a model where PI-PLC is localized to subdomains within the plasma membrane to restrict $PtdInsP_2$ distribution (Fig. 3). The accumulation of $PtdInsP_2$ specifically at the growing apex could then affect a host of factors already characterized as being key elements of the growth machinery (Figs. 3a, b; Drøbak and Watkins 1994; Kovar et al. 2001; van Rheenen and Jalink 2002; Wu et al. 2002; Chen et al. 2003; Dowd et al. 2006).

While the "$PtdInsP_2$ restriction" model for PI-PLC action seems an important insight into the role of plant PI-PLCs in the regulation of plant growth, it is important to note that in pollen, the spatial distribution of PI-PLC in the membrane also appears to regulate the production of at least one of its $PtdInsP_2$ turnover products, DAG (Fig. 3a, c; Helling et al. 2006). Again, although the precise role of DAG remains unclear, rapid phosphorylation to form PA via DGK could reflect a regulatory activity (see chapter, "Diacylglycerol Kinase"). PA can also be produced from phosholipids directly, via activation of phospholipase D (PLD; Testerink and Munnik 2005; see chapter, "Plant Phospholipase D"), another well characterized important element of tip growth control (Potocký et al. 2003; Monteiro et al. 2005). Rapid interconversion of DAG to PA may provide an elegant way for the plant to integrate between PI-PLC and PLD-related regulatory systems. These PLC and PLD activities acting to degrade phospholipids in the membrane will undoubtedly be balanced by the activities of enzymes synthesizing their substrates. Thus, plants deficient in one of the 4-phosphate 5-kinase (PIP5K) genes, the enzyme that catalyzes the conversion of PtdIns4P into $PtdInsP_2$ showed a loss of polarity at the tip of root hairs (Stenzel et al. 2008; Kusano et al. 2008; see chapter, "PIP-Kinases as Key Regulators of Plant Function"). When this gene was over-expressed in pollen tubes, they had an increased level of $PtdInsP_2$ at the tip indicating a key role in determining the amount of $PtdinsP_2$ residing in the membrane (Fig. 3; Stenzel et al. 2008). The interplay between PLD, PI-PLC, DGK, and 4-phosphate 5-kinase once again highlights the importance of appreciating how dynamic the phospholipid regulatory network is likely to be when interpreting the role of each element. Clearly, we are only just at the beginning of understanding the complexities of the phospholipases in the regulation of growth, but we can predict that the use of very well-defined growth response systems such as tip growth in root hairs and pollen tubes will continue to be important tools in revealing the intricacies of how PI-PLC activity relates to the modulation of cell expansion.

4 Conclusions and Perspectives

Over the past decade, great strides have been made in understanding both PC-PLC and PI-PLCs based mechanisms. Critically, plant investigators have managed to clone both PC-PLC and PI-PLCs genes setting the stage for an in-depth functional analysis of these enzymes. However, despite their cloning, the role of the PC-PLCs in plants remains enigmatic. One key problem is our lack of knowledge about the subcellular localization and potential stimulus-related relocalization of these enzymes. Protein fusions to GFP or immunolocalization with specific antibodies raised against the individual PC-PLCs would provide key tools to help solve this problem. The role of NPC4 in phosphate starvation undoubtedly represents just the "tip of the iceberg" as far as the physiological roles of this class of enzymes and we can anticipate that a key role in maintenance of the plasma membrane is likely to emerge in the future.

Great progress in elucidating the mode of action of mammalian PI-PLCs came by analyzing interacting and regulatory partners, but, at present, there are no reports demonstrating protein:protein interactions for the plant PI-PLCs. Finding the regulators/interactors for these enzymes remains one of the major challenges for researchers in this field.

Acknowledgments The authors thank Dr. S. Swanson for critical reading of the manuscript. Funding from the USDA is gratefully acknowledged.

References

Abo-El-Saad M, Wu R (1995) A rice membrane calcium-dependent protein kinase is induced by gibberellin. Plant Physiol 108:787–793

Albi E, Lazzarini R, Viola Magni M (2008) Phosphatidylcholine/ sphingomyelin metabolism crosstalk inside the nucleus. Biochem J 410:381–389

Andersson MX, Larsson KE, Tjellstrom H, Liljenberg C, Sandelius AS (2005) Phosphate-limited Oat: the plamsma membrane and the tonoplast as major targets for phospholipid-to-glycolipid replacement and stimulation of phospholipases in the plasma membrane. J Biol Chem 280:27578–27586

Apone F, Alyeshmerni N, Wiens K, Chalmers D, Chrispeels MJ, Colucci G (2003) The G-protein-coupled receptor GCR1 regulates DNA synthesis through activation of phosphatidylinositol-specific phospholipase C. Plant Physiol 133:571–579

Blatt MR, Thiel G, Trentham DR (1990) Reversible inactivation of K+ channels of Vicia stomatal guard cells following the photolysis of caged inositol 1, 4, 5-trisphosphate. Nature 346:766–769

Braiman A, Barda-Saad M, Sommers CL, Samelson LE (2006) Recruitment and activation of PLCgamma1 in T cells: a new insight into old domains. EMBO J 25:774–784

Charron D, Pingret JL, Chabaud M, Journet EP, Barker DG (2004) Pharmacological evidence that multiple phospholipid signaling pathways link Rhizobium nodulation factor perception in *Medicago truncatula* root hairs to intracellular responses, including Ca^{2+} spiking and specific ENOD gene expression. Plant Physiol 136:3582–3593

Chen CY, Cheung AY, Wu HM (2003) Actin-depolymerizing factor mediates Rac/Rop GTPase-regulated pollen tube growth. Plant Cell 15:237–249

Coursol S, Pierre JN, Vidal J (2000) Role of the phosphoinositide pathway in the light-dependent C4 phosphoenolpyruvate carboxylase phosphorylation cascade in *Digitaria sanguinalis* protoplasts. Biochem Soc Trans 6:821–823

de Jong CF, Laxalt AM, Bargmann BOR, de Wit PJGM, Joosten MHAJ, Munnik T (2004) Phosphatidic acid accumulation is an early response in the Cf-4/Avr4 interaction. Plant J 39:1–12 den Hartog M, Verhoef N, Munnik T (2003) Nod factor and elicitors activate different phospholipid signaling pathways in suspension-cultured alfalfa cells. Plant Physiol 132:311–317

Deswal R, Chowdhary GK, Sopory SK (2004) Purification and characterization of a PMA-stimulated kinase and identification of PMA-induced phosphorylation of a polypeptide that is dephosphorylated by low temperature in *Brassica juncea*. Biochem Biophys Res Commun 322:420–427

Dowd PE, Coursol S, Skirpan AL, T-h K, Gilroy S (2006) Petunia phospholipase c1 is involved in pollen tube growth. Plant Cell 18:1438–1453

Drøbak BK (1992) The plant phosphoinositide system. Biochem J 288:697–712

Drøbak BK, Watkins PAC (1994) Inositol (1, 4, 5) trisphosphate production in plant cells: stimulation by the venom peptides, Melittin and Mastoparan. Biochem Biophys Res Commun 205:739–745

Engstrom EM, Ehrhardt DW, Mitra RM, Long SR (2002) Pharmacological analysis of nod factor-induced calcium spiking in *Medicago truncatula*. Evidence for the requirement of type IIA calcium pumps and phosphoinositide signaling. Plant Physiol 128:1390–1401

Exton JH (1994) Phosphatidylcholine breakdown and signal transduction. Biochim Biophys Acta 1212:26–42

Fantuzzi L, Spadaro F, Purifcato C, Cecchetti S, Podo F, Belardelli F, Gessani S, Ramoni C (2008) Phosphatidylcholine-specific phospholipase C activation is required for CCR5-dependent, NF-kB-driven CCL2 secretion elicited in response to HIV-1 gp120 in human primary macrophages. Blood 111:3355–3363

Franklin-Tong VE, Drøbak BK, Allan AC, Watkins PAC, Trewavas AJ (1996) Growth of pollen tubes of Papaver rhoeas is regulated by a slow-moving calcium wave propagated by inositol 1, 4, 5-trisphosphate. Plant Cell 8:1305–1321

Gaude N, Nakamura Y, Scheible WR, Ohta H, Dörmann P (2008) Phospholipase C5 (NPC5) is involved in galactolipid accumulation during phosphate limitation in leaves of Arabidopsis. Plant J 56(1):28–39

Gilroy S, Read ND, Trewavas AJ (1990) Elevation of cytoplasmic calcium by caged calcium or caged inositol triphosphate initiates stomatal closure. Nature 346:769–771

Helling D, Possart A, Cottier S, Klahre U, Kost B (2006) Pollen tube tip growth depends on plasma membrane polarization mediated by tobacco PLC3 activity and endocytic membrane recycling. Plant Cell 18:3519–3534

Helsper JPFG, Heemskerk JWM, Veerkamp JH (1987) Cytosolic and particulate phosphotidylinositol phospholipase C activities in pollen tubes of *Lilium longiflorum*. Physiol Plant 71:120–126

Holdaway-Clarke TL, Hepler PK (2003) Control of pollen tube growth: role of ion gradients and fluxes. New Phytol 159:539–563

Hunt L, Mills LN, Pical C, Leckie CP, Aitken FL, Kopka J, Mueller-Roeber B, McAinsh MR, Hetherington AM, Gray JE (2003) Phospholipase C is required for the control of stomatal aperture by ABA. Plant J 34:47–55

Jacob T, Ritchie S, Assmann SM, Gilroy S (1999) Abscisic acid signal transduction in guard cells is mediated by phospholipase D activity. Proc Natl Acad Sci U S A 96:12192–12197

Jones NP, Katan M (2007) Role of phospholipase Cgamma1 in cell spreading requires association with a beta-Pix/GIT1-containing complex, leading to activation of Cdc42 and Rac1. Mol Cell Biol 16:5790–5805

Katan M (2005) New insights into the families of PLC enzymes: looking back and going forward. Biochem J 391:e7–e9

Kates M (1955) Hydrolysis of lecithin by plant plastid enzymes. Can J Biochem Physiol 33:575–589

Kim YJ, Kim JE, Lee JH, Lee MH, Jung HW, Bahk YY, Hwang BK, Hwang I, Kim WT (2004) The Vr-PLC3 gene encodes a putative plasmamembrane-localized phosphoinositide-specific phospholipase C whose expression is induced by abiotic stress in mung bean (*Vigna radiata* L.). FEBS Lett 556:127–136

Kopka J, Pical C, Gray JE, Muller-Rober B (1998) Molecular and enzymatic characterization of three phosphoinositide-specific phospholipase C isoforms from potato. Plant Physiol 116:239–250
Kost B, Lemichez E, Spielhofer P, Hong Y, Tolias K, Carpenter C, Chua N-H (1999) Rac homologues and compartmentalized phosphatidylinositol 4, 5-bisphosphate act in a common pathway to regulate polar pollen tube growth. J Cell Biol 145:317–330
Kovar DR, Drøbak BK, Collings DA, Staiger CJ (2001) The characterization of ligand-specific maize (*Zea mays*) profilin mutants. Biochem J 358:49–59
Krinke O, Novotná Z, Valentová O, Martinec J (2006) Inositol trisphosphate receptor in higher plants: is it real? J Exp Bot 58:361–376
Kusano H, Testerink C, Vermeer JE, Tsuge T, Shimada H, Oka A, Munnik T, Aoyama T (2008) The Arabidopsis phosphatidylinositol phosphate 5-kinase PIP5K3 is a key regulator of root hair tip growth. Plant Cell 2:367–380
Lee Y, Assmann SM (1991) Diacylglycerols induce both ion pumping in patch-clamped guard-cell protoplasts and opening of intact stomata. Proc Natl Acad Sci U S A 88:2127–2131
Lemtiri-Chlieh F, MacRobbie EAC, Webb AAR, Manison NF, Brownlee C, Skepper JN, Chen J, Prestwich GD, Brearley CA (2003) Inositol hexakisphosphate mobilizes an endomembrane store of calcium in guard cells. Proc Natl Acad Sci U S A 100:10091–10095
Malho R (1998) Role of 1, 4, 5-inositol triphosphate-induced Ca2+ release in pollen tube orientation. Sex Plant Reprod 11:231–235
Mills LN, Hunt L, Leckie CP, Aitken FL, Wentworth M, McAinsh MR, Gray JE, Hetherington AM (2004) The effects of manipulating phospholipase C on guard cell ABA-signaling. J Exp Bot 55:199–204
Monteiro D, Liu Q, Lisboa S, Scherer GEF, Quader H, Malho R (2005) Phosphoinositides and phosphatidic acid regulate pollen tube growth and reorientation through modulation of $[Ca^{2+}]c$ and membrane secretion. J Exp Bot 56:1665–1674
Mueller-Roeber B, Pical C (2002) Inositol phospholipid metabolism in Arabidopsis. characterized and putative isoforms of inositol phospholipid kinase and phosphoinositide-specific phospholipase C. Plant Physiol 130:22–46
Munnik T, Musgrave A, de Vrije T (1994) Rapid turnover of polyphosphoinositides in carnation flower petals. Planta 193:89–98
Munnik T, Irvine RF, Musgrave A (1998) Phospholipid signaling in plants. Biochim Biophys Acta 1389:222–272
Nakamura Y, Awai K, Masuda T, Yoshioka Y, K-i T, Ohta H (2005) A novel phosphatidylcholine-hydrolyzing phospholipase C induced by phosphate starvation in Arabidopsis. J Biol Chem 280:7469–7476
Nanmori T, Taguchi W, Kinugasa M, Oji Y, Sahara S, Fukami Y, Kikkawa U (1994) Purification and characterization of protein kinase C from a higher plant, *Brassica campestris* L. Biochem Biophys Res Commun 203:311–318
Nomura H, Komori T, Kobori M, Nakahira Y, Shiina T (2008) Evidence for chloroplast control of external Ca^{2+}-induced cytosolic Ca^{2+} transients and stomatal closure. Plant J 53:988–998
Otterhag L, Sommarin M, Pical C (2001) N-terminal EF-hand-like domain is required for phosphoinositide-specific phospholipase C activity in *Arabidopsis thaliana*. FEBS Lett 497:165–170
Pan YY, Wang X, Ma LG, Sun DY (2005) Characterization of phosphatidylinositol-specific phospholipase C (PI-PLC) from *Lillum daviddi* pollen. Plant Cell Physiol 10:1657–1665
Perera IY, Heilmann I, Chang SC, Boss WF, Kaufman PB (2001) A role for inositol 1, 4, 5-trisphosphate in gravitropic signaling and the retention of cold-perceived gravistimulation of oat shoot pulvini. Plant Physiol 125:1499–1507
Pical C, Kopka J, Mueller-Roeber B, Hetherington AM, Gray JE (1997) Isolation of two cDNA clones for phosphoinositide-specific phospholipase C from epidermal peels (Accession No. X95877) and guard cells (Accession No. Y11931) of *Nicotiana rustica*. Plant Physiol 114:748
Potocký M, Eliás M, Profotová B, Novotná Z, Valentová O, Zárský V (2003) Phosphatidic acid produced by phospholipase D is required for tobacco pollen tube growth. Planta 217:122–130

Scherer GFE, Paul RU, Holk A, Martinec J (2002) Down-regulation by elicitors of phosphatidylcholine-hydrolyzing phospholipase C and up-regulation of phospholipase A in plant cells. Biochem Biophys Res Commun 293:766–770

Shi J, Gonzales RA, Bhattacharyya MK (1995) Characterisation of a plasma membrane associated phosphoinositide-specific phospholipase C from soybean. Plant J 8:381–390

Spadaro F, Cecchetti S, Sanchez M, Ausiello CM, Podo F, Ramoni C (2006) Expression and role of phosphatidylcholine-specific phospholipase C in human NK and T lymphocyte subsets. Eur J Immunol 36:3277–3287

Stenzel I, Ischebeck T, König S, Hołubowska A, Sporysz M, Hause B, Heilmann I (2008) The type B phosphatidylinositol-4-phosphate 5-kinase 3 is essential for root hair formation in *Arabidopsis thaliana*. Plant Cell 1:124–141

Swann K, Saunders CM, Rogers NT, Lai FA (2006) PLCzeta(zeta): a sperm protein that triggers Ca^{2+} oscillations and egg activation in mammals. Semin Cell Dev Biol 17:264–273

Tang RH, Han S, Zheng H, Cook CW, Choi CS, Woerner TE, Jackson RB, Pei ZM (2007) Coupling diurnal cytosolic Ca^{2+} oscillations to the CAS-IP_3 pathway in *Arabidopsis*. Science 315:1423–1426

Testerink C, Munnik T (2005) Phosphatidic acid: a multifunctional stress signaling lipid in plants. Trends Plant Sci 10:368–379

Vainonen JP, Sakuragi Y, Stael S, Tikkanen M, Allahverdiyeva Y, Paakkarinen V, Aro E, Suorsa M, Scheller HV, Vener AV, Aro EM (2008) Light regulation of CaS, a novel phosphoprotein in the thylakoid membrane of *Arabidopsis thaliana*. FEBS J 275:1767–1777

van Leeuwen W, Vermeer JE, Gadella TW Jr, Munnik T (2007) Visualization of phosphatidylinositol 4, 5-bisphosphate in the plasma membrane of suspension-cultured tobacco BY-2 cells and whole Arabidopsis seedlings. Plant J 52:1014–1026

van Rheenen J, Jalink K (2002) Agonist-induced PIP_2 hydrolysis inhibits cortical actin dynamics: regulation at a global but not at a micrometer scale. Mol Biol Cell 13:3257–3267

Wang X (2004) Lipid signaling. Curr Opin Plant Biol 7:329–336

Wu L, Bauer CS, Zhen XG, Xie C, Yang J (2002) Dual regulation of voltage-gated calcium channels by PtdIns(4,5)P_2. Nature 419:947–952

Yoon GM, Dowd PE, Gilroy S, McCubbin AG (2006) Calcium-dependent protein kinase isoforms in petunia have distinct functions in pollen tube growth, including regulating polarity. Plant Cell 18:867–878

Plant Phospholipase D

Wenhua Zhang, Xiaobo Wan, Yueyun Hong, Weiqi Li, and Xuemin Wang

Abstract Phospholipase D (PLD) is involved in different plant processes, ranging from responses to abiotic and biotic stresses to plant development and seed quality. The PLD family consists of multiple members that have distinguishable biochemical and regulatory properties. The differential activation of different PLDs regulates the temporal and spatial production of the lipid messenger, phosphatidic acid (PA), and the selective hydrolysis of membrane lipids. PLD and PA may regulate plant functions through their effects on signal transduction, cytoskeletal reorganization, vesicular trafficking, membrane remodeling, and/or lipid degradation, and the modes of action may differ depending on the specific PLDs and stimulations. The molecular heterogeneity of PLDs plays important roles in their different functions.

1 Introduction

One major advance in modern biochemistry and cell biology in recent years has been the realization that phospholipids are rich sources for generating mediators and are active participants in cell regulation. Activation of phospholipases (PL) is often an early event in the cascade of signal transduction. Phospholipase D (PLD) (EC 3.1.4.4) constitutes a major phospholipase family in plants that hydrolyzes phospholipids to produce phosphatidic acid (PA) and a free head group, such as choline, ethanolamine, or serine. This activity is widespread and readily detectable

W. Zhang (✉)
College of Life Science, State Key Laboratory of Crop Genetics and Germplasm Enhancement, Nanjing Agricultural University, Nanijing 210095, P R, China
e-mail: whzhang@njau.edu.cn

X. Wang (✉)
Department of Biology, University of Missouri, St. Louis and Donald Danforth Plant Science Center, St. Louis, MO 63121, USA
e-mail: wangxue@umsl.edu

T. Munnik (ed.), *Lipid Signaling in Plants*, Plant Cell Monographs 16,
DOI 10.1007/978-3-642-03873-0_3, © Springer-Verlag Berlin Heidelberg 2010

in various plant tissues. PLD was first discovered in plants, about 60 years ago. Early studies have provided valuable information on the biochemical properties and occurrence of the enzyme. For example, the discovery that PLD can transfer the phosphatidyl moiety to a primary alcohol has been useful in current studies concerning the activity of PLD and PA functions in cells.

The application of molecular approaches, albeit later than other families of phospholipases, has propelled the PLD research to a new era. The first cloning of PLD from castor bean was a catalyst for cloning PLDs from different sources, including yeast and mammals (Wang et al. 1994). This was followed by identification and analysis of multiple PLDs, revealing biochemical, molecular, and functional heterogeneity of the PLD family. Over the past 10 years, considerable progress has been made toward understanding the function of PLDs in processes of plant growth, development, and stress response (Bargmann and Munnik 2006; Wang et al. 2006). In particular, the role of PLD in producing PA as a class of lipid messengers has received increasing attention. Recent results have led to the proposition that "PA (is) promoted to manager" in mammalian cells (Hancock 2007). The notion of PA as cellular manager is applicable to the regulation of not only G proteins (Zhao et al. 2007), but also protein phosphorylation, transcription, and translation in glycerolipid biosynthesis, cell proliferation, and growth. PLD and PA are involved in a wide variety of plant processes, ranging from responses to hormones, drought, salinity, cold, freezing, nutrient deficiencies, and pathogens, to root growth, senescence, reproduction, and seed quality. They affect these processes through roles in cell signaling, membrane trafficking, and cytoskeletal rearrangement, as well as lipid remodeling and membrane degradation. The exciting developments in PLD research have been the subject of several recent reviews (Wang 2000, 2005, 2002; Testerink and Munnik 2005; Zhang et al. 2005; Bargmann and Munnik 2006; Wang et al. 2006). The signaling function of PA has also been reviewed recently (Testerink and Munnik 2005; Wang et al. 2006), which revealed that PA is not only generated via PLD but also via diacylglycerol kinase (see also chapters "Diacylglycerol Kinase," "Phosphatidic Acid – An Electrostatic/Hydrogen- Bond Switch?," "Nitric Oxide and Phosphatidic Acid Signaling in Plants"). Here, the recent progress in PLD research in plants is outlined.

2 The Expanding PLD Family and Domain Structures

The PLD family in *Arabidopsis thaliana* is most extensively characterized. Twelve PLD genes have been identified (Elias et al. 2002; Qin and Wang 2002), and subdivided into six classes, three αs, two βs, three γs, one δ, one ε, and two ζs, based on gene architecture, sequence similarity, domain structure, biochemical properties, and the order of cDNA cloning (Wang 2005; Wang et al. 2006). Based on the overall protein domain structure, PLDs have been grouped into C2-PLDs and PX/PH-PLDs (Fig. 1; Wang 2005). Ten of the 12 *Arabidopsis* PLDs have the N-terminal C2 domain, which is a Ca^{2+}-dependent phospholipid binding

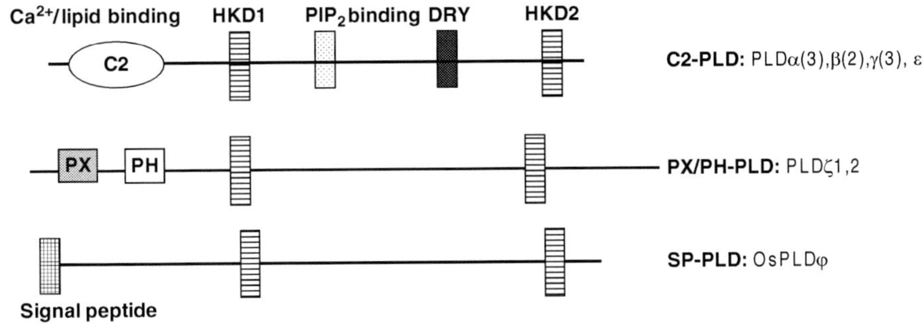

Fig. 1 Main domain structures of the plant PLD family. The common plant PLDs consist of two distinct groups: the C2-PLDs and the PX/PH-PLDs. A putative PLD, OsPLDϕ was found in rice as predicated from genomic sequences. Individual PLDs can differ in key amino acid residues in these regulatory motifs such as C2, PIP_2-binding, and DRY. OsPLDφ has a signal peptide instead of C2 or PX/PH domain at N-terminus. *C2*, Ca^{2+}/phopsholipid binding domain; *PH*, Pleckstrin homology domain; *PX*, phox homology domain. The duplicated HKD motifs are involved in catalysis

structure for activity regulation (Qin and Wang 2002). The Ca^{2+}-binding enhances the affinity of substrates to the enzyme (Wang 2005). Most *PLDs* cloned from other plant species, including castor bean (Wang et al. 1994), rice (Ueki et al. 1995), maize (Ueki et al. 1995), tomato (Laxalt et al. 2001), tobacco (Lein and Saalbach 2001), and strawberry (Yuan et al. 2005), are C2 PLDs. In comparison, two PLDζs have PX (phox homology) and PH (pleckstrin homology) domains at the N-terminus similar to mammalian PLDs (Qin and Wang 2002). All these PLDs contain two HKD (H × K × × × ×D) motifs that are required for PLD catalysis. The two HKD motifs are separated by approximately 320 amino acids in the primary structure, but have been suggested to interact with each other to form an active site in plant PLDs (Fig. 1; Wang 2000).

Genome-wide analysis of rice has identified 17 PLDs: 14 PLDs have the C2 domain, whereas two PLDs have the PX/PH domains (Li and Xue 2007). In addition, the analysis suggests that rice has one unique PLD, PLDφ (termed SP-PLD), which has a signal peptide instead of the C2 or PX/PH domains at the N-terminus (Fig. 1; Li and Xue 2007). Such a SP-PLD-like sequence has not been identified in *Arabidopsis*. The enzymatic identity of SP-PLD as a PLD still needs to be confirmed, and the function of this PLD is also unknown. Recently, a novel PLD, mitochondrial PLD (MitoPLD), was identified in mammals (Choi et al. 2006). Unlike PLD1 and PLD2, MitoPLD has only one single HKD catalytic motif (Wang et al. 2006; Choi et al. 2006). MitoPLD is a divergent and ancestral family member most similar to bacterial cardiolipin synthase. It is associated with the external face of the mitochondria. MitoPLD hydrolyzes cardiolipin to PA and promotes transmitochondrial membrane adherence (Choi et al. 2006).

Several observations can be made with the limited comparison of PLDs from different organisms: (1) The C2-PLDs are the major forms of plant PLDs; they are unique to plants and not found in animals. (2) The two PX/PH-PLDs are evolutionarily conserved in plants and mammals. It would be of interest to determine whether their specific cellular functions are also conserved. (3) In addition to the above two types, plants may have other types of PLDs that may vary from species to species. (4) The PLD family in plants has more members and more diverse domain structures than that in other organisms (Wang et al. 2006). This is in contrast to other types of phospholipases. For instance, mammalian cells have more types of PLC and PLA_2 than plants.

In addition, the number of PLD enzymes is greater than that of the genes identified because alternative splicing variants for PLDδ and PLDγ 2 have been reported (Qin et al. 2006). Some PLDs contain a G protein-interacting, DRY motif and/or a polybasic polyphosphoinositide-binding motif (Zhao and Wang 2004). Moreover, individual PLDs can differ in key amino acid residues in the various domains and motifs. In some cases, these differences have been shown to underlie a structural basis for their different biochemical and regulatory properties (Wang et al. 2006). The presence of these different regulatory motifs that are involved in interactions with Ca^{2+}, phosphoinositides, and G proteins provides insights to the different modes of activation and regulations of PLDs.

3 Different Biochemical Properties and Activity Regulation Among PLDs

One of the most noted biochemical properties associated with plant PLD from early studies was its need of high millimolar concentrations of Ca^{2+} and detergent for activity in vitro. It is now known that in *Arabidopsis* this "traditional PLD activity" comes primarily from PLDα1, and that most of PLDs, such as β, γ, δ, and ζ are not active in the presence of the high levels of Ca^{2+} (Wang et al. 2006; Wang 2004). PLDβ1, PLDγ 1, and PLDγ 2 are most active at micromolar concentrations of Ca^{2+}, whereas ζ1 does not need Ca^{2+} for activity at all. In addition, PLDα1 is active at micromolar concentrations, near-physiological range of Ca^{2+}, when substrates are presented as a mixture of membrane lipids at an acidic pH range (pH 4.5–5.5) (Pappan and Wang 1999). In plants, cytosolic free Ca^{2+} ($[Ca^{2+}]_{cyt}$) oscillates in response to extracellular stimuli and developmental signals. The transient and local $[Ca^{2+}]_{cyt}$ could activate different PLDs at various locations in cells. A positive correlation between increased $[Ca^{2+}]_{cyt}$ and PLD activity was reported when the Ca^{2+} levels in cells were perturbed with Ca^{2+}-ATPase inhibitors and calmodulin antagonists (de Vrije and Munnik 1997).

Recent studies have shed light on the mechanisms of Ca^{2+} regulation of PLD activity. Using recombinant proteins expressed in *E. coli*, putative Ca^{2+}-binding C2 domains have been characterized in PLDα and PLDβ, with three putative Ca^{2+}-binding sites in PLDβ (Zheng et al. 2000). The Ca^{2+} binding induces conformational changes of the C2 domain and promotes binding of the C2 domain to PC (Zheng et al. 2000). The PLD C2 domains also bind PIP_2 (Fig. 1), and the C2–PIP_2 interaction is inhibited by submillimolar levels of Ca^{2+} (Zheng et al. 2002).

The reverse relationship between Ca^{2+} and PIP_2 binding of the C2 domain appears to be in conflict with the previous findings that PLDβ needs both Ca^{2+} and PIP_2 for its activation (Pappan et al. 1998). To explore further the mechanisms for the regulation of PLDβ by the cofactors, a C2 domain-deleted PLDβ1 (PLDβcat) was constructed, and the deletion of C2 domain decreases, but does not abolish Ca^{2+} binding to the enzyme when compared with the full-length PLDβ1 (Pappan et al. 2004). The result indicates the presence of other Ca^{2+} binding site(s) in the catalytic region of the enzyme, probably between the two HKD motifs. The Ca^{2+}–PLDβcat interaction increases its affinity to its activator PIP_2, but not to its substrate PC (Pappan et al. 2004). These findings are in direct contrast to those that found Ca^{2+} binding to the C2 domain leads to the stimulation of PC binding and the reduction of PIP_2 binding in the full PLDβ (Zheng et al. 2000). A membrane-scooting model has been proposed to explain the regulation of PLDβ activity by changing cellular Ca^{2+} concentrations (Fig. 2; Pappan et al. 2004).

Another key regulator of PLD activity is PIP_2, but the requirements for PIP_2 vary among different PLDs (Pappan et al. 1997; Qin et al. 1997; Pappan et al. 1998; Zheng et al. 2000; Wang and Wang 2001; Zheng et al. 2002; Pappan et al. 2004; Qin et al. 2002a; Qin et al. 2006). PIP_2 is required for the activity of PLDβ, γ, and ζ; however, PLDα1 and δ are quite active without PIP_2. Instead, PLDδ is stimulated by oleic acid at a broad range of Ca^{2+} concentrations (μM–mM) (Wang and Wang 2001). PLDs have at least two types of PIP_2-binding sites. The study of PLDβ shows that PIP_2 binds directly to the C2 and catalytic domains and induces a

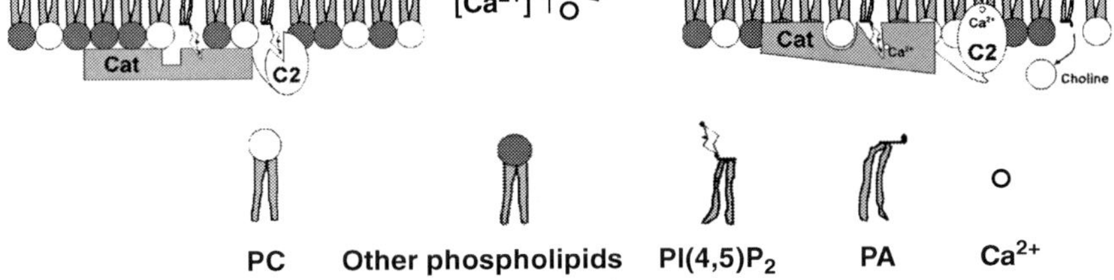

Fig. 2 Membrane-scooting model depicting the regulation of PLDβ1 activity by changing cellular Ca^{2+} concentrations and PIP_2 binding (adapted from Pappan et al. 2004). At the resting, low $[Ca^{2+}]_{cyt}$, the C2 domain interact with anionic lipids, including scare PIP_2. When cells are stimulated, $[Ca^{2+}]_{cyt}$ increase, the Ca^{2+}-C2 binding induces a conformational change that alters the relative affinity of the C2 domain in favor of PC binding over PIP_2. The C2 domain releases PIP_2 and bonds to PC. Meanwhile, Ca^{2+} binds to the catalytic region and increases the affinity of this region to a now more abundant PIP_2. The PIP_2 binding induces a conformational change in the catalytic region, which increases PLD activity by increasing its affinity to its substrate, PC

conformational change in the catalytic region, which increases PLD activity by increasing its affinity for its substrate, PC (Fig. 2; Pappan et al. 2004). It should be noted that the catalytic requirement for PIP_2 cannot be applied to all PLDs, particularly PLDαs (Wang 2000; Qin et al. 1997).

In addition, a recent study reveals that the two highly similar PLDγ 1 and γ 2 may differ from the effect that PIP_2 and detergent Triton X-100 have on their activities (Qin et al. 2006). While PLDγ 2 and γ 1 both require PIP_2 for activity, Triton X-100 totally abolishes PLDγ 2 activity and prohibits any stimulation effect from PIP_2. The detergent greatly activates PLDγ 1 activity and serves only as a neutral diluent in the substrate vesicles. PLDγ 2 misses one of the basic PIP_2-interacting residues, which may weaken the binding of PIP_2 to PLDγ 2 (Qin et al. 2006). These differences in biochemical properties suggest that the highly homologous PLDγ s are subjected to unique regulations and may have distinguishable functions.

4 PLD Interaction with and Regulation by Proteins

In addition to Ca^{2+}, PIP_2, and free fatty acids, the activity of PLDs is modulated by their interaction with different proteins. PLD in plants has been found to interact with heterotrimeric G-proteins (Lein and Saalbach 2001; Zhao and Wang 2004; Mishra et al. 2006), tubulin (Gardiner et al. 2001), actin (Kusner et al. 2003), protease, and potentially protein kinases and phosphatases (Novotna et al. 2003). Several studies on PLD/G–protein interaction have provided valuable clues to the function of the interaction. Using potential G-protein activators and inhibitors, such as toxins, nonhydrolyzable guanine nucleotide analogs, and alcohols, the role of G-proteins in regulating plant PLD has been suggested (Munnik et al. 1995; Ritchie and Gilroy 2000). In barley aleurone cells, a PLDα-like activity has been suggested to be associated with a G-protein on the plasma membrane to mediate ABA signaling (Ritchie and Gilroy 2000). Coincubation of a recombinant tobacco PLDα with Gα decreased the PLD activity (Lein and Saalbach 2001). Direct interaction between PLDα1 and α subunit of heterotrimeric G-proteins (GPA1) has been characterized using protein coprecipitation expressed in either *E. coli* or *Arabidopsis*. PLDα1 has a Gα binding motif (EKF) that is analogous to the DRY motif present in G-protein coupled receptors (GPCRs) (Zhao and Wang 2004).

The identification of PLDα1 interaction with Gα raises intriguing questions regarding PLD regulation by potentially G-protein-coupled receptors. In mammalian cells, PLD is regulated by various G-proteins and also by a large number of cell surface receptors and GPCRs (Oude Weernink et al. 2007). For instance, muscarimic acetylcholine receptors (mAChRs) in neuroblastoma cells activate Rho-kinase, which phosphorylates LIM-kinase (LIM, Lin-11/Isl-1/Mec-3). LIM-kinase-phosphorylated cofilin, the pivotal actin regulatory protein, directly and specifically interacts with PLD1 and stimulates its activity (Han et al. 2007 and references therein). In oat seedlings, PLD activity is regulated by light.

PLD activity was found to be higher in etiolated oat seedlings than in green seedlings. White and red light inhibits PLD activity in etiolated seedlings, while far-red light eliminates the red-light-induced decrease in PLD activity. These results imply a possible involvement of phytochrome photoreceptor in regulating PLD activity (Kabachevskaya et al. 2007).

In addition, many negative regulators of PLD activity have been identified in mammals, including fodrin (Lukowski et al. 1996), α-actinin (Park et al. 2000), actin (Lee et al. 2001), tubullin (Chae et al. 2005), gelsolin (Steed et al. 1996), amphiphysin (Lee et al. 2000), aldolase (Kim et al. 2002), α-/β-synuclein (John et al. 1998), synaptojanin (Chung et al. 1997), clathrin assembly protein 3 (Lee et al. 1997), and collapsin response mediator protein-2 (Lee et al. 2002). In plants, much less is known about negative regulation of PLD activity.

5 Subcellular Distribution and Expression Patterns

PLDα1 occurs in both soluble and membrane fractions, and the relative distribution changes in response to stress and growth stages (Fan et al. 1999). Both PLDα1 and PLDγ can be detected in the fractions of the plasma membrane, intracellular membranes, clathrin-coated vesicles, and mitochondria. Most of membrane-associated PLDα1 is cofractionated with the plasma membrane, whereas most PLDγ is associated with intracellular membranes (Fan et al. 1999). PLDδ is associated exclusively with the plasma membrane (Wang and Wang 2001) and proposed to be the PLD that binds the microtubule cytoskeleton (Gardiner et al. 2001). PLDα3 is predominantly localized at the plasma membrane (Hong et al. 2008a). The different distributions of PLDs are expected to play an important role in the spatial regulation of membrane lipid hydrolysis and PA production in specific subcellular compartments.

Individual PLDs exhibit different and overlapping patterns of expression based on the results from northern blotting (Fan et al. 1999), real-time PCR (Li et al. 2006a), and Genevestigator (www.genevestigator.ethz.ch) (Fig. 3). *PLDα1* is most abundant in plant tissues, *PLDδ* the second, and *PLDγ* , *PLDβ*, and *PLDζ* are of moderate level, whereas levels for *PLDε*, *PLDα2*, and *PLDα3* are rather low (Fig. 3a). For *PLDα1*, the level of expression is high in all tissues except pollen. In contrast, *PLDδ*, *PLDα2*, *PLDβ1*, *PLDβ2*, and *PLDε* show a high level of expression in pollen relative to other tissues (Fig. 3a). The function implication of the differential expression of different PLDs in sexual reproduction has yet to be studied.

Under stress conditions, PLDs show different expression patterns (Fig. 3b). Osmotic stress induces the expression of *PLDα1* and δ, as well as *PLDζs*. High salinity, which causes hyperosmotic and hyperionic stresses, increases the expression of *PLDζs*, *PLDγ1*, *γ3*, and reduces *PLDγ2*, but does not affect *PLDα1* expression (Fig. 3b). *PLDα1* expression is induced by cold treatment, which is consistent with its activation by cold stress and involvement in cold tolerance

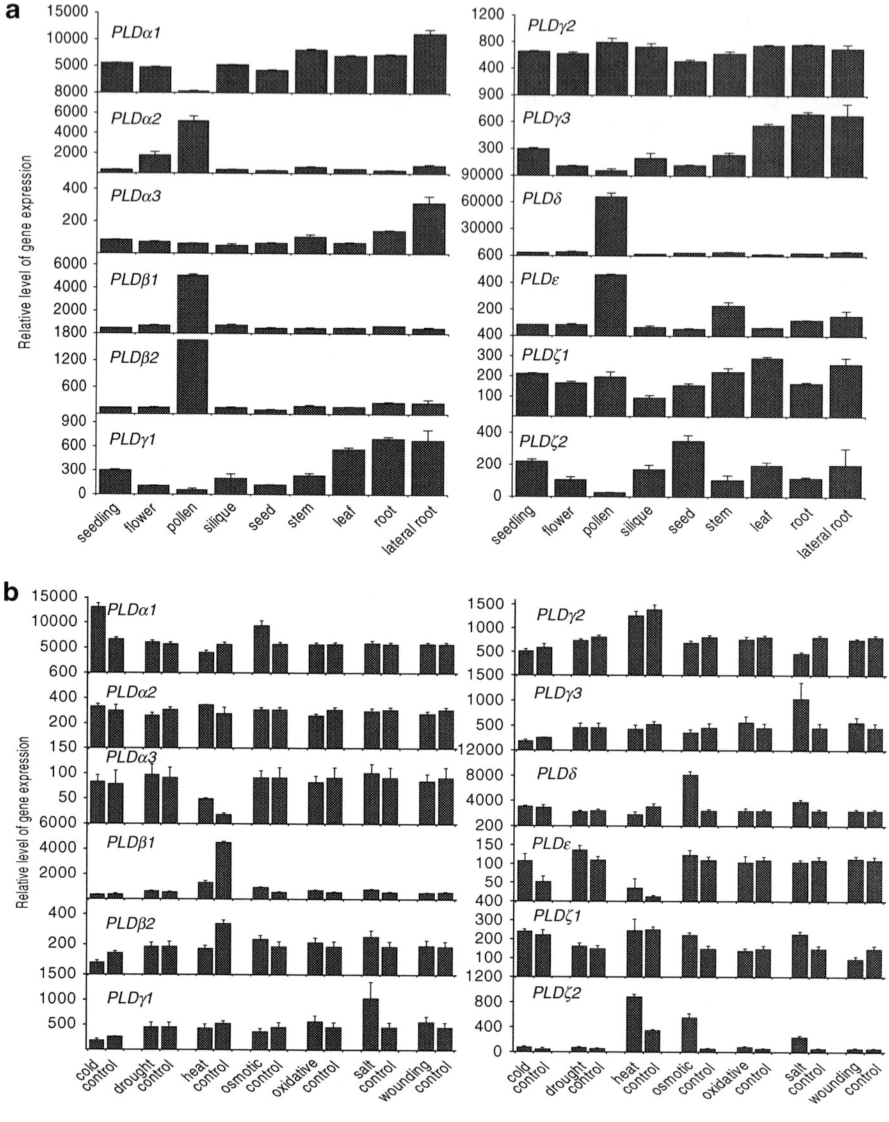

Fig. 3 Relative level of PLD gene expression. (**a**) PLD gene expression in plant tissues. (**b**) PLD gene expression under stressed conditions. Data are from GENEVESTIGATOR (https://www.genevestigator.ethz.ch)

(Welti et al. 2002). In contrast, *PLD*δ expression is not affected by oxidative stress (Fig. 3b), but its activity is stimulated by H_2O_2 and is important to oxidative tolerance (Zhang et al. 2003). Therefore, the induction of PLD expression can sometimes be uncoupled with their functions under specific stress. Posttranscriptional and posttranslational regulation of PLDs, such as the interaction with different effectors as described above and phosphorylation (Novotna et al. 2003), play important roles in regulation of PLD activation and functions.

6 Cellular and Physiological Functions

The multiple regulatory domains and molecular interactions indicate that PLD activity is strictly controlled in cells (Wang et al. 2006). The differences of PLDs in substrate preferences, subcellular distribution, and the spatial and temporal gene expression, suggest that different PLDs play diverse roles in response to stimuli. PLD and its product PA have been implicated in multiple plant growth and developmental processes, such as seed germination, seedling growth, pollen tube germination and elongation, and leaf senescence (Munnik 2001; Wang 2002, 2005; Testerink and Munnik 2005; Bargmann and Munnik 2006; Wang et al. 2006). Increasing reports indicate that PLDs and PA play pivotal roles in signaling plant responses to various biotic and abiotic cues, such as drought, freezing, oxidative assault, wounding, pathogen, and the stress hormones abscisic acid and jasmonic acid (Wang et al. 2000; Zhang et al. 2003, 2005; Li et al. 2004). The following discussion will be focused on the recent results on the role of PLDs in plant response to hormones, abiotic stresses, and plant–microbial interactions.

6.1 PLD and PA in ABA Effects on Stomatal Movements and Seed Germination

The two major effects of ABA on plants include (1) the decrease of water loss by promoting stomatal closure, and (2) the regulation of seed dormancy and germination. PLD is involved in both of these ABA-mediated physiological processes. Stomata close during drought stress. This change is crucial for maintaining hydration status in leaves and for plant survival. In *Vicia faba* guard cell protoplasts, application of 1-butanol, which diverts part of PLD-produced PA to phosphatidylbutanol, results in the impediment of the ABA responses, whereas, the addition of PA partially mimics the effect of ABA (Jacob et al. 1999). In *Arabidopsis*, antisense suppression or gene knockout of *PLDα1* inhibits stomatal closure induced by ABA, resulting in a higher rate of transpirational water loss than in wild type plants (Sang et al. 2001; Zhang et al. 2004). These data suggest that abundant PLD, *PLDα1,* is involved in ABA responses.

Further studies reveal a connection between PLD and PA with G proteins and protein phosphatase 2Cs in regulating the ABA response in stomata. *Arabidopsis* has a single canonical Gα gene, *GPA1*, which encodes the Gα subunit of heterotrimeric G protein (Wang et al. 2001). *GPA1*-null and *PLDα1*-null *Arabidopsis* plants share some similar phenotypes in ABA and water-deficit responses, such as retardation of stomatal movement and increasing water loss (Sang et al. 2001; Wang et al. 2001), implying that PLDα1 and Gα may interact in ABA signal pathway. Molecular and biochemical evidence shows that PLDα1 binds to Gα through a DRY motif, and mutations in the DRY motif disrupt the Gα–PLDα1 binding (Zhao and Wang 2004). PLDα1 binds to Gα in a GDP-bound state but not in a GTP-bound state. The PLDα1–Gα interaction decreases PLDα activity. This activity inhibition is reduced with the activation of Gα1 by GTP addition (Zhao and Wang 2004). The decrease in the inhibition is correlated with the dissociation of the PLDα1–Gα complex. These results have led to the hypothesis that PLDα1 is activated in vivo with the activation of G-protein by environmental stimuli via a receptor (Perfus-Barbeoch et al. 2004). However, the binding of PLDα1 to Gα stimulates the intrinsic GTPase activity that converts active $G\alpha_{\text{-GTP}}$ to inactive $G\alpha_{\text{-GDP}}$ (Zhao and Wang 2004; Mishra et al. 2006). These results suggest that the interaction reciprocally modulates the activities of PLDα1 and Gα.

Protein phosphatases are important mediators in ABA signaling. PLDα1-derived PA has been found to bind to ABI1, a protein phosphatase 2C (PP2C) that is a negative regulator in ABA response (Himmelbach et al. 2002). Site-specific mutational analyses have identified that arginine 73 in ABI1 is essential for the PA–ABI1 binding (Zhang et al. 2004). The PA–ABI1 binding leads to the inhibition of PP2C activity and the tethering to plasma membrane. The membrane tethering prevents the translocation of ABI1 from cytosol to the nucleus, where it binds to and activates *ATHB6*, a transcriptional factor that negatively regulates ABA responses (Himmelbach et al. 2002; Zhang et al. 2004).

To further investigate the functional significance of PA–ABI1 interaction in plants, wild type *ABI1* ($ABI1_{WT}$) and mutant *ABI1* ($ABI1_{R73A}$) were introduced into an ABI1–null (*abi1–KO*) mutant *Arabidopsis*. $ABI1_{WT}$ and *abi1–KO* plants show a normal response to ABA, as do wild type. However, *abi1–KO* plants carrying the $ABI1_{R73A}$ transgene are insensitive to the promotion of stomatal closure, but display normal sensitivity to the ABA inhibition of stomatal opening (Mishra et al. 2006). These lines of genetic and physiological evidence indicate that PA binding to ABI1 is required for ABA promotion of stomatal closure but not for ABA inhibition of stomatal opening. In contrast, the disruption of the PLDα1–GPA1 interaction using transgenic methods renders plants hypersensitive to ABA in inhibiting stomatal opening because both the GPA1 and PLDα1 functions are less inhibited. PLDα1 and GPA1 are positive regulators in ABA inhibition of stomatal opening (Mishra et al. 2006; Coursol et al. 2003). So the disruption of the PLDα1–GPA1 interaction releases and activates GPA1, which inhibits the K^+ inward channel, thereby promoting ABA inhibition of stomatal opening (Coursol et al. 2003; Pei et al. 2000), but it does not affect the ABA-induced stomatal closure (Mishra et al. 2006). Therefore, PLDα1 regulates both ABA-mediated stomatal closing and opening through bifurcating interactions with ABI1 and Gα (Fig. 4; Mishra et al. 2006).

ABA is an important regulator of seed dormancy and germination, and PLD activation in the aleurone is essential for this process. An earlier report showed that ABA activated PLD activity transiently in barley aleurone protoplasts . The application of PA to the protoplasts induced an ABA-like inhibition of α-amylase

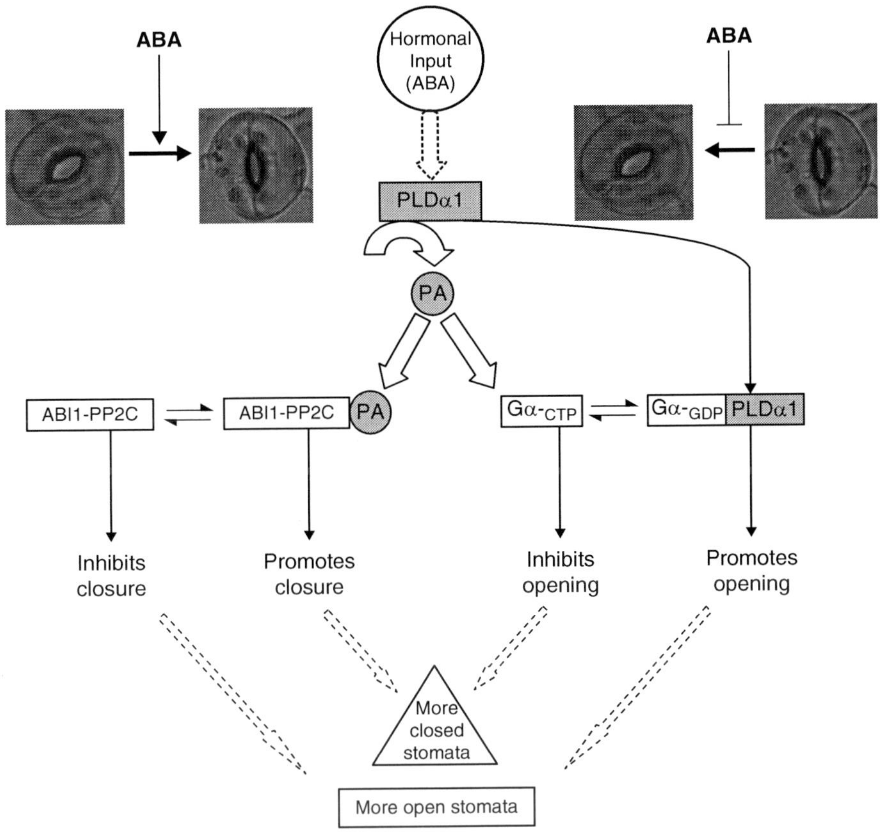

Fig. 4 A model depicting PLD-mediated bifurcating pathways directing ABA effects on stomatal closure and opening in *Arabidopsis* (data from Mishra et al. 2006). On ABA-induced stomatal closure, PLDα1-derived PA binds to ABI1, promoting ABA effect on stomatal closure. On ABA inhibition of stomatal opening, PLDα1 binds to GPA1 and regulates its function by promoting the conversion of GTP-bound Gα ($G\alpha_{-GTP}$) to GDP-bound Gα ($G\alpha_{-GDP}$) and by producing PA that acts upstream of Gα to inhibit stomatal opening. Note that this model is not comprehensive and only includes some of the signaling components implicated in the ABA signaling cascades. *Arrows with solid lines* indicate established links and *arrows with dashed lines* denote putative links

production, which is activated by another plant hormone, gibberellic acid (GA) (Ritchie and Gilroy 1998). One probable reason for α-amylase inhibition may be associated with the induction of proteins ASI (amylase subtilisin inhibitor) and RAB (responsive to ABA) by PA (Ritchie and Gilroy 1998). Because the synthesis and secretion of this hydrolytic enzyme are needed to mobilize endosperm reserves for germination, ABA activation of PLD is an important step for its regulation of seed dormancy.

A recent study indicates that PLDβ1 is a component in ABA-regulated seed germination in cereal plant rice (Li and Xue 2007). Transgenic studies have shown that the suppressed expression of rice *PLDβ1* results in reduced sensitivity to exogenous ABA during seed germination. One possible mechanism of the PLDβ1 action is that the PLD-produced PA activates the expression of *SAPK8* and *SAPK10* (genes of osmotic stress/ABA–activated protein kinases) (Kobayashi et al. 2004). PA reduces the expression of *GAmyb,* which encodes a protein that binds to the GA response element of α-amylase promoters, directly or through SAPK functions to inhibit seed germination (Li and Xue 2007).

LPP (lipid phosphate phosphatase) catalyzes the sequential conversion of diacylglycerol pyrophosphate (DGPP; see chapters, "Phosphatidic Acid Phosphatases in Seed Plants," "Diacylglycerol Pyrophosphate, a Novel Plant Signaling Lipid") to PA, or PA to DAG, regulating seed germination. T-DNA insertional mutants (*lpp2-1* and *lpp2-2*) show an increase in PA levels and are hypersensitive to ABA during germination. Double mutant analyses with transcription regulators, ABI3 and ABI4, suggest that AtLPP2 acts as a negative regulator upstream of ABI4 in ABA signaling during germination (Katagiri et al. 2005). The results suggest that PA plays a signaling role in ABA-regulated seed germination and stomatal movement.

6.2 PLD in Auxin Transport and Growth Responses

Auxin is involved in many developmental processes including root elongation, gravity response, shoot architecture, and vein pattern formation (Rahman et al. 2007; Zhang et al. 2007; Sieburth et al. 2006). The deficiency of *PLDζ2* gene in *Arabidopsis* leads to decreased sensitivity to auxin, reduced root gravitropism, suppressed auxin-dependent hypocotyl elongation at 29°C, and suppressed expression of auxin-responsive genes (Li and Xue 2007). It has been well documented that one important characteristic of auxin is the effect it has on polar transport from the site of synthesis to the site of action through influx and efflux carrier proteins. It is also known that the influx of auxin is mediated by the influx carrier AUX1, a presumptive auxin permease exhibiting basal plasma membrane localization (Swarup et al 2001). However, auxin efflux is mediated by several efflux carriers, of which PIN1 is essential for IAA acropetal transport in root tissues (Galweiler et al. 1998), and that PIN-FORMED2 (PIN2) is involved in auxin redistribution in root gravitropism (Muller et al. 1998; Shin et al. 2005; Abas et al. 2006).

Early work showed that PLD and PA were involved in vesicle trafficking (reviewed in Wang 2005). By using a genetic approach, both PLDζ1 and PLDζ2 have been implicated in regulating the vesicle trafficking (Li and Xue 2007; Li et al. 1996a). PLDζ2 and PA regulate positively the vesicle trafficking and PIN cycling (Li and Xue 2007). Because the cycling of PIN proteins between membrane and endosomal compartments is important to auxin transport and is dependent on vesicle trafficking, a positive effect of PLDζ2 on auxin responses may be derived from its regulation of the vesicle trafficking (Li and Xue 2007). However, how PLD and PA are involved in vesicular trafficking and auxin responses requires further investigation.

6.3 PLD in Ethylene and Cytokinin Response

Ethylene is a hormone that is essential for plant growth and development, especially for senescence, along with responses to biotic and abiotic stresses. Several important components of ethylene signal transduction cascades have been identified. ETR1 (Ethylene Resistant 1) encodes a histidine kinase-like receptor protein that mediates ethylene perception. The binding of ethylene to ETR1 causes the Raf-like kinase protein CTR1 (Constitutive Triple Response 1) to be inactive, releasing the repression of EIN2 (transmembrane protein). This leads to the upregulation of expression of downstream genes by the transcription factor, EIN3 (Taiz and Zeiger 2002). PLDα is an important mediator in ethylene- and ABA-promoted senescence in detached leaves (Fan et al. 1997). Ethylene-treated leaves showed a stimulated PLDα expression, while PLDα-deficient transgenic plants showed a slower rate of senescence than WT plants (Fan et al. 1997). The recent identification of CTR1 as a PA target connects the relation between PLD and ethylene-induced senescence. Biochemical evidence shows that PA binds CTR1 through a PA-binding region in the kinase domain, leading to the inhibition of its kinase activity. In addition, the binding blocks the interaction between CTR1 and ETR1 (Testerink et al. 2007).

Cytokinins are plant hormones that have an opposite effect on ethylene in plant senescence. Treatment of plants with 1-butanol partially blocks cytokinin-induced GUS activity and reduces the accumulation of *ARR5* gene transcripts, indicating a role of PLD in cytokinin signaling (Romanov et al. 2002). The above results suggest that PLD-mediated senescence is complex. It will be interesting to test whether, and/or how, PLDs regulate plant senescence by mediating ethylene and cytokinin signal cascades *in planta*.

6.4 PLD in Plant–Pathogen Interactions

When plants are challenged with pathogens, various intracellular signaling systems are activated. These include an increase in cytosolic Ca^{2+}, protein phosphorylation, depolarization of the plasma membrane, and generation of reactive oxygen species (ROS) (Nurnberger and Scheel 2001). Earlier studies in rice indicated that PLD is involved in the interaction between plant and bacterial pathogen (Young et al. 1996). Infiltration of *Arabidopsis* with *P. syringae*, with or without AvrRpm1 protein, induces transcriptional activation of *PLDα*, β, γ (de Torres Zabela et al. 2002). In addition, the induced expression of bacterial (*P. syringae*) Avr protein, AvrRpm, caused biphasic accumulations of PA (Andersson et al. 2006). The first wave of PA accumulation was correlated with PLC-mediated PIP (phosphatidylinositol phosphate) degradation with subsequent phosphorylation of DAG by DAGK. The second wave of PA was due to activation of a PLD (Andersson et al. 2006). When tobacco cells, expressing the tomato Cf-4, extracellular-LLR disease-resistance protein, were treated with the corresponding avirulent *Cladosporum fulvum* protein Avr4, PA levels dramatically increased, which in this case was suggested to be mainly derived via DAG kinase (DGK; see chapters, "Diacylglycerol Kinase," "Nitric Oxide and Phosphatidic Acid Signaling in Plants") and PLC-mediated PIP turnover (de Jong et al. 2004).

In bacterial Avr-R interaction, the accumulation of PA induces the expression of the defense gene (*PR1*) and a hypersensitive response. Ca^{2+} influx is required for the activation of PLD, but not for PLC-mediated PIP degradation (Andersson et al. 2006). Inclusion of NADPH oxidase inhibitor diphenyleneiodonium (DPI) has little effect on the Avr4- or AvrRpm1- dependent PIP degradation and PA accumulation (de Jong et al. 2004; Andersson et al. 2006), suggesting that phospholipase signaling pathways act upstream or independent of ROS.

To characterize which PLD is involved in pathogen tolerance, a gene-specific RNAi construct was used to knock down *LePLDβ1* in suspension-cultured tomato cells (Bargmann and Munnik 2006). This *PLD* was selected because it was induced by the fungal elicitor xylanase (Laxalt et al. 2001). Suppression of *LePLDβ1* resulted in a strong decrease in xylanase-induced PLD activity, and a disproportionately oxidative burst (Bargmann and Munnik 2006). In addition, imaging of LePLDβ1–GFP indicates that LePLDβ1–GFP is translocated from cytosol to the punctated structures close to the plasma membrane after xylanase treatment. In contrast, the LePLDα1–GFP remains cytosolic after the xylanase treatment (Bargmann and Munnik 2006). It would be of interest to determine how the activation and relocalization of *LePLDβ1* would affect plant response to pathogens.

6.5 Different Roles of PLDs in Freezing Tolerance

Changes in the PLD activity in plant response to low temperatures were reported decades ago (Yoshida and Sakai 1974). Recent studies have provided information about different functions of individual PLDs in plants coping with cold temperatures (Welti et al. 2002; Li et al. 2004; Wang et al. 2006). PLDα1 plays a negative role in freezing tolerance. Suppression of PLDα1 renders *Arabidopsis* more tolerance than wild type plants (Welti et al. 2002; Rajashekar et al. 2006). The freezing tolerance behavior of PLDα1-deficient plants was similar in both high-survival rate and low-membrane damage. Compared to normal growth conditions, cold acclimation does not induce changes in PA levels, whereas freezing induces a 5-fold increase in PA levels (Welti et al. 2002). The 50% increase in PA during freezing is derived mostly from PLDα1 hydrolysis of PC. Antisense suppression of *PLDα*1 causes a substantial decrease in freezing-induced production of PA (Welti et al. 2002).

During postfreezing recovery, the changes in membrane lipids, especially the individual molecular species of PA, are distinctly different from that during freezing (Li et al. 2008). The level of 34:4 PA during postfreezing recovery is significantly higher than that during freezing, while other PA species remain unchanged and, in some cases, are decreased. 34:4 PA is likely produced by PLDα1-mediated hydrolysis of plastidic lipid 34:4 PG (Li et al. 2008), and the plastidic PG may become accessible to PLDα1 due to damages of plastidic membranes, such as decompartmentalization by freezing. Thus, PLDα1 may be involved in the degradation of plastidic lipids during postfreezing recovery.

Under subzero temperatures, the formation of a nonbilayer phase, hexagonal II phase, is believed to be the main cause of membrane injuries. PC is a bilayer-stabilizing lipid, whereas PA associated with Ca^{2+} favors the formation of hexagonal II phase. Compared to wild type plants, PA level is lower and PC level is higher in *PLDα1*-deficient plants; thus, the propensity of hexagonal II phase is lower, which could result in less damage to membranes and make plants more tolerant to freezing (Welti et al. 2002). A decrease in the degradation of plastidic PG to PA in *PLDα1*-deficient plants may help the recovery of chloroplast membrane from freezing injury (Li et al. 2008).

In contrast, PLDδ shows a positive role in freezing tolerance. *PLDδ*-knockout *Arabidopsis* is more sensitive to freezing, while *PLDδ*-overexpressing *Arabidopsis* is more tolerant to freezing than wild type (Li et al. 2004). Two observations suggest that the function of PLDδ in freezing is complex. One observation is that the alteration of freezing tolerance in *PLDδ*-altered plants is dependent on cold acclimation. Without cold acclimation, the freezing tolerance in the PLDδ-altered plants is the same as that of wild type plants. A second observation is that freezing-induced ion leakage in the transgenic plants is at the same level as that in wild type plants (Li et al. 2004). These results suggest that the function of PLDδ in freezing stress is associated with other factors required for cold acclimation and that suppression of PLDδ does not alter membrane damage during freezing. Comparative profiling of lipids reveals that PLDδ has only subtle effects on membrane lipid composition. When freezing occurs, the amount of PA in *PLDδ*-knockout and overexpressing plants is about 80% and 125%, respectively, of wild type (Li et al. 2004). During postfreezing recovery, however, the loss of PLDδ was associated with an increase in PA production. The increased amount of PA formed includes the 34:4 species, which is likely to originate from plastidic 34:4 PG (Li et al. 2008).

The positive role of PLDδ in freezing tolerance is postulated to result from its interaction with microtubule cytoskeleton (Li et al. 2008). Knockout and overexpression of *PLDδ* could restrict and enhance, respectively, the capacity of microtubules to reorganize the organelles recovering from the freezing, thus altering the freezing sensitivity (Li et al. 2004). The sensitivity to oxidative stress (Zhang et al. 2003; Li et al. 2004) during postfreezing recovery in *PLDδ*-knockout *Arabidopsis* could make the plants sensitive to freezing.

The opposite roles of PLDα1 and PLDδ in freezing response in *Arabidopsis* could reflect their distinctively different biochemical properties, subcellular associations, and gene expression patterns (Qin and Wang 2002). PLDδ alterations do not affect the expression of CBF-regulated genes (Li et al. 2004; Jaglo-Ottosen et al. 1998), nor do cold-induced increases in the levels of compatible osmolytes, proline, or soluble sugars, which are known to play a role in plant freezing tolerance. These results suggest that PLDs and associated membrane lipid hydrolysis are important components mediating plant response to freezing stress.

6.6 PLDs in Plant Response to Phosphorus Deficiency

Root elongation and root hair formation are important in nutrient absorption. Phosphorus is an essential macronutrient for plant growth, development, and reproduction. It plays an important role in regulation of various enzymes acting as a constituent of components, such as membrane phospholipids and nucleic acids (Schachtman et al. 1998). Under phosphate limited conditions, plants hydrolyze phospholipids, resulting in an increase in internal phosphate availability. In order to maintain the function and structure of membrane systems, the hydrolyzed phospholipids are replaced by nonphosphorus lipids, such as galactolipids and sulfolipids.

PLDζ1 and *ζ2* are involved in plant response to phosphorus deficiency that play metabolic and signaling roles, depending on the severity of Pi deficiency. Under moderate phosphorus limitation, *PLDζ1* and *PLDζ2* double knockout mutants display shorter primary roots than wild type, indicating that PLDζs promote primary root growth (Li et al. 2006a). During severe phosphorus starvation, these PLDs hydrolyze phospholipids, particularly PC to supply phosphorus and diacylglycerol moieties for galactolipid synthesis (Li et al. 2006b). The expression of *PLDζ2* increases greatly during Pi starvation in *Arabidopsis* (Cruz-Ramirez et al. 2006; Li et al. 2006b). In *pldζ2* plants, phospholipid hydrolysis as well as their capacity to accumulate galactolipids are diminished when Pi starvation occurs (Cruz-Ramirez et al. 2006; Li et al. 2006b). The accumulation of digalactosyldiacylglycerol (DGDG) in the roots of Pi-limited *pldζ2* plants is reduced while PC and PE accumulate in Pi-starved *pldζ2* roots when compared with roots in wild type (Cruz-Ramirez et al. 2006; Li et al. 2006b). Therefore, PLDζ2 is involved in hydrolyzing PC and PE to produce DAG for DGDG synthesis that allows Pi to maintain other biochemical and physiological processes requiring phosphate (Cruz-Ramirez et al. 2006; Li et al. 2006b). Taken together, these results indicate that *PLDζ1* and *PLDζ2* play a role in regulating root development in response to nutrient limitation.

6.7 PLDs in Plant Response to Salt and Hyperosmotic Stress

High salinity causes both ionic and hyperosmotic stress (Hasegawa et al. 2000). NaCl rapidly activates PLD activity and PA accumulation in alfalfa, tomato, and *Chlamydomonas moewusii* cells (Munnik et al. 2000). Some of the PA is converted to DGPP through a PA kinase. Besides NaCl, other osmotica like KCl, glycerol, mannitol, and sucrose, with concentrations equivalent to NaCl, showed similar activation of PLD and increases in PA (Munnik et al. 2000). These results suggested that hyperosmotic stress plays a major role in the activation of PLD.

The expression of *AtPLDδ* is induced rapidly by NaCl and dehydration treatment, but not by ABA or cold treatment (Katagiri et al 2001). However, when *PLDα* is constitutively expressed during these stress treatments, *PLDβ* and *PLDγ* are not significantly induced (Katagiri et al. 2001). These results suggest that *AtPLDδ* is a PLD gene that is upregulated by hyperosmotic stress, including salt stress. The changes in PLD activity, expression, and PA formation under these conditions suggest a role for PLD in response to salinity and other hyperosmotic stresses.

A recent study shows that manipulation of PLDα3 alters plant response to hyperosmotic conditions (Hong et al. 2008a). *pldα3-1* seeds are more susceptible to salt stress, as indicated by delayed germination, lower germination rate, retarded seedlings and root growth. In contrast, *PLDα3*-OE seeds displayed more resistance to salt with enhanced germination rates and seedling growth. In addition, the alterations of PLDα3 also change the plant's response to water deficiency, indicating that PLDα3 plays a role in hyperosmotic stress, rather than specifically to salt. Furthermore, the results indicate that PLDα3 plays a role in modulating plant growth and development under hyperosmotic stresses and suggests a possible connection between PLD-based signaling and some of the key regulators, such as ABA, flower-timing regulators, and TOR (target of rapamycin)-signaling, in flowering promotion and the hyperosmotic response (Hong et al. 2008a). Further studies on the potential interactions of PLD with the regulators will help to better understand the mechanisms by which plants respond to salinity and water deficiency.

Nitric oxide (NO), an endogenous signaling molecule in animals and plants, mediates responses to abiotic and biotic stresses (see also chapter, "Nitric Oxide and Phosphatidic Acid Signaling in Plants"). Like PLD, NO is involved in salt stress response in plants (Zhang et al. 2006). In maize leaves, NO acts as a signal molecule in increasing the activities of vacuolar H^+-ATPase and H^+-PPase, which provide the driving force for Na^+/H^+ exchange under NaCl treatment. A treatment of 1-butanol, which diverts part of PA to phosphatidylbutanol, impairs the NO-mediated H^+-pump activation (Zhang et al. 2006). The result could mean that PA production is involved in mediating the NO effect on NaCl response. In *V. faba* guard cells, PA levels increase transiently with a NO treatment. This increase is generated via PLD or PLC (Distefano et al. 2008; see chapter, "Nitric Oxide and Phosphatidic Acid Signaling in Plants"). These results imply that NO could trigger PLD activation and PA production. However, PA can regulate NO production by increasing the inducible nitric oxide synthase (iNOS) in animal cells (Lim et al. 2003). Whether PLD/PA regulates NO production is yet to be investigated in plants (see also chapter, "Nitric Oxide and Phosphatidic Acid Signaling in Plants").

7 Multiple Functions of PLD and Implications in Biotechnology

Genetic manipulation of *PLDα1*, *PLDα3*, *PLDβ*, *PLDδ*, and *PLDζs* have resulted in alterations in cellular and physiological processes in response to different stimuli, including ABA, auxin, ROS, freezing, P_i deficiency, and elicitors (Zhang et al. 2003; Li et al. 2004; Bargmann and Munnik 2006; Cruz-Ramirez et al. 2006; Li et al. 2006a, b; Mishra et al. 2006). The identification of molecular targets of

PLD and PA further support a notion that PLD and PA are involved in signal transduction processes (Wang 2000; Munnik 2001; Testerink and Munnik 2005; Bargmann and Munnik 2006; Wang et al. 2006). However, as a major family of membrane lipid-hydrolyzing enzymes, PLD activity has been implicated in membrane degradation. PLDs may have dual functions at different stages in response to stimuli. Unbridled membrane hydrolysis would be detrimental to cell integrity. For example, overexpression of *PLDα1* in tobacco results in dual changes in water loss (Hong et al. 2008b). In the early stage of drought, increased *PLDα1* expression reduces water loss that is associated with a rapid stomatal closure (Sang et al. 2001; Zhang et al. 2004). However, with prolonged drought stress, the PLD-overexpressing plants show more damage, including increasing membrane lipid oxidation and ionic leakage (Hong et al. 2008b).

High levels of PLDα1 in seeds promote seed aging and deterioration (Devaiah et al. 2007). The formation of PA by PLD is thought to initiate a chain reaction to disrupting cell membranes and storage lipids in seeds (List et al. 1992). The ablation of PLDα1 abolishes aging-induced increase in PA and decreases the levels of lipid peroxidation, which may underlie the basis for increased resistance to aging in the PLDα1-deficient seeds. The PLDα1-mediated hydrolysis of membrane lipids could destabilize oil bodies that consist of triacylglycerol, coated by a phospholipid monolayer (Devaiah et al. 2007). In addition, the potential activation of NADPH oxidase by PA derived from PLDα1 (Sang et al. 2001) could produce ROS, which accelerates the lipid peroxidation and membrane damage (Devaiah et al. 2007).

By contrast, ablation of PLDδ leads to less tolerance to aging, which implies that PLDδ plays a positive role in ameliorating the effects of seed aging (Devaiah et al. 2007). This may be related to its effect on mitigating ROS damages (Zhang et al. 2003). Wild type cells showed more tolerance to ROS than PLDδ-deficient cells (Zhang et al. 2003). Furthermore, overexpression of *PLDδ* makes *Arabidopsis* plants tolerant to freezing, partially because overexpressed plants have a higher ability to tolerate ROS induced by freezing (Li et al. 2004). As stated above, knocking down of *PLDα*1 results in the increase in freezing tolerance in *Arabidopsis* (Welti et al. 2002).

In plants, there is a delicate balance between signaling and catabolic function of PLDs, which is important to plant growth, development, and response to internal and environmental stimuli (Hong et al. 2008b). Temporal and spatial control of the expression of different PLDs in plants can be explored for improving plant growth, performance, and product quality.

8 Perspectives

Great progress has been made in past years towards understanding the biochemistry, molecular biology, and metabolic and physiological functions of PLD. Recent advances from the PLD studies have provided valuable insights to the understanding of lipid-mediated signaling in regulating plant growth and stress responses.

However, the current understanding of PLDs and lipid regulation of plant functions is still limited. The PLD family in plants is complex, and the cellular and physiological functions for most PLDs are yet to be uncovered. The potential overlapping effects of multiple PLDs on plant morphology makes the functional study rather challenging and requires detailed analysis of metabolic alterations and/or identification of growth conditions. Using mutant plants deficient in more than one PLD will help to understand the functional studies. In addition, little is known about the upstream regulatory components that connect signal perceptions and the activation of different PLDs. Addressing these questions will shed light on the mechanism of PLD activation and also the specific process in which specific PLDs are involved. The recent identification of specific targets of PLD and/or PA in the cell is an exciting development, and many PLD and PA targets are yet to be identified. It is possible that PLD and PA, together with other membrane lipid-based signaling, provide important linkages between cellular regulators, such as phytohormones, oxidative stress, G proteins, protein phosphatase, and kinases in plant growth and stress responses. Current knowledge from *Arabidopsis* PLDs indicates that the molecular heterogeneity of PLDs plays important roles in their different functions. Future studies of different PLDs in crop plants may provide opportunities to integrate the basic knowledge in lipid-mediated signaling with potential applications to crop production and quality.

Acknowledgments The work was supported by the National Science Foundation of China Research Grants (30625027, 90817014), the Chinese National Key Basic Research Project (2006CB100100), and grants from the Ministry of Education of China (20060307019, 111 Project) to W Zhang. Much of the work in the XW laboratory was supported by grants from U.S. Department of Agriculture (2005-35818-15253; 2007-35318-18397) and the National Science Foundation (IOS-0454866; IOS-0818740). We thank Marti Shafer for critically reading the manuscript.

References

Abas L, Benjamins R, Malenica N, Paciorek T, Wisniewska J, Jeanette C, Moulinier-Anzola JC, Sieberer T, Friml J, Luschnig C (2006) Intracellular trafficking and proteolysis of the *Arabidopsis* auxin-efflux facilitator PIN2 are involved in root gravitropism. Nat Cell Biol 8:249–256

Andersson MX, Kourtchenko O, Dangl JL, Mackey D, Ellerstrom M (2006) Phospholipase-dependent signaling during the AvrRpm1- and AvrRpt2-induced disease resistance responses in *Arabidopsis thaliana*. Plant J 47:947–959

Bargmann BO, Munnik T (2006) The role of phospholipase D in plant stress responses. Curr Opin Plant Biol 9:515–522

Chae YC, Lee S, Lee HY, Heo K, Kim JH, Kim JH, Suh PG, Ryu SH (2005) Inhibition of muscarinic receptor-linked phospholipase D activation by association with tubulin. J Biol Chem 280:3723–3730

Choi SY, Huang P, Jenkins GM, Chan DC, Schiller J, Frohman MA (2006) A common lipid links Mfn-mediated mitochondrial fusion and SNARE-regulated exocytosis. Nat Cell Biol 8:1255–1262

Chung JK, Sekiya F, Kang HS, Lee C, Han JS, Kim SR, Bae YS, Morris AJ, Rhee SG (1997) Synaptojanin inhibition of phospholipase D activity by hydrolysis of phosphatidylinositol 4, 5-bisphosphate. J Biol Chem 272:15980–15985

Coursol S, Fan L-M, Stuff HL, Spiegel S, Gilroy A, Assmann SM (2003) Sphingolipid signaling in *Arabidopsis* guard cells involves heterotrimeric G proteins. Nature 423:651–654

Cruz-Ramirez A, Oropeza-Aburto A, Razo-Hernandez F, Ramirez-Chavez E, Herrera-Estrella L (2006) Phospholipase DZ2 plays an important role in extraplastidic galactolipid biosynthesis and phosphate recycling in *Arabidopsis* roots. Proc Natl Acad Sci U S A 103:6765–6770

de Jong CF, Laxalt AM, Bargmann BO, de Wit PJ, Joosten MH, Munnik T (2004) Phosphatidic acid accumulation is an early response in the Cf-4/Avr4 interaction. Plant J 39:1–12

de Torres Zabela M, Fernandez-Delmond I, Niittyla T, Sanchez P, Grant M (2002) Differential expression of genes encoding *Arabidopsis* phospholipases after challenge with virulent or avirulent *Pseudomonas* isolates. Mol Plant Microbe Interact 15:808–816

de Vrije T, Munnik T (1997) Activation of phospholipase D by calmodulin antagonists and mastoparan in carnation petals. J Exp Bot 48:1631–1637

Devaiah SP, Pan X, Hong Y, Roth M, Welti R, Wang X (2007) Enhancing seed quality and viability by suppressing phospholipase D in *Arabidopsis*. Plant J 50:950–957

Distefano AM, Garcia-Mata C, Lamattina L, Laxalt AM (2008) Nitric oxide-induced phosphatidic acid accumulation: a role for phospholipases C and D in stomatal closure. Plant Cell Environ 31:187–194

Elias M, Potocky M, Cvrckova F, Zarsky V (2002) Molecular diversity of phospholipase D in angiosperms. BMC Genom 3:2

Fan L, Zheng S, Wang X (1997) Antisense suppression of phospholipase Da retards abscisic acid- and ethylene-promoted senescence of postharvest. Plant Cell 9:2183–2196

Fan L, Zheng S, Cui D, Wang X (1999) Subcellular distribution and tissue expression of phospholipase Dα, β, and γ in Arabidopsis. Plant Physiol 119:1371–1378

Galweiler L, Guan C, Muller A, Wisman E, Mendgen K, Yephremov A, Palme K (1998) Regulation of polar auxin transport by AtPIN1 in *Arabidopsis* vascular tissue. Science 282:2226–2230

Gardiner JC, Harper JDI, Weerakoon ND, Collings DA, Ritchie S, Gilroy S, Cyr RJ, Marc J (2001) A 90-kD phospholipase D from tobacco binds to microtubules and the plasma membrane. Plant Cell 13:2143–2158

Han L, Stope MS, de Jesús ML, Weernink PAO, Urban M, Wieland T, Rosskopf D, Mizuno K, Jakobs KH, Schmidt M (2007) Direct stimulation of receptor-controlled phospholipase D1 by phospho-cofilin. EMBO J 26:4189–4202

Hancock JF (2007) PA promoted to manager. Nat Cell Biol 9:615–617

Hasegawa PM, Bressan RA, Zhu JK, Bohnert HJ (2000) Plant cellular and molecular responses to high salinity. Annu Rev Plant Mol Biol 51:463–499

Himmelbach A, Hoffmann T, Leube M, Hohener B, Grill E (2002) Homeodomain protein ATHB6 is a target of the protein phosphatase ABI1 and regulates hormone responses in *Arabidopsis*. EMBO J 21:3029–3038

Hong YY, Pan XQ, Welti R, Wang XM (2008a) Phospholipase Dα3 is Involved in the Hyper-osmotic Response in *Arabidopsis*. Plant Cell 20:803–816

Hong Y, Zheng S, Wang X (2008b) Dual functions of phospholipase Dα1 in plant response to drought. Mol Plant 1:262–269

Jacob T, Ritchie S, Assmann SM, Gilroy S (1999) Abscisic acid signal transduction in guard cells is mediated by phospholipase D activity. Proc Nat Acad Sci USA 96:12192–12197

Jaglo-Ottosen KR, Gilmour SJ, Zarka DG, Schabengerger O, Thomashow MF (1998) *Arabidopsis* CBF1 overexpression induces COR genes and enhances freezing tolerance. Science 280:104–106

John M, Jenco AR, Daniels B, Morris AJ (1998) Regulation of phospholipase D2: selective inhibition of mammalian phospholipase D isoenzymes by α- and β-synucleins. Biochemistry 37:4901–4909

Kabachevskaya AM, Liakhnovich GV, Kisel MA, Volotovski ID (2007) Red/far-red light modulates phospholipase D activity in oat seedlings: relation of enzyme photosensitivity to photosynthesis. J Plant Physiol 164:108–110
Katagiri T, Takahashi S, Shinozaki K (2001) Involvement of a novel *Arabidopsis* phospholipase D, AtPLDdelta, in dehydration-inducible accumulation of phosphatidic acid in stress signaling. Plant J 26:595–605
Katagiri T, Ishiyama K, Kato T, Tabata S, Kobayashi M, Shinozaki K (2005) An important role of phosphatidic acid in ABA signaling during germination in *Arabidopsis thaliana*. Plant J 43:107–117
Kim JH, Lee S, Kim JH, Lee TG, Hirata M, Suh PG, Ryu SH (2002) Phospholipase D2 directly interacts with aldolase via its PH domain. Biochemistry 41:3414–3421
Kobayashi Y, Yamamoto S, Minami H, Kagaya Y, Hattori T (2004) Differential activation of the rice sucrose nonfermenting1–related protein kinase2 family by hyperosmotic stress and abscisic acid. Plant Cell 16:1163–1177
Kusner DJ, Barton JA, Qin C, Wang X, Lyer SS (2003) Evolutionary conservation of physical and functional interactions between phospholipase D and actin. Arch Biochem Biophys 412:231–241
Laxalt AM, ter Riet B, Verdonk JC, Parigi L, Tameling WI, Vossen J, Haring M, Musgrave A, Munnik T (2001) Characterization of five tomato phospholipase D cDNAs: rapid and specific expression of LePLDbeta1 on elicitation with xylanase. Plant J 26:237–247
Lee C, Kang HS, Chung JK, Sekiya F, Kim JR, Han JS, Kim SR, Bae YS, Morris AJ, Rhee SG (1997) Inhibition of phospholipase D by clathrin assembly protein 3 (AP3). J Biol Chem 272:15986–15992
Lee C, Kim SR, Chung JK, Frohman MA, Kilimann MW, Rhee SG (2000) Inhibition of phospholipase D by amphiphysins. J Biol Chem 275:18751–18758
Lee S, Park JB, Kim JH, Kim Y, Kim JH, Shin KJ, Lee JS, Ha SH, Suh PG, Ryu SH (2001) Actin directly interacts with phospholipase D, inhibiting its activity. J Biol Chem 276:28252–28260
Lee S, Kim JH, Lee CS, Kim JH, Kim Y, Heo K, Ihara Y, Goshima Y, Suh PG, Ryu SH (2002) Collapsin response mediator protein-2 inhibits neuronal phospholipase D2 activity by direct interaction. J Biol Chem 277:6542–6549
Lein W, Saalbach G (2001) Cloning and direct G-protein regulation of phospholipase D from tobacco. Biochim Biophys Acta 1530:172–183
Li G, Xue HW (2007) *Arabidopsis* PLDζ2 regulates vesicle trafficking and is required for auxin response. Plant Cell 19:281–295
Li W, Li M, Zhang W, Welti R, Wang X (2004) The plasma membrane-bound phospholipase Dδ enhances freezing tolerance in *Arabidopsis thaliana*. Nat Biotechnol 22:427–433
Li M, Qin C, Welti R, Wang X (2006a) Double knockouts of phospholipases Dζ1 andζ2 in *Arabidopsis* affect root elongation during phosphate-limited growth but do not affect root hair patterning. Plant Physiol 140:761–770
Li M, Welti R, Wang X (2006b) Quantitative profiling of *Arabidopsis* polar glycerolipids in response to phosphorus starvation. Roles of phospholipases D ζ1 and Dζ2 in phosphatidylcholine hydrolysis and digalactosyldiacylglycerol accumulation in phosphorus-starved plants. Plant Physiol 142:750–761
Li W, Wang R, Li M, Li L, Wang C, Welti R, Wang X (2008) Differential degradation of extraplastidic and plastidic lipids during freezing and post-freezing recovery in *Arabidopsis thaliana*. J Biol Chem 283:461–468
Lim HK, Choi YA, Park W, Lee T, Ryu SH, Kim SY, Kim JR, Kim JH, Baek SH (2003) Phosphatidic acid regulates systemic inflammatory responses by modulating the Akt-mammalian target of rapamycin-p70 S6 kinase 1 pathway. J Biol Chem 278:45117–45127
List GR, Mounts TL, Lanser AC (1992) Factors promoting the formation of nonhydratable soybean phosphatides. J Am Oil Chem Soc 69:443–446
Lukowski S, Lecomte MC, Mira JP, Marin P, Gautero H, Russo-Marie F, Geny B (1996) Inhibition of phospholipase D activity by fodrin. J Biol Chem 271:24164–24171

Mishra G, Zhang W, Deng F, Zhao J, Wang X (2006) A bifurcating pathway directs abscisic acid effects on stomatal closure and opening in *Arabidopsis*. Science 312:264–266

Muller A, Guan C, Galweiler L, Tanzler P, Huijser P, Marchant A, Parry G, Bennett M, Wisman E, Palme K (1998) AtPIN2 defines a locus of *Arabidopsis* for root gravitropism control. EMBO J 17:6903–6911

Munnik T (2001) Phosphatidic acid: an emerging plant lipid second messenger. Trends Plant Sci 6:227–233

Munnik T, Arisz SA, De Vrije T, Musgrave A (1995) G Protein activation stimulates phospholipase D signaling in plants. Plant Cell 7:2197–2210

Munnik T, Meijer HJG, Bt R, Hirt HH, Frank W, Bartels D, Musgrave A (2000) Hyperosmotic stress stimulates phospholipase D activity and elevates the levels of phosphatidic acid and diacylglycerol pyrophosphate. Plant J 22:147–154

Novotna Z, Linek J, Hynek R, Martinec J, Potocky M, Valentova O (2003) Plant PIP_2-dependent phospholipase D activity is regulated by phosphorylation. FEBS Lett 554:50–54

Nurnberger T, Scheel D (2001) Signal transmission in the plant immune response. Trends Plant Sci 6:372–379

Oude Weernink PA, Han L, Jakobs KH, Schmidt M (2007) Dynamic phospholipid signaling by G protein-coupled receptors. Biochim Biophys Act 1768:888–900

Pappan K, Wang X (1999) Plant phospholipase Dα is an acidic phospholipase active at near-physiological Ca^{2+} concentrations. Arch Biochem Biophys 368:347–353

Pappan K, Zheng S, Wang X (1997) Identification and characterization of a novel plant phospholipase D that requires polyphosphoinositides and submicromolar calcium for activity in *Arabidopsis*. J Biol Chem 272:7048–7054

Pappan K, Austin-Brown S, Chapman KD, Wang X (1998) Substrate selectivities and lipid modulation of phospholipase Dα, β and γ from plants. Arch Biochem Biophys 353: 131–140

Pappan K, Zheng L, Krishnamoorthi R, Wang X (2004) Evidence for and characterization of Ca^2+ binding to the catalytic region of *Arabidopsis thaliana* phospholipase Dβ. J Biol Chem 279:47833–47839

Park JB, Kim JH, Kim Y, Ha SH, Kim JH, Yoo JS, Du G, Frohman MA, Suh PG, Ryu SH (2000) Cardiac phospholipase D2 localizes to sarcolemmal membranes and is inhibited by α-actinin in an ADP-ribosylation factor-reversible manner. J Biol Chem 275:21295–21301

Pei ZM, Murata Y, Benning G, Thomine S, Klusener B, Allen GJ, Grill E, Schroeder JI (2000) Calcium channels activated by hydrogen peroxide mediate abscisic acid signaling in guard cells. Nature 406:731–734

Perfus-Barbeoch L, Jones AM, Assmann SM (2004) Plant heterotrimeric G protein function: insights from *Arabidopsis* and rice mutants. Curr Opin Plant Biol 7:719–731

Qin C, Wang X (2002) The *Arabidopsis* phospholipase D family. Characterization of a calcium-independent and phosphatidylcholine-selective PLDζ 1 with distinct regulatory domains. Plant Physiol 128:1057–1068

Qin W, Pappan K, Wang X (1997) Molecular heterogeneity of phospholipase D (PLD): Cloning of PLDγ and regulation of plant PLDα, and β by polyphosphoinositides and calcium. J Biol Chem 272:28267–28273

Qin C, Li M, Qin W, Bahn SC, Wang C, Wang X (2006) Expression and characterization of Arabidopsis phospholipase Dγ 2. Biochim Biophys Acta 1761:1450–1458

Rahman A, Bannigan A, Sulaman W, Pechter P, Blancaflor EB, Baskin TI (2007) Auxin, actin and growth of the *Arabidopsis thaliana* primary root. Plant J 50:514–528

Rajashekar CB, Zhou HE, Zhang Y, Li W, Wang X (2006) Suppression of phospholipase Dalpha1 induces freezing tolerance in *Arabidopsis*: response of cold-responsive genes and osmolyte accumulation. J Plant Physiol 163:916–926

Ritchie S, Gilroy S (1998) Abscisic acid signal transduction in the barley aleurone is mediated by phospholipase D activity. Proc Nat Acad Sci USA 95:2697–2702

Ritchie S, Gilroy S (2000) Abscisic acid stimulation of phospholipase D in the barley aleurone ils G-protein-mediated and localized to the plasma membrane. Plant Physiol 124:693–702

Romanov GA, Kieber JJ, Schmulling T (2002) A rapid cytokinin response assay in *Arabidopsis* indicates a role for phospholipase D in cytokinin signaling. FEBS Lett 515:39–43

Sang Y, Zheng S, Li W, Huang B, Wang X (2001) Regulation of plant water loss by manipulating the expression of phospholipase Dα. Plant J 28:135–144

Schachtman D, Reid RJ, Ayling SM (1998) Phosphorus uptake by plants: from soil to cell. Plant Physiol 116:447–453

Shin H, Shin HS, Guo Z, Blancaflor EB, Masson PH, Chen R (2005) Complex regulation of *Arabidopsis* AGR1/PIN2-mediated root gravitropic response and basipetal auxin transport by cantharidin-sensitive protein phosphatases. Plant J 42:188–200

Sieburth LE, Muday GK, King EJ, Benton G, Kim S, Metcalf KE, Meyers L, Seamen E, Van Norman JM (2006) SCARFACE encodes an ARF-GAP that is required for normal auxin efflux and vein patterning in *Arabidopsis*. Plant Cell 18:1396–1411

Steed PM, Nagar S, Wennogle LP (1996) Phospholipase D regulation by a physical interaction with the actin-binding protein gelsolin. Biochemistry 35:5229–5237

Swarup R, Friml J, Marchant A, Ljung K, Sandberg G, Palme K, Bennett MJ (2001) Localization of the auxin permease AUX1 suggests two functionally distinct hormone transport pathways operate in the *Arabidopsis* root apex. Genes Dev 15:2648–2653

Taiz L, Zeiger E (2002) Plant physiology, 3rd edn. Sinauer, MA, USA, pp 532–538

Testerink C, Munnik T (2005) Phosphatidic acid: a multifunctional stress signaling lipid in plants. Trends Plant Sci 10:368–375

Testerink C, Larsen PB, van der Does D, van Himbergen JA, Munnik T (2007) Phosphatidic acid binds to and inhibits the activity of *Arabidopsis* CTR1. J Exp Bot 58:3905–3914

Ueki J, Morioka S, Komari T, Kumashiro T (1995) Purification and characterization of phospholipase D (PLD) from rice (*Oryza sativa* L.) and cloning of cDNA for PLD from rice and maize (*Zea mays* L.). Plant Cell Physiol 36:903–914

Wang X (2000) Multiple forms of phospholipase D in plants: the gene family, catalytic and regulatory properties, and cellular functions. Prog Lipid Res 9:109–149

Wang X (2002) Phospholipase D in hormonal and stress signaling. Curr Opin Plant Biol 5:408–414

Wang X (2004) Lipid signaling. Curr Opin Plant Biol 7:329–336

Wang X (2005) Regulatory functions of phospholipase D and phosphatidic acid in plant growth, development, and stress responses. Plant Physiol 139:566–573

Wang C, Wang X (2001) A novel phospholipase D of *Arabidopsis* that is activated by oleic acid and associated with the plasma membrane. Plant Physiol 127:1102–1112

Wang X, Xu L, Zheng L (1994) Cloning and expression of phosphatidylcholine-hydrolyzing phospholipase D from *Ricinus communis* L. J Biol Chem 269:20312–20317

Wang C, Zien CA, Afitlhile M, Welti R, Hildebrand DF, Wang X (2000) Involvement of phospholipase D in wound-induced accumulation of jasmonic acid in *Arabidopsis*. Plant Cell 12:2237–2246

Wang XQ, Ullah H, Jones AM, Assmann SM (2001) G protein regulation of ion channels and abscisic signaling in *Arabidopsis* guard cells. Science 292:2070–2072

Wang X, Devaiah SP, Zhang W, Welti R (2006) Signaling functions of phosphatidic acid. Prog Lipid Res 45:250–278

Welti R, Li W, Li M, Sang Y, Biesiada H, Zhou HE, Rajashekar CB, Williams TD, Wang X (2002) Profiling membrane lipids in plant stress responses. Role of phospholipase Dα in freezing-induced lipid changes in *Arabidopsis*. J Biol Chem 277:31994–32002

Yoshida S, Sakai A (1974) Phospholipid degradation in frozen plant cells associated with freezing injury. Plant Physiol 53:509–511

Young SA, Wang X, Leach JE (1996) Changes in the plasma membrane distribution of rice phospholipase D during resistant interactions with *Xanthomonas oryzae* pv *oryzae*. Plant Cell 8:1079–1090

Yuan H, Chen L, Paliyath G, Sullivan A, Murr DP (2005) Characterization of microsomal and mitochondrial phospholipase D activities and cloning of a phospholipase Dα cDNA from strawberry fruits. Plant Physiol Biochem 43:535–547

Zhang W, Wang C, Qin C, Wood T, Olafsdottir G, Welti R, Wang X (2003) The oleate-stimulated phospholipase D, PLDδ, and phosphatidic acid decrease H_2O_2-induced cell death in *Arabidopsis*. Plant Cell 15:2285–2295

Zhang W, Qin C, Zhao J, Wang X (2004) Phospholipase Dα1-derived phosphatidic acid interacts with ABI1 phosphatase 2C and regulates abscisic acid signaling. Proc Nat Acad Sci USA 101:9508–9513

Zhang W, Yu L, Zhang Y, Wang X (2005) Phospholipase D in the signaling networks of plant response to abscisic acid and reactive oxygen species. Biochim Biophys Acta 1736:1–9

Zhang Y, Wang L, Liu Y, Zhang Q, Wei Q, Zhang W (2006) Nitric oxide enhances salt tolerance in maize seedlings through increasing activities of proton-pump and Na^+/H^+ antiport in the tonoplast. Planta 224:545–555

Zhang ZB, Yang G, Arana F, Chen Z, Li Y, Xia HJ (2007) *Arabidopsis* inositol polyphosphate 6-/3-kinase (AtIpk2β) is involved in axillary shoot branching via auxin signaling. Plant Physiol 144:942–951

Zhao J, Wang X (2004) *Arabidopsis* phospholipase Dalpha1 interacts with the heterotrimeric G-protein alpha-subunit through a motif analogous to the DRY motif in G-protein-coupled receptors. J Biol Chem 279:1794–1800

Zhao C, Du G, Skowronek K, Frohman MA, Bar-Sagi D (2007) Phospholipase D2-generated phosphatidic acid couples EGFR stimulation to Ras activation by Sos. Nat Cell Biol 9:706–712

Zheng L, Krishnamoorthi R, Zolkiewski M, Wang X (2000) Distinct Ca^{2+} binding properties of novel C_2 domains of plant phospholipase Dα and β. J Biol Chem 275:19700–19706

Zheng L, Shan J, Krishnamoorthi R, Wang X (2002) Activation of plant phospholipase Dβ by phosphatidylinositol 4, 5-bisphosphate: characterization of binding site and mode of action. Biochemistry 41:4546–4553

Part II
Kinases

Phosphatidylinositol 4-Phosphate is Required for Tip Growth in *Arabidopsis thaliana*

Amy L. Szumlanski and Erik Nielsen

Abstract Polyphosphoinositides (PPIs) are important signaling molecules for a variety of cellular processes including intracellular membrane trafficking. In plants, recent studies have led to an increased understanding of the role of PPIs in a number of cellular processes. Here, we discuss data concerning phosphatidylinositol 4-phosphate (PtdIns4P) function in tip growing cells. Plants lacking members of phosphatidylinositol 4-OH kinase (PI4K) or PtdIns4P phosphatase families display defects in growth and morphology of root hairs and pollen tubes. Imaging of PtdIns4P localization in root hairs and pollen tubes revealed that PtdIns4P is present on internal compartments and is enriched at the plasma membrane at the tips of growing cells. These data indicate that regulation of PtdIns4P formation and turnover is important for tip growth. Once viewed as an intermediate for synthesis of other PPIs and as a substrate for production of signaling molecules, PtdIns4P is emerging as an important PPI in its own right.

1 Introduction

1.1 PPIs in Membrane Trafficking

One important aspect of cell growth is the mechanism by which various intracellular compartments maintain separate identities while simultaneously exchanging membranes. Additionally, cellular signaling cascades often initiate within cellular membranes but propagate through aqueous luminal or cytosolic compartments. One class of molecules that has important roles in these processes is the polyphosphoinositides (PPIs). Their relative low abundance, rapid turnover, and variety of

E. Nielsen (✉)
Department of Molecular, Cellular, and Developmental Biology, University of Michigan, 830 North University Avenue, Ann Arbor, MI 48109, USA
e-mail: nielsene@umich.edu

T. Munnik (ed.), *Lipid Signaling in Plants*, Plant Cell Monographs 16,
DOI 10.1007/978-3-642-03873-0_4, © Springer-Verlag Berlin Heidelberg 2010

stereoisomers make these lipids versatile and dynamic signaling molecules (for reviews see Corvera et al. 1999; Stevenson et al. 2000; Meijer and Munnik 2003; De Matteis and Godi 2004; Thole and Nielsen 2008; Munnik and Testerink 2009).

As lipids, PPIs are embedded in cellular membranes and can serve as a means to mark or differentiate intracellular membranes or membrane domains. Addition of phosphate to sites on the inositol head group allows for creation of seven possible PPIs, five of which have been detected in plants (Munnik and Testerink 2009). Recognition of PPIs by lipid binding domains can lead to the selective recruitment of subsets of cytosolic proteins to a particular membrane domain. Additionally, PPIs can be cleaved by a variety of phosphatases and phospholipases, leading to generation of secondary messengers.

1.2 PtdIns4P Metabolism

The most abundant PPI in plants is phosphatidylinositol 4-phosphate (PtdIns4P), accounting for approximately 80% of total PPIs (Munnik et al. 1994; Meijer et al. 2001; Meijer and Munnik 2003). By contrast, in animal cells, the level of PtdIns4P is often less than or equal to the amount of phosphatidylinositol 4,5-bisphosphate (PtdIns(4,5)P_2; Adel-Latif et al. 1985; Cunningham et al. 1995). PtdIns4P is generated by phosphorylation of phosphatidylinositol (PI) on the D-4 position of the inositol head group by phosphatidylinositol 4-OH kinases (PI4Ks). PtdIns4P can be converted back to PI by PtdIns4P phosphatases. In yeast and animal cells, PtdIns4P can also be generated through the partial dephosphorylation of PtdIns(4,5)P_2, although whether this occurs in plants remains unknown (Balla and Balla 2006).

Initially, the role of PtdIns4P in plants was viewed to be similar to the well-characterized function of PtdIns4P in animal systems, serving mainly as a precursor to PtdIns(4,5)P_2, which is a substrate for phospholipase C (PLC). In animals, PLC-based turnover plays a central role in phospholipid signaling by generating the secondary messengers diacylglyerol (DAG) and inositol 1,4,5-triphosphate (InsP_3). However, recent examination in both plants and animal systems indicates that PtdIns4P has regulatory functions independent of PtdIns(4,5)P_2 and PLC (see D'Angelo et al. 2008; Thole and Nielsen 2008; Vermeer et al. 2009). The relatively high abundance of PtdIns4P in plant cells as compared to PtdIns(4,5)P_2, and studies of enzymes responsible for synthesis and turnover of PtdIns4P support the idea that PtdIns4P is an important signaling molecule in its own right during growth and development in plants (Vermeer et al. 2009).

2 The PI4K Family in *A. thaliana*

Examination of the *Arabidopsis thaliana* genome reveals twelve predicted PI4K family members (Fig. 1; Stevenson et al. 2000; Müeller-Roeber and Pical 2002). PI4Ks are characterized by a predicted catalytic domain, and are divided into three

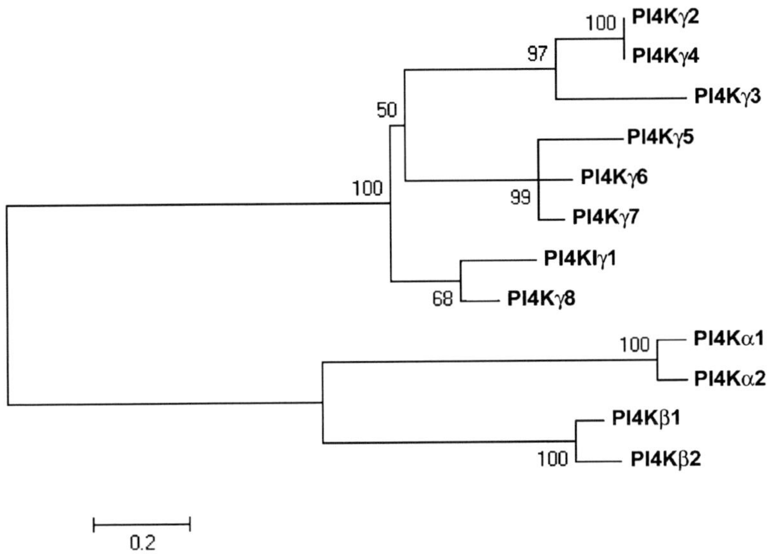

Fig. 1 *A. thaliana* has Twelve Predicted PI4K Proteins. Protein sequences of the predicted AtPI4K family members were obtained from The *Arabidopsis* Information Resource (www.arabidopsis.org). Full-length protein sequences were aligned using CLUSTALX PC V.181 (Thompson et al. 1997). Unrooted phylogenetic trees of full-length PI4K family members were created by MEGA V.21 (Kumar et al. 2001), using the *p*-distance method with gaps treated by pairwise deletions and a 500 bootstrap replicate

groups: alpha (α), beta (β), and gamma (γ), based on additional domains (Fig. 2). PI4Ks are also classified as type-II or type-III kinases by their sensitivities to inhibitors, with type-II kinases inhibited by adenosine and type-III kinases inhibited by wortmannin (Balla and Balla 2006). The α- and β-groups are type-III kinases, while members of the γ-group are type II-kinases. The α- and β-groups, with the exception of AtPI4Kα2, have a lipid kinase unique (LKU) domain, also termed as helical domain (Müeller-Roeber and Pical 2002). In yeast, the LKU domain of the PI4Kβ, Pik1p, binds frequenin, a Ca^{2+} binding protein (Huttner et al. 2003). Similarly, the LKU domain of AtPI4Kβ1 interacts with the frequenin-like Ca^{2+} sensor AtCBL1, indicating that Ca^{2+} regulation of β class PI4Ks may be conserved between these organisms (Preuss et al. 2006). AtPI4Kα1 contains a domain which displays some similarity to pleckstrin homology (PH) domains, while AtPI4Kα2 is truncated and contains only the catalytic domain (Müeller-Roeber and Pical 2002). PH domains bind phospholipids and often, but not always, selectively bind certain phospholipids (Lemmon and Ferguson 2000; Yu et al. 2004). The PH domain of AtPI4Kα1 binds PtdIns4P, phosphatidic acid (PA), and PtdIns(4,5)P_2, is important for protein localization, and may have a role in regulating the activity of AtPI4Kα1 (Stevenson et al. 1998; Stevenson-Paulik et al. 2003). The β-group members have a novel homology (NH) domain, and also contain an N-terminal LKU domain (Xue et al. 1999). NH domains function in protein/protein interaction. The NH domain of

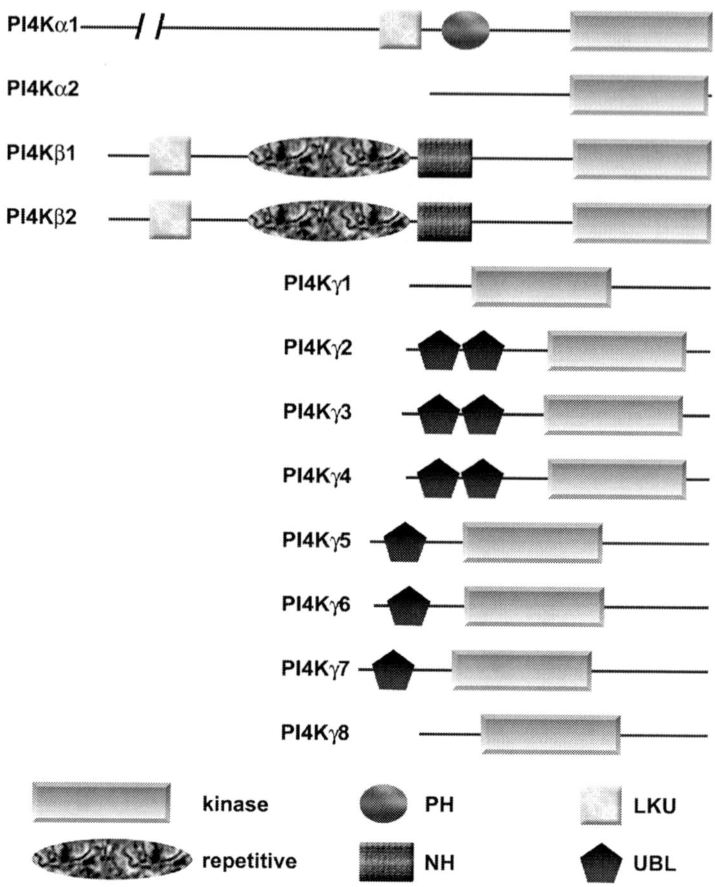

Fig. 2 Protein Domain Structure of PI4K-Family Members in *A. thaliana*. The Simple Modular Architecture Research Tool (SMART) was used to predict locations of kinase domains (Schultz et al. 1998). Information concerning the size and location of the PH, NH, UBL, repetitive, and LKU domains was obtained from (Xue et al. 1999; Müeller-Roeber and Pical 2002; Galvão et al. 2008). A truncated image of AtPI4Kα1 is presented due large protein size and lack of predicted domains in the n-terminal portion of AtPI4Kα. A space flanked by two slanted lines indicates the location of the truncation

AtPI4Kβ1 has been demonstrated to interact with members of the *A. thaliana* RabA family of small GTPase proteins (Preuss et al. 2006; Szumlanski and Nielsen 2009). In plants, PI4Kβ proteins have a repetitive section also termed as plant PI4K charged (PPC) domain (Lou et al. 2006). The repetitive domain is important for membrane localization of plant β-group members, likely through its ability to bind membrane lipids, including PA and PtdIns4P (Lou et al. 2006). Little is known about the function of plant PI4Kγ proteins. Unlike their yeast and animal homologues, most plant γ-group members contain one or two ubiquitin-like (UBL) domains (Fig. 2), which interact with proteins in the ubiquitin pathway in vitro (Müeller-Roeber and Pical 2002; Galvão et al. 2008).

Members of yeast and animal PI4K families have been studied extensively. Yeast has three PI4K family members, while mammals have five (reviewed in Balla and Balla 2006). In yeast, two proteins, Stt4p and Pik1p, are similar to the α- and β-groups, respectively. Stt4p and Pik1p are essential proteins and fulfill nonredundant functions in secretion (Audhya et al. 2000). The third yeast PI4K, Lsp6p, also termed Pik2p, is a γ-group member. Loss of Lsp6p is not lethal and leads to only moderate alterations in PtdIns4P levels (Han et al. 2002; Shelton et al. 2003). Two mammalian α-group members are isoforms of a single gene, PI4KIIIα, with isoform 1 containing just the c-terminal portion of the protein. Mammalian PI4KIIIβ belongs to the β-group, while PI4KIIα and PI4KIIβ belong to the γ-group (reviewed in Balla and Balla 2006). Mammalian PI4Ks regulate vesicle trafficking on several intracellular compartments (reviewed in Balla and Balla 2006). Expansion of the PI4K family in *A. thaliana* as compared to yeast and mammals makes it tempting to speculate that PI4Ks may fulfill specialized regulation of membrane trafficking in plants.

While all 12 *A. thaliana* PI4K family members are classified as PI4K proteins, so far, only two family members, AtPI4Kα1 and AtPI4Kβ1, have been demonstrated to have lipid kinase activity in vitro (Stevenson et al. 1998; Xue et al. 1999). Recent data show that AtPI4Kγ4 and AtPI4Kγ7 have protein kinase activity in vitro (Galvão et al. 2008), raising the possibility that AtPI4Kγ4 and AtPI4Kγ7 may have protein kinase activity in vivo, in addition to their predicted lipid kinase activity.

2.1 AtPI4Kβ-Group Members are Required for Tip Growth

Tip growth is a highly polarized form of cell expansion exhibited by root hairs and pollen tubes, in which growth is restricted to the apex of the cell. Properly controlled expansion of tip growing cells requires tip-focused delivery of secretory vesicles to an apical plasma membrane domain. Root hairs and pollen tubes are surface-exposed cells, and, in the case of root hairs, are nonessential cell types, making root hairs and pollen tubes convenient systems for studying membrane trafficking.

The Rab family of small GTPase proteins is important for various stages of membrane trafficking, such as vesicle formation and vesicle motility (Molendijk et al. 2004). Several Rab family members are required for tip growth. AtRabA4b, is important for polarized expansion of root hairs, while the closely-related AtRabA4d is necessary for pollen tube tip growth (Preuss et al. 2004; Szumlanski and Nielsen 2009). Additionally, over-expression of mutant forms of tobacco homologues of *A. thaliana*, RabA1 and RabB families, leads to abnormal pollen tube growth (Cheung et al. 2002; de Graaf et al. 2005).

A connection between Rab GTPase proteins and PtdIns4P came to light when AtPI4Kβ1 was identified in a screen for proteins which interact specifically with active AtRabA4b (Preuss et al. 2006). This interaction was specific to the β-group and AtPI4Kβ1 colocalized with AtRabA4b at the tips of root hair cells, indicating that AtPI4Kβ1 may be important during root hair growth (Preuss et al. 2006). AtPI4Kβ1 also interacted with active AtRabA4d, supporting the idea that AtPI4Kβ1 may be important for tip growth of both pollen tubes and root hairs (Szumlanski and Nielsen 2009). Plants lacking AtPI4Kβ1 and AtPI4Kβ2 had defects in both root hair and pollen tube morphology (Preuss et al. 2006; Szumlanski and Nielsen 2009). However, normal polarized growth was disrupted, and not abolished in these double mutant plants, implying that AtPI4Kβ1 and AtPI4Kβ2 may serve a regulatory function during polarized membrane trafficking in tip growing cell types.

3 PtdIns4P Phosphatases in *A. thaliana*

While PtdIns4P generation by PI4Ks is clearly important during cell expansion, turnover of PtdIns4P by phosphatases is also integral to proper regulation of membrane trafficking during this process. Mammals and yeast have two groups of PPI phosphatases; one consists of Sac1p-like and Fig4-like proteins, while the other consists of synaptojanin proteins, which contain both Sac1 PI phosphatase and type II phosphatase catalytic domains (reviewed in Hughes et al. 2000a). The *A. thaliana* genome has nine predicted PI phosphatases (Fig. 3a; see chapter, "Signaling and the Polyphosphoinositide Phosphatases from Plants"). Members of the *A. thaliana* SAC (AtSAC) family can be classified into three groups based on their protein domains and by similarity to yeast phosphatase proteins (Fig. 3b). Five members, AtSAC1–5, are predicted to have only the SAC domain and show similarity to the yeast phosphatidylinositol 3,5-bisphosphate [PtdIns(3,5)P_2] phosphatase, Fig4p. AtSAC6–8 have two predicted transmembrane domains and are most similar to the yeast PtdIns4P phosphatase Sac1p. AtSAC9 contains a WW domain, which is a protein/protein interaction motif and this class of phosphatase may be unique to plants (Ilsley et al. 2002; Zhong and Ye 2003). Currently, there are no predicted synaptojanin-like proteins in *A. thaliana*. This absence is noteworthy as the synaptojanin proteins have important roles in coordinating vesicle trafficking in yeast and mammalian cells.

The phosphatase activities of most members of the *A. thaliana* SAC domain family remain uncharacterized. In *S. cerevisiae*, Sac1p was initially identified as a suppressor of actin defects (Novick et al. 1989), and was later demonstrated to have PtdIns4P phosphatase activity (Hughes et al. 2000b). Three *A. thaliana* SAC family members, AtSAC6-8, are able to rescue the yeast *sac1* mutant, showing these phosphatases can display PtdIns4P phosphatase activity. However, as yeast Sac1p is able to dephosphorylate multiple PPIs in vitro (Hughes et al. 2000b; Despres et al. 2003), the possibility that AtSAC6-8 may act on other PPI isoforms other than PtdIns4P in plants cannot be excluded. Of the *A. thaliana* Sac1-like proteins, only AtSAC7, also known as *ROOT HAIR DEFECTIVE4* (RHD4), has been shown to

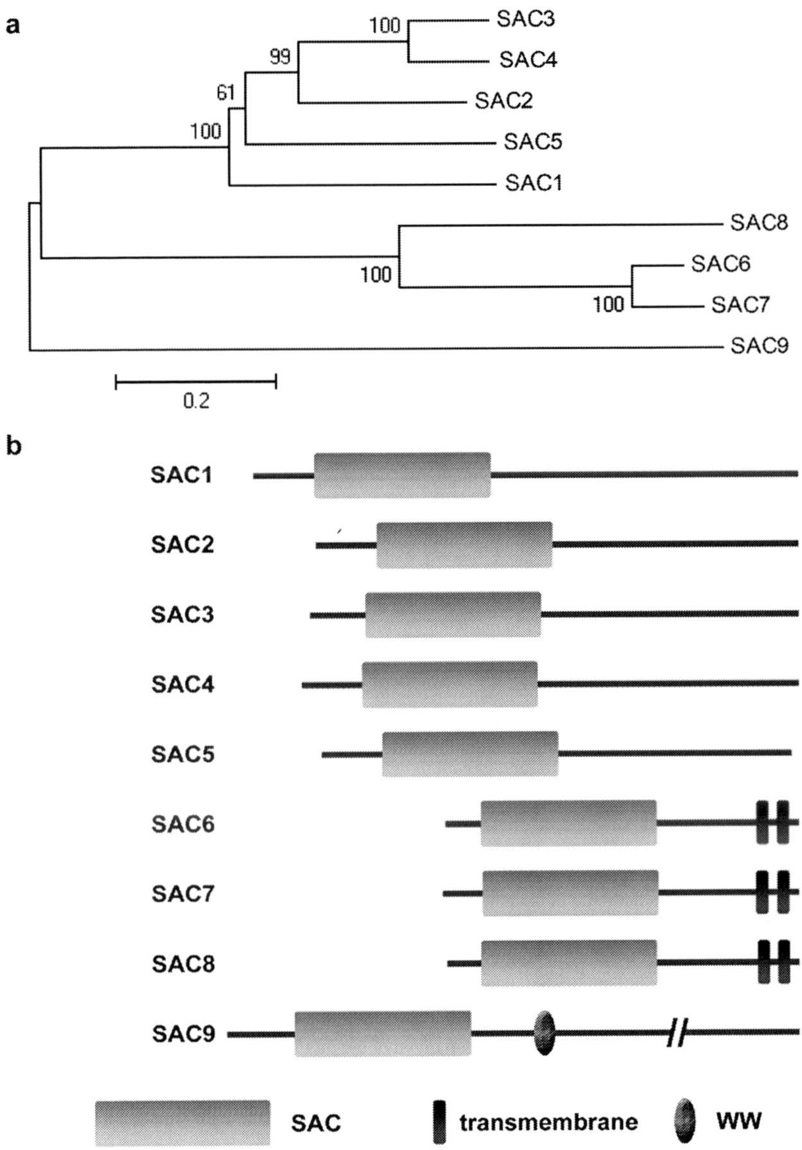

Fig. 3 *A. thaliana* has Nine Predicted SAC Domain-Containing Proteins. (a). Protein sequences of the predicted AtSAC family members were obtained from The *Arabidopsis* Information Resource (www.arabidopsis.org). Full-length protein sequences were aligned using CLUSTALX PC V.181 (Thompson et al. 1997). Unrooted phylogenetic trees of full-length AtSAC family members were created by MEGA V.21 (Kumar et al. 2001), using the *p*-distance method with gaps treated by pairwise deletions and a 500 bootstrap replicate. (b). Protein domain structures of AtSAC domain-containing family members are indicated by shapes of various patterns. The Simple Modular Architecture Research Tool (SMART) was used to predict locations of SAC, transmembrane and WW domains (Schultz et al. 1998). A truncated image of AtSAC9 is presented due to large protein size and lack of predicted domains in the c-terminal portion of AtSAC9. A space flanked by two slanted lines indicates the location of the truncation

have phosphatase activity against PtdIns4P both in vivo and in vitro (Thole et al. 2008). Loss of AtSAC7 leads to short, bulged root hairs which often form a bulge and then reinitiate tip growth (Thole et al. 2008), implying that turnover of PtdIns4P is important for maintenance of root hair tip growth. Given the similarity of root hairs and pollen tubes, it is likely that some AtSAC family members function to regulate PtdIns4P turnover in pollen tubes. Also, promoter GUS analysis showed that three family members, AtSAC6-8, are expressed in pollen, and AtSAC6 may be pollen specific (Despres et al. 2003).

Of the *A. thaliana* Fig4-like proteins, AtSAC1, also termed *fragile fiber7* (*fra7*), has in vitro phosphatase activity towards PI(3,5)P_2 (Zhong et al. 2005). Loss of AtSAC1 leads to a decrease in overall cell wall thicknesses in stems and a reduction of secondary cell wall material (Zhong et al. 2005). While it is not yet known how AtSAC1 phosphatase activity contributes to cell wall deposition, these results show that the AtSAC family is necessary for normal deposition of secondary cell walls in these cells. Phosphatase activity of AtSAC9, which is unique to plants, has not been tested in vitro. However, loss of AtSAC9 leads to a small increase of in vivo PtdIns(4,5)P_2 levels (Williams et al. 2005), indicating that AtSAC9 may normally function in PtdIns(4,5)P_2 turnover. In summary, of the characterized AtSAC family members, only one, AtSAC7, has been directly demonstrated to use PtdIns4P as a substrate. However, the large family size and relatively moderate phenotypes of single mutants (Williams et al. 2005; Zhong et al. 2005; Thole et al. 2008) implies that there are likely multiple family members able to utilize PtdIns4P in vivo.

4 Imaging PtdIns4P in Tip-Growing Cells

While the presence and abundance of PtdIns4P have been well established, current research has focused on determining its localization and dynamics in living cells. Recent data support the theory that cells contain different pools of PtdIns4P, which are created by different PI4Ks (Balla et al. 2005), putting forth the idea that generation and subsequent turnover of PtdIns4P are highly regulated.

Fusions between fluorescent proteins and lipid-binding domains have become useful tools for studying PPI localization (see also chapter, "Imaging Lipids in Living Plants"). Among others, PH domains can be used for this purpose as they can have specificity for particular PPI isoforms (Lemmon and Ferguson 2000; Yu et al. 2004), allowing for the detection of specific lipids in living cells. The PH domain of human FAPP1 protein (FAPP1), which binds PtdIns4P (Dowler et al. 2000), was recently used to monitor subcellular localization of PtdIns4P in several plant species. These studies showed that enhanced yellow fluorescent protein (EYFP) fused to the PH domain of FAPP1 was present on internal compartments, identified as Golgi membranes, and the plasma membrane (Thole et al. 2008; Vermeer et al. 2009). EYFP–FAPP1 was enriched in plasma membranes at the tips of growing root hairs of both *Medicago truncatula* and *A. thaliana*

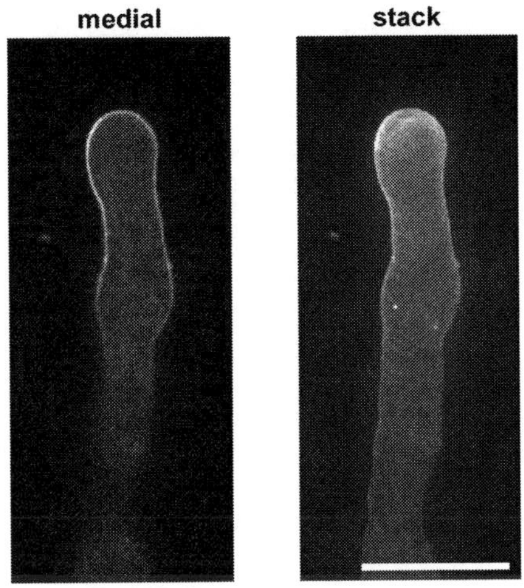

Fig. 4 The PtdIns4P-Binding Protein FAPP1 is Tip-Localized in *A. thaliana* Pollen Tubes. EYFP-FAPP1 localization in wild-type pollen tubes was imaged using confocal microscopy at ×100 (NA=1.46). A maximum projection and medial confocal image are shown (bar=25 μm)

(Thole et al. 2008; Vermeer et al. 2009). EYFP–FAPP1 displays a similar pattern in growing *A. thaliana* pollen tubes (Fig. 4), supporting the idea that presence of PtdIns4P at the apex of tip growing cells is important for polarized cell expansion. Root hairs of *rhd4* plants, which have an increased level of PtdIns4P, display a loss of membrane accumulated PtdIns4P and show defects in polarized root hair growth (Thole et al. 2008).

While tip-growing cells expand specifically at the most apical region of the cell, the physical location of the tip changes over time. Therefore, generation and removal of tip-localized PtdIns4P must be highly regulated and dynamic. A working model (Fig. 5a) can be envisioned in which PtdIns4P is generated on secretory vesicles targeted to the tips of growing cells by lipid kinases, such as PI4Kβ1. Fusion of these secretory compartments with the apical cell membrane would lead to enrichment of PtdIns4P in an apical plasma membrane domain. Accumulation of PtdIns4P on these secretory vesicles prior to successful targeting to the cytoplasm underlying the growing tip of the cell would be blocked by activity of phosphatases, such as RHD4, which revert PtdIns4P back to PI. Evidence for RHD4 function in restricting PtdIns4P accumulation on secretory vesicles is supported by its subcellular distribution in tip growing cells. RHD4 is localized on internal membrane compartments in a tip-focused manner in both root hairs and pollen tubes, but does not show complete colocalization with membrane compartments labeled by AtRabA4b or AtRabA4d (Thole et al. 2008; Szumlanski and Nielsen 2009). Instead, RHD4 is often absent from the most apical population of AtRabA4b-or AtRabA4d-labeled compartments. Absence of RHD4 from a subset of vesicles could allow for creation of two distinct vesicle populations at the tip of the cell; one with RHD4 present, which therefore could have lower levels of PtdIns4P, and one which lacks RHD4 and therefore, could have higher levels of PtdIns4P. A second possible means of restricting PtdIns4P to the cell apex could be by combined actions of phosphatidylinositol-4-phosphate 5-kinase (PIP5K) proteins, which can

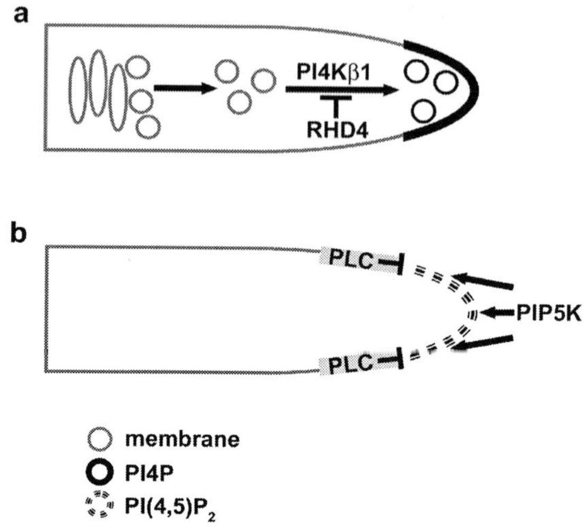

Fig. 5 Model of PtdIns4P Generation and Turnover in Tip Growing Cells. (**a**) PtdIns4P is generated on internal membranes by PI4Ks. These internal membranes fuse with the tip of the growing cell leading to enrichment of PtdIns4P on the apical cell membrane. Generation of PtdIns4P on internal membranes is countered by phosphatase proteins such as RHD4. (**b**) Tip-localized PtdIns4P is converted to tip-localized PtdIns(4,5)P_2 by PIP5Ks. PtdIns(4,5)P_2 is restricted to the apical cell membrane by PLC, which cleaves PtdIns-(4,5)P_2 to DAG and $InsP_3$

phosphorylate PtdIns4P to form PtdIns(4,5)P_2, and phospholipase C (PLC), which can cleave PtdIns(4,5)P_2 to form DAG and $InsP_3$ (Fig. 5b). PtdIns(4,5)P_2 is localized to the apical cell membrane of root hairs and pollen tubes (Kost et al. 1999; Vincent et al. 2005; Dowd et al. 2006; van Leeuwen et al. 2007). Curiously, AtPIP5K3 is localized to the apical cell membrane of growing root hairs (Kusano et al. 2008), while AtPIP5K4 and AtPIP5K5 are localized to flanking plasma membrane domains near the apex of growing pollen tubes (Ischebeck et al. 2008; Sousa et al. 2008). However, PtdIns(4,5)P_2 has not been observed to accumulate in the lateral plasma membrane of tip growing cells. One possible explanation is that PtdIns(4,5)P_2 is rapidly turned over in these lateral regions by PLC, which localizes to the lateral plasma membrane very close to the growing tip (Dowd et al. 2006; Helling et al. 2006). Clearly, PtdIns4P generation and PtdIns4P turnover by these various enzymes must be carefully coordinated to maintain distinct distributions of PtdIns4P and PtdIns(4,5)P_2 during tip growth.

References

Adel-Latif AA, Smith JP, Akhtar RA (1985) Polyphosphoinositides and muscarinic cholinergic and β-adrenergic receptors in iris smooth muscle. In: Bleasdale JE, Eichberg J, Hauser G (eds) Inositol and phosphoinositides. Humana, Clifton, pp 275–298

Audhya A, Foti M, Emr SD (2000) Distinct roles for the yeast phosphatidylinositol 4-kinases, Stt4p and Pik1p, in secretion, cell growth, and organelle membrane dynamics. Mol Biol Cell 11:2673–2689

Balla A, Balla T (2006) Phosphatidylinositol 4-kinases: old enzymes with emerging functions. Trends Cell Biol 16:351–361

Balla A, Tuymetova G, Tsiomenko A, Várnai P, Balla T (2005) A plasma membrane pool of phosphatidylinositol 4-phosphate is generated by phosphatidylinositol 4-kinase type-III alpha: Studies with the PH domains of the oxysterol binding protein FAPP1. Mol Biol Cell 16:1282–1295

Cheung AY, Chen CY, Glaven RH, de Graaf BH, Vidali L, Hepler PK, Wu HM (2002) Rab2 GTPase regulates vesicle trafficking between the endoplasmic reticulum and the Golgi bodies and is important to pollen tube growth. Plant Cell 14:945–962

Corvera S, D'Arrigo A, Stenmark H (1999) Phosphoinositides in membrane traffic. Curr Opin Cell Biol 11:460–465

Cunningham E, Thomas GM, Ball A, Hiles I, Cockcroft S (1995) Phosphatidylinositol transfer protein dictates the rate of inositol trisphosphate production by promoting the synthesis of PIP_2. Curr Biol 5:775–785

D'Angelo G, Vicinanza M, Di Campli A, De Matteis MA (2008) The multiple roles of PtdIns(4)P – not just the precursor of PtdIns(4, 5)P2. J Cell Sci 121:1955–1963

de Graaf BH, Cheung AY, Andreyeva T, Levasseur K, Kieliszewski M, Wu HM (2005) Rab11 GTPase-regulated membrane trafficking is crucial for tip-focused pollen tube growth in tobacco. Plant Cell 17:2564–2579

De Matteis MA, Godi A (2004) PI-loting membrane traffic. Nat Cell Biol 6:487–492

Despres B, Bouissonnie F, Wu H, Gomord V, Guilleminot J, Grellet F, Berger F, Delseny M, Devic M (2003) Three *SAC1*-like genes show overlapping patterns of expression in *Arabidopsis* but are remarkably silent during embryo development. Plant J 34:293–306

Dowd PE, Coursol S, Skirpan AL, Kao TH, Gilroy S (2006) Petunia phospholipase c1 is involved in pollen tube growth. Plant Cell 18:1438–1453

Dowler S, Currie RA, Cambell DG, Deak M, Kular GS, Downes CP, Alessi DR (2000) Identification of pleckstrin-homology-domain-containing proteins with novel phosphoinositide-binding specificities. Biochem J 351:19–31

Galvão RM, Kota U, Soderblom EJ, Goshe MB, Boss WF (2008) Characterization of a new family of protein kinases from Arabidopsis containing phosphoinositide 3/4-kinase and ubiquitin-like domains. Biochem J 409:117–127

Han GS, Audhya A, Markley DJ, Emr SD, Carman GM (2002) The *Saccharomyces cerevisiae* LSB6 gene encodes phosphatidylinositol 4-kinase activity. J Biol Chem 277: 47709–47718

Helling D, Possart A, Cottier S, Klahre U, Kost B (2006) Pollen tube tip growth depends on plasma membrane polarization mediated by tobacco PLC3 activity and endocytic membrane recycling. Plant Cell 18:3519–3534

Hughes WE, Cooke FT, Parker PJ (2000a) Sac phosphatase domain proteins. Biochem J 350:337–352

Hughes WE, Woscholski R, Cooke FT, Patrick RS, Dove SK, McDonald NQ, Parker PJ (2000b) SAC1 encodes a regulated lipid phosphoinositide phosphatase, defects in which can be suppressed by the homologous Inp52p and Inp53p phosphatases. J Biol Chem 275:801–808

Huttner IG, Strahl T, Osawa M, King DS, Ames JB, Thorner J (2003) Molecular interactions of yeast frequenin (Frq1) with the phosphatidylinositol 4-kinase isoform Pik1. J Biol Chem 278:4862–4874

Ilsley JL, Sudol M, Winder SJ (2002) The WW domain: linking cell signaling to the membrane cytoskeleton. Cell Signal 14:183–189

Ischebeck T, Stenzel I, Heilmann I (2008) Type B phosphatidylinositol-4-phosphate 5-kinases mediate *Arabidopsis* and *Nicotiana tabacum* pollen tube growth by regulating apical pectin secretion. Plant Cell 20:3312–3330

Kost B, Lemichez E, Spielhofer P, Hong Y, Tolias K, Carpenter C, Chua NH (1999) Rac homologues and compartmentalized phosphatidylinositol 4, 5-bisphosphate act in a common pathway to regulate polar pollen tube growth. J Cell Biol 145:317–330

Kumar S, Tamura K, Jakobsen IB, Nei M (2001) MEGA2: molecular evolutionary genetics analysis software. Bioinformatics 17:1244–1245

Kusano H, Testerink C, Vermeer JEM, Tsuge T, Shimada H, Oka A, Munnik T, Aoyama T (2008) The *Arabidopsis* phosphatidylinositol phosphate 5-kinase PIP5K3 is a key regulator of root hair tip growth. Plant Cell 20:367–380

Lemmon MA, Ferguson KM (2000) Signal-dependant membrane targeting by pleckstrin homology (PH) domains. Biochem J 350:1–18

Lou Y, Ma H, Lin W, Chu Z, Müeller-Roeber B, Xu Z, Xue H (2006) The highly charged region of plant β-type phosphatidylinositol 4-kinase is involved in membrane targeting and phospholipid binding. Plant Mol Biol 60:729–746

Meijer HJ, Munnik T (2003) Phospholipid-based signaling in plants. Annu Rev Plant Biol 54:265–302

Meijer HJ, Berrie CP, Lurisci C, Divecha N, Musgrave A, Munnik T (2001) Identification of a new phosphoinositide in plants, phosphatidylinositol 5-monophosphate (PtdIns5P) and its accumulation upon osmotic stress. Biochem J 360:491–498

Molendijk AJ, Ruperti B, Palme K (2004) Small GTPases in vesicle trafficking. Curr Opin Plant Biol 7:694–700

Müeller-Roeber B, Pical C (2002) Inositol phospholipid metabolism in *Arabidopsis*. Characterized and putative isoforms of inositol phospholipid kinase and phosphoinositide-specific phospholipase C. Plant Physiol 130:22–46

Munnik T, Testerink C (2009) Plant phospholipid signaling – 'in a nutshell'. J Lipid Res 50: S260–S265

Munnik T, Irvine RF, Musgrave A (1994) Rapid turnover of a phosphatidylinositol 3-phosphate in the green alga *Chlamydomonas eugametos*: signs of a phosphatidylinositide 3-kinase signaling pathway in the lower plants? Biochem J 298:269–273

Novick P, Osmond BC, Botstein D (1989) Suppressors of yeast actin mutations. Genetics 121:659–674

Preuss ML, Serna J, Falbel TG, Bednarek SY, Nielsen E (2004) The *Arabidopsis* Rab GTPase RabA4b localizes to the tips of growing root hair cells. Plant Cell 16:1589–1603

Preuss ML, Schmitz AJ, Thole JM, Bonner HK, Otegui MS, Nielsen E (2006) A role for the RabA4b effector protein PI-4Kbeta1 in polarized expansion of root hair cells in *Arabidopsis thaliana*. J Cell Biol 172:991–998

Schultz J, Milpetz F, Bork P, Ponting CP (1998) SMART, a simple modular architecture research tool: identification of signaling domains. PNAS 95:5857–5864

Shelton SN, Barylko B, Binns DD, Horazdovsky BF, Albanesi JP, Goodman JM (2003) *Saccharomyces cerevisiae* contains a type II phosphoinositide 4-kinase. Biochem J 371:533–540

Sousa E, Kost B, Malhó R (2008) *Arabidopsis* phosphatidylinositol-4-monophosphate 5-kinase 4 regulates pollen tube growth and polarity by modulating membrane recycling. Plant Cell 20:3050–3064

Stevenson JM, Perera IY, Boss WF (1998) A phosphatidylinositol 4-kinase pleckstrin homology domain that binds phosphatidylinositol 4-monophosphate. J Biol Chem 273:22761–22767

Stevenson JM, Perera IY, Heilmann I, Persson S, Boss WF (2000) Inositol signaling and plant growth. Trends Plant Sci 5:252–258

Stevenson-Paulik J, Love J, Boss WF (2003) Differential regulation of two *Arabidopsis* type III phosphatidylinositol 4-kinase isoforms. A regulatory role for the pleckstrin homology domain. Plant Physiol 132:1053–1064

Szumlanski AL, Nielsen E (2009) The Rab GTPase RabA4d regulates pollen tube tip growth in *Arabidopsis thaliana*. Plant Cell 21:526–544

Thole JM, Nielsen E (2008) Phosphoinositides in plants: novel functions in membrane trafficking. Curr Opin Plant Biol 11:620–631

Thole JM, Vermeer JEM, Zhang Y, Gadella TW Jr, Nielsen E (2008) *ROOT HAIR DEFECTIVE4* encodes a phosphatidylinositol-4-phosphate phosphatase required for proper root hair development in *Arabidopsis thaliana*. Plant Cell 20:381–395

Thompson JD, Gibson TJ, Plewniak F, Jeanmougin F, Higgins DG (1997) The CLUSTAL_X windows interface: flexible strategies for multiple sequence alignment aided by quality analysis tools. Nucleic Acids Res 25:4876–4882 van Leeuwen W, Vermeer JEM, Gadella TWJ Jr, Munnik T (2007) Visualization of phosphatidylinositol 4, 5-bisphosphate in the plasma membrane of suspension-cultured tobacco BY-2 cells and whole Arabidopsis seedlings. Plant J 52:1014–1026

Vermeer JEM, Thole JM, Goedhart J, Nielsen E, Munnik T, Gadella TW Jr (2009) Imaging phosphatidylinositol 4-phosphate dynamics in living plant cells. Plant J 57:356–372

Vincent P, Chua M, Nogue F, Fairbrother A, Mekeel H, Xu Y, Allen N, Bibikova TN, Gilroy S, Bankaitis VA (2005) A Sec14p-nodulin domain phosphatidylinositol transfer protein polarizes membrane growth of *Arabidopsis thaliana* root hairs. J Cell Biol 168:801–812

Williams ME, Torabinejad J, Cohick E, Parker K, Drake EJ, Thompson JE, Hortter M, DeWald DB (2005) Mutations in the Arabidopsis phosphoinositide phosphatase gene SAC9 lead to overaccumulation of PtdIns(4, 5)P2 and constitutive expression of the stress-response pathway. Plant Physiol 138:686–700

Xue H, Pical C, Brearly C, Elge S, Müeller-Roeber B (1999) A plant 126-kDa phosphatidylinositol 4-kinase with a novel repeat structure. J Biol Chem 274:5738–5745

Yu JW, Mendrola JM, Audhya A, Singh S, Keleti D, DeWald DB, Murray D, Emr SD, Lemmon MA (2004) Genome-wide analysis of membrane targeting by *S. cerevisiae* pleckstrin homology domains. Mol Cell 13:677–688

Zhong R, Ye Z (2003) The SAC domain-containing protein gene family in *Arabidopsis*. Plant Physiol 132:544–555

Zhong R, Burk DH, Nairn CJ, Wood-Jones A, Morrison III WH, Ye Z (2005) Mutation of SAC1, an *Arabidopsis* SAC domain phosphoinositide phosphatase, causes alterations in cell morphogenesis, cell wall synthesis, and actin organization. Plant Cell 17:1449–1466

PIP-Kinases as Key Regulators of Plant Function

Till Ischebeck and Ingo Heilmann

Abstract The phosphoinositide pathway arose early in evolution and is conserved in eukaryotic organisms of all kingdoms. Phosphoinositides derive from phosphatidylinositol (PtdIns) by sequential phosphorylation of the lipid head group and can interact with various protein partners to regulate their biochemical activity or subcellular localization. This chapter is concerned with PIP-kinases, the enzymes producing PtdIns-bisphosphates with regulatory function, such as the well-characterized PtdIns(4,5)P_2 or the less-well understood PtdIns(3,4)P_2 and PtdIns (3,5)P_2. In recent years, the study of plant PIP-kinases has revealed new targets for perturbing plant phosphoinositide metabolism and enabled functional investigations on the roles of PtdIns-bisphosphates, including involvement in the regulation of cell polarity, guard cell function, defense responses, and stress signaling. Together with new techniques for lipid analysis and new tools for the visualization of specific lipids, these recent advances have opened the field to exciting discoveries, indicating central roles for PIP-kinases and phosphoinositide signaling in plant function and development.

1 Phosphatidylinositol-Bisphosphates and Their Function in Eukaryotic Cells

Phosphorylated derivatives of phosphatidylinositol (PtdIns) perform a multitude of regulatory roles in the physiology of eukaryotic cells. Phosphatidylinositol 4,5-bisphosphate [PtdIns(4,5)P_2] is the best-studied phosphoinositide and can, both in plants and animals, perform cellular functions either as an intact lipid ligand for protein partners (Drobak et al. 1994; Lemmon 2003; Balla 2005) or as a

I. Heilmann (✉)
Department of Plant Biochemistry, Albrecht-von-Haller-Institute for Plant Sciences, Georg-August-University Göttingen, Justus-von-Liebig-Weg 11, 37077, Göttingen, Germany
e-mail: iheilma@uni-goettingen.de

T. Munnik (ed.), *Lipid Signaling in Plants*, Plant Cell Monographs 16,
DOI 10.1007/978-3-642-03873-0_5, © Springer-Verlag Berlin Heidelberg 2010

substrate for phospholipase C (PLC), which hydrolyzes the lipid into inositol 1,4,5-trisphosphate ($InsP_3$) and diacylglycerol (DAG) (Berridge 1983; Meijer and Munnik 2003). PtdIns(4,5)P_2 functions reported from various eukaryotic systems have been extensively reviewed (Stevenson et al. 2000; Meijer and Munnik 2003; Balla 2006; Santarius et al. 2006), and examples from both the plant and animal models include the regulation of ion-channel or ATPase activity (Cote et al. 1996; Suh and Hille 2005), cytoskeletal dynamics (Lemmon et al. 2002; Doughman et al. 2003; Wasteneys and Galway 2003; Wenk and De Camilli 2004), and hormonal and stress signaling (Stevenson et al. 2000; Mueller-Roeber and Pical 2002; Meijer and Munnik 2003; Balla 2006; Santarius et al. 2006). A recent report demonstrated association of PtdIns(4,5)P_2 with clathrin-coated vesicles in plant cells (König et al. 2008), although so far the involvement of PtdIns(4,5)P_2 in vesicle trafficking has only been conclusively demonstrated for animal cells (Roth 2004; Wenk and De Camilli 2004).

Besides PtdIns(4,5)P_2, a number of reports suggest the formation of PtdIns(3,4) P_2 and PtdIns(3,5)P_2 in plants (Dove et al. 1997; Meijer et al. 1999; Westergren et al. 2001), however, considerably less information is available on the physiological relevance of these isomeric PtdIns-bisphosphates than on PtdIns(4,5)P_2.

2 PIP-Kinases Generate Regulatory Phosphatidylinositol-Bisphosphates

The different PtdIns-bisphosphates are produced by phosphorylation of PtdIns-monophosphates catalyzed by PIP-kinases (Fig. 1). The enzymes share a PIP-kinase domain that is well-conserved among isoforms and between enzymes from different organisms. Based on their characteristic substrate specificities, mammalian PIP-kinases have been categorized into three types. Type I enzymes convert PtdIns4P into PtdIns(4,5)P_2, whereas type II enzymes generate the same product by phosphorylation of the D4-position of PtdIns5P (Fig. 1a). Types I and II will also utilize PtdIns3P in vitro (Hinchliffe et al. 1998); however, it is unclear whether these activities have physiological relevance in vivo (McEwen et al. 1999). Type III PIP-kinases preferentially phosphorylate PtdIns3P in the D5-position, yielding PtdIns (3,5)P_2 (Cooke et al. 1998; Gary et al. 1998; McEwen et al. 1999) (Fig. 1b).

From the known crystal-structure of the human PIPKIIβ (Rao et al. 1998), it can be concluded that PIP-kinases are structurally related to protein kinases; however, there is only limited similarity at the sequence level (Rao et al. 1998). One outcome of the structural studies conducted was the finding that PIP-kinases have special adaptations, facilitating the interaction of the soluble enzymes with membrane surfaces holding their substrate lipids. For instance, the membrane-interfacial aspect of the human PIPK IIβ is flattened and does not hold large, obstructing residues that could sterically hinder protein–lipid interaction (Burden et al. 1999). A PIP-kinase domain named according to its structural equivalent found in protein kinases (Rao et al. 1998), is the "activation loop," a stretch of amino acids in close

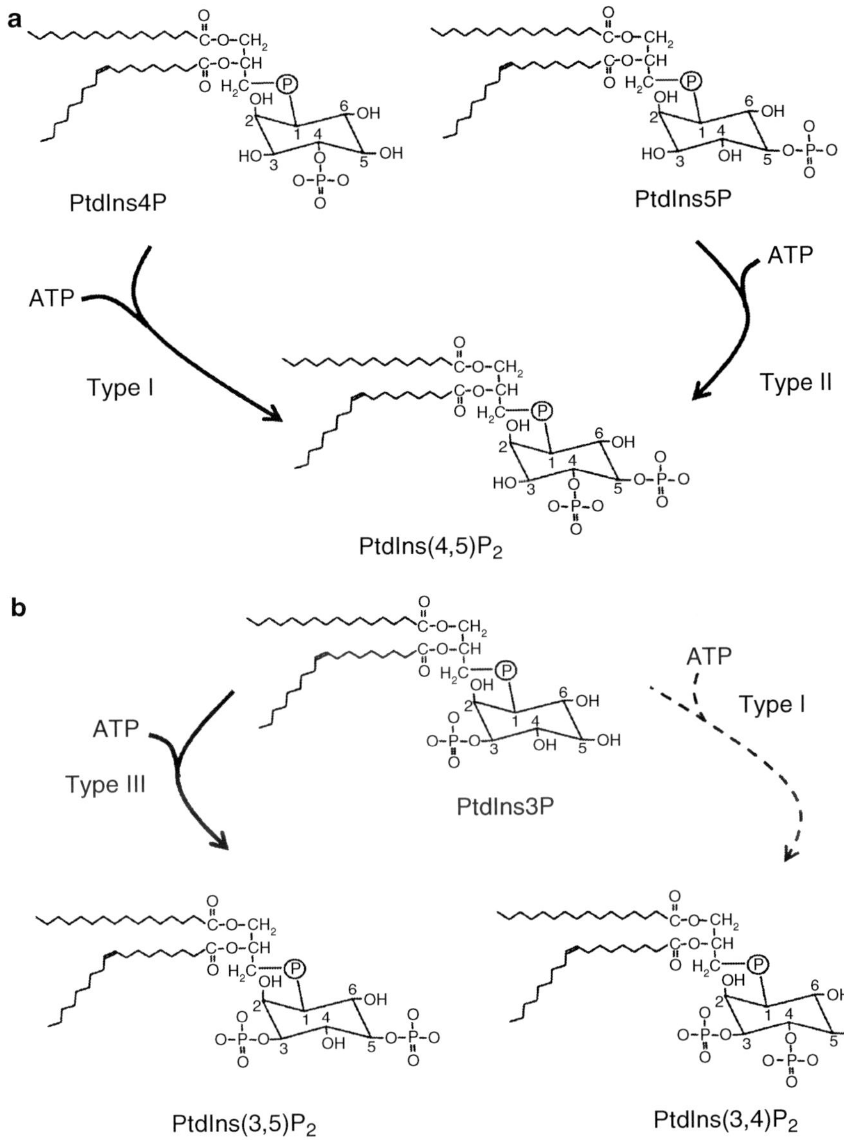

Fig. 1 Generation of phosphatidylinositol-bisphosphates by PIP-kinases. The phosphatidylinositol-monophosphates, PtdIns3P, PtdIns4P and PtdIns5P, are phosphorylated by PIP-kinases to different bisphosphates under consumption of ATP. Depending on their substrate preference, three types of PIP-kinases are distinguished. (**a**) Formation of PtdIns(4,5)P_2. Type I enzymes phosphorylate PtdIns4P in the D5-position of the inositol-ring to PtdIns(4,5)P_2. Type II enzymes form the same product by phosphorylation of PtdIns5P in the D4-position. (**b**) Formation of PtdIns(3,5)P_2 and PtdIns(3,4)P_2. Type III enzymes phosphorylate PtdIns3P in the D5-position, yielding PtdIns(3,5) P_2. As a side activity, class I enzymes also phosphorylate PtdIns3P to (possibly) PtdIns(3,4)P_2. Only type I and type III PIP-kinases have been reported for plants

proximity to the active site, which was demonstrated to be a determinant for substrate specificity of human PIP-kinases (Kunz et al. 2000, 2002).

3 PIP-Kinase Gene Families in *Arabidopsis thaliana*

A survey of the *Arabidopsis* genome for genes related to phosphoinositide signaling revealed the presence of multiple sequences encoding putative proteins with similarity to mammalian PIP-kinases (Mueller-Roeber and Pical 2002). Ramifications for the evolution of the respective gene families were reviewed earlier (Mueller-Roeber and Pical 2002) and shall not be reiterated here. Based on the sequence determinants for substrate-specificity (Kunz et al. 2000, 2002), in *Arabidopsis,* a large gene family of 11 members encoding possible type I PIP-kinases (PIP5K1–11) can be distinguished from a smaller family of 4 putative type III PIP-kinases (Fab1a–d). In *Arabidopsis* and mammals, PIP-kinases are encoded by multiple genes, whereas yeast, contains only one PIP-kinase type I (Mss4p) and one type III-enzyme (Fab1p). The relevance of this difference in isoenzyme patterns between multicellular organisms and yeast is not clear. The presence of so many isoforms of *Arabidopsis* PIP-kinases, in contrast to the PI-kinases (four PI4-kinases and one PI3-kinase) may indicate functional diversification of PIP-kinases, possibly correlated to the establishment of different PtdInsP$_2$-pools.

3.1 Type I PIP-Kinases

Several *Arabidopsis* type I enzymes have been experimentally demonstrated to exhibit PI4P 5-kinase activity in vitro (Mikami et al. 1998; Elge et al. 2001; Perera et al. 2005; Nagano et al., 2008; Stenzel et al. 2008). While actual and putative *Arabidopsis* PIP-kinase sequences share typical catalytic and dimerization domains with their mammalian counterparts, a subfamily of plant PIP-kinases carries additional characteristic domains (Fig. 2). As mentioned above, the plant type I PIP-kinases can also phosphorylate PtdIns3P in vitro (Westergren et al. 2001; Perera et al. 2005; Stenzel et al. 2008), but the physiological relevance of this side-activity is not clear. The formation of PtdIns(3,4,5)P$_3$ in insect cells expressing *Arabidopsis* PIP5K1 has been reported (Elge et al. 2001), but again it is not clear whether the enzyme is immediately responsible for the production of that lipid or whether it formed a precursor that was further metabolized by other enzymes. So far, PtdIns (3,4,5)P$_3$ has never been detected in plant cells.

According to their domain-structure, plant type I PIP-kinases can be classified into type A or B (Mueller-Roeber and Pical 2002). The smaller type A enzymes (isoforms 10 and 11) have molecular masses of 46 and 49 kDa, respectively, and exhibit domain structures similar to those described in mammals (Fig. 2). In contrast, type B kinases (isoforms 1 to 9) are larger, between 81 and 92 kDa, and

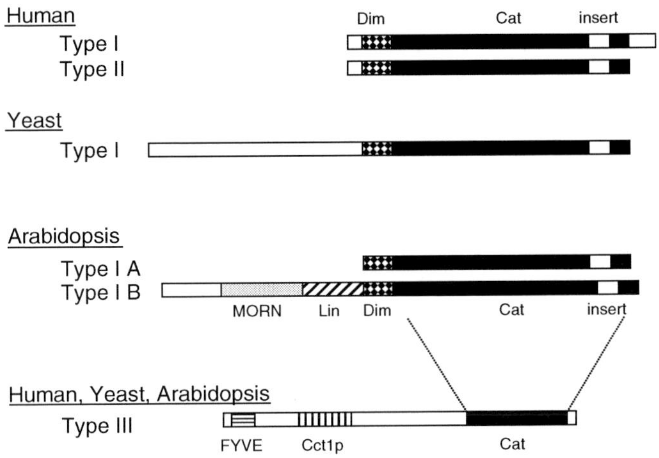

Fig. 2 Domain-structure of PIP-kinases. PIP-kinases share a well-conserved catalytic domain (Cat) that is interrupted by a variable stretch of sequence (*insert*). The dimerization domain (Dim) is thought to form the interface for dimerization of monomers. Mammalian and human PIP-kinases do not contain additional protein domains. Yeast type I enzymes contain a large N-terminal extension with regulatory functions in addition to the conserved catalytic domain. Plant type I PIP-kinases are grouped into subfamilies A and B. The smaller type I A enzymes have a domain structure similar to that of mammalian and human type I PIP-kinases. In contrast, type I B enzymes contain a large N-terminal extension that is not similar to that of yeast enzymes. In plant type I B PIP-kinases, a domain containing several membrane-occupation and recognition nexus (MORN)-repeats is conserved that may have functions in lipid binding and subcellular targeting. The MORN-repeat-domain is connected to the catalytic portion of the enzyme via a linker-domain (Lin), which has been shown for *Arabidopsis* PIP5K1 to bind phosphatidic acid and actin. Type III PIP-kinases of mammalian, yeast or plant origin are much larger and can contain a FYVE-domain capable of binding to PtdIns3P. Another conserved domain of type III PIP-kinases is the Cct1p-homology domain of unknown function

unique to plants in containing a large additional N-terminal extension which includes several membrane-occupation-and-recognition-nexus (MORN) repeats (Mueller-Roeber and Pical 2002). MORN-repeats are found in both animal and plant proteins and can mediate protein-membrane contacts, e.g., in the *Arabidopsis*, ARC3 protein involved in plastidial fission (Shimada et al. 2004; Maple et al. 2007), the toxoplasma protein MORN1 involved in cell division (Gubbels et al. 2006), or junctophilins, which mediate endomembrane-to-plasma membrane contacts in mammalian cells (Takeshima et al. 2000). For the *Arabidopsis* type I PIP-kinase 3, it has been demonstrated that the N-terminus, including the MORN-repeat domain, is not required for enzyme activity, but is necessary for physiological function and complementation of the *pip5k3*-mutant phenotype (Stenzel et al. 2008). A recent report indicates that the N-terminus of *Arabidopsis* PI4P 5-kinase isoform 1 (PIP5K1) has multiple regulatory effects on enzyme activity and may guide subcellular localization of the enzyme (Im et al. 2007). PI4P 5-kinases from different organisms have been reported to associate with different subcellular locations, including the plasma membrane of plants, yeast, and mammals

(Perera et al. 1999; Heilmann et al. 2001; Kobayashi et al. 2005; Santarius et al. 2006), the nucleus of yeast and mammals (Ciruela et al. 2000; Audhya and Emr 2003; Santarius et al. 2006), the actin cytoskeleton of plants, yeast, and mammals (Desrivieres et al. 1998; Doughman et al. 2003; Davis et al. 2007), and endomembranes of plants and mammals (Whatmore et al. 1996; Heilmann et al. 1999; Im et al. 2007). Correct subcellular localization may, thus, be important for physiological functionality of PIP-kinases (Stenzel et al. 2008), and the multiple subcellular locations of PIP-kinase activity reported for plant cells may indicate multiple possibilities for physiological functions in plants.

3.2 Type II PIP-Kinases

Based on the sequence determinants for substrate specificity (Kunz et al. 2000, 2002), the *Arabidopsis* genome seems to lack type II PIP 4-OH-kinases that convert PtdIns5P to PtdIns(4,5)P_2 (Mueller-Roeber and Pical 2002). Consistent with this prediction, all *Arabidopsis* candidate proteins recombinantly analyzed so far for catalytic activity or their substrate and regio-specificity, including PIP5K isoforms 1, 2, 3, 4, 7, 8, and 9, did not show more than trace activity against PtdIns5P (Elge et al. 2001; Westergren et al. 2001; Perera et al. 2005; Kusano et al. 2008; Stenzel et al. 2008). Thus, type II PIP-kinases appear to be either absent or may have been missed. It must also be noted that no enzyme activity phosphorylating PtdIns to PtdIns5P has been discovered to date, and that PtdIns5P detected in plants (Meijer et al. 2001) may, therefore, be derived from dephosphorylation of PtdIns-bisphosphates.

3.3 Type III PIP-Kinases

In comparison to the type I-enzymes described above, considerably less is known about the 4 putative *Arabidopsis* type III PIP-kinases, with similarity to the yeast Fab1p-protein. The deduced type III-PIP-kinases from *Arabidopsis* represent putative enzymes of approximately 200 kDa (Fig. 2), two of which contain FYVE-domains (termed after its occurrence in the proteins Fab-1, YGL023, Vps27, and EEA1) capable of binding the proposed catalytic substrate, PtdIns3P, while the other two putative isoforms lack recognizable FYVE-domains. All the four sequences share a conserved domain with significant similarity to the catalytic core of PIP-kinases and the Cct1p-homology-domain with similarity to chaperonin components (Mueller-Roeber and Pical 2002). The yeast type III PIP-kinase, Fab1p, is functionally involved in endomembrane trafficking and vacuolar dynamics (Cooke et al. 1998). As no characterization of recombinant *Arabidopsis* type III PIP-kinases has been reported, the discussion of plant Fab1p-like type III PIP-kinases must remain speculative at this point, as was previously noted (Mueller-Roeber and Pical 2002). Nonetheless, plants do make PtdIns(3,5)P_2 (Meijer et al. 1999; Zonia and Munnik 2004).

4 PIP-Kinases Have Distinct Organ-Specific Expression Patterns

Candidate functions for some PIP-kinase isoforms may be inferred by the tissue-specific expression patterns of the genes in question, and the collection of transcript array-data accessible through the Genevestigator (Zimmermann et al. 2004) or AtGenexpress (Kilian et al. 2007) portals provides a powerful tool for the first approximation of such predictions. According to data deposited in the Genevestigator database, some PIP-kinases exhibit characteristic expression patterns with expression restricted to certain organs or cells. For instance, expression of type I isoforms 5, 6, 10, or 11 appears to occur predominantly or even exclusively in pollen, that of isoform 4 in roots and pollen, and that of isoform 3 exclusively in roots. The other type I-isoforms 1, 2, 7, 8, and 9 are expressed at moderate levels in all plant organs. In part, these predicted expression patterns have been corroborated by experimental evidence. Specifically, isoforms 1 (Mikami et al. 1998; Elge et al. 2001), 3 (Kusano et al. 2008; Stenzel et al. 2008), 10 (Perera et al. 2005), and 2, 3, 7, 8, and 9 (Stenzel et al. 2008) were studied using promoter GUS-fusions and histochemical staining and/or RT-PCR-approaches. According to Genevestigator, type III PIP-kinases exhibit moderate levels of expression throughout all plant organs. From the expression patterns, it may be concluded that some PIP-kinase isoforms are required in all organs, whereas other isoforms may fulfill tissue- or cell-type-specific functions. It is important to note that expression of several isoforms can be induced by certain stresses that also induce increases in phosphoinositide metabolites, for instance osmotic stress (Mikami et al. 1998).

5 Physiological Functions of PIP-Kinases

The identification of genes encoding PIP-kinase isoforms and the characterization of the corresponding T-DNA insertion mutants has yielded the first clear phenotypes attributable to particular PIP-kinase isoforms. The following sections will give an overview on relevant recent discoveries.

5.1 PIP-Kinases and Guard Cell Function

While the Genevestigator prediction of organ-specific expression patterns did not suggest expression of the *Arabidopsis* type I PIP-kinase isoform 4, a recent report indicates a role for this enzyme in the control of guard cell function, highlighting the limitations of large-scale high-throughput databases for predictions about transcripts of low abundance or spatially restricted occurrence. The characterization of T-DNA insertion mutants for the PIP5K4 gene revealed that the mutant plants were compromised in guard cell functionality, and stomatal opening was significantly reduced (Lee et al. 2007). Signaling processes controlling guard cell opening and closure involve the interplay of numerous ion channels, including proton-ATPases and channels for Ca^{2+}, K^{+}, and various anions in the plasma membrane and the tonoplast (Fan et al. 2004). The proposed mechanism by which PtdIns(4,5)P_2 may contribute to guard cell function includes the control of slow anion channels in the guard cell-plasma membrane by that lipid (Lee et al. 2007). Because no regulatory effects on guard cell function were found for PtdIns (3,4)P_2 or PtdIns(3,5)P_2, the data presented by Lee and coworkers highlight the specific functions of PtdInsP_2 headgroup isomers. The data are in line with previous reports from various eukaryotic model systems indicating that ion channels and ATPases are regulated by their lipid ligand, PtdIns(4,5)P_2 (Shyng et al. 2000; Hilgemann et al. 2001; Cho et al. 2005; Suh and Hille 2005).

5.2 PIP-Kinases, Cell Polarity and Polar Growth of Plant Cells

Phosphoinositides and PtdIns(4,5)P_2 have been implicated in the control of cell polarity and polar growth of eukaryotic cells (Golub and Caroni 2005; Lacalle et al. 2007). Previous biochemical evidence from gravistimulated maize and oat plants has suggested vectorial distribution of PIP-kinase and PtdIns(4,5)P_2 in tissues altering their polar growth axis during gravitropism (Perera et al. 1999, 2001). The use of fluorescence-tagged protein modules with specific lipid-binding properties (Varnai and Balla 1998) has greatly aided the study of subcellular phosphoinositide distribution. Using this technique, a microdomain of PtdIns(4,5)P_2 was described in vivo in the plasma membrane of the apical tips of growing pollen tubes (Kost et al. 1999; Dowd et al. 2006; Helling et al. 2006), a highly polar cell type characterized by localized tip-growth. In root hairs, a cell type structurally similar to pollen tubes, the presence of a PtdIns(4,5)P_2-microdomain in the plasma membrane of the apical region of hair cells was demonstrated at about the same time using immunofluorescence-detection (Braun et al. 1999), which has meanwhile been confirmed by the GFP-based biosensor approach (Van Leeuwen et al. 2007) . The localized accumulation of PtdIns(4,5)P_2 in the tip plasma membrane of pollen tubes and root hairs suggests action of a PIP-kinase at this site that is responsible for PtdIns(4,5)P_2-biosynthesis. Based on the Genevestigator prediction that *Arabidopsis* type I PIP-kinase 3 is exclusively expressed in roots, it was demonstrated that T-DNA-insertion mutants deficient in this isoform were compromised in root hair formation (Kusano et al. 2008; Stenzel et al. 2008). Catalytic activity and PtdIns (4,5)P_2 formation were required for complementation of the mutant phenotype and fluorescence-tagged *Arabidopsis* PIP5K3 localized to the extreme periphery of the growing apices of root hair cells (Kusano et al. 2008; Stenzel et al. 2008). Interestingly, root hair-specific overexpression of PIP5K3 resulted in a gradual loss of cellular polarity, indicating that a fine balance of PtdIns(4,5)P_2 is necessary for cellular polarity (Stenzel et al. 2008). Together the data suggest that PtdIns(4,5)P_2 production by this particular enzyme is essential for polar tip-growth of root hair cells, and that other type I PIP-kinase isoforms also present in root hair cells cannot functionally complement the loss of PIP5K3. Previous work by the Nielsen group (Preuss et al. 2006) has shown that exocytotic vesicles, destined to fuse to the apical plasma membrane of root hairs, associate with PI4-kinase and are coated with PtdIns4P, the preferred substrate for PIP5K3. It remains to be seen whether there is a mechanistic link between PtdIns(4,5)P_2 production and vesicle-to-plasma membrane fusion at the root hair apex. By analogy to the root hair system, it is likely that other type I PIP-kinases will be found to be essential components of the machinery driving polar tip growth in pollen tube-cells or involved in polar growth of plant cells in other contexts.

6 PIP-Kinases and Their Signaling Environment

In order to establish the delicate balance of PtdInsP_2-pools with regulatory effects on metabolism or growth, the action of PIP-kinases must be tightly controlled. Mechanisms regulating PIP-kinase activity may include intrinsic properties of the PIP-kinase proteins as well as factors recruiting PIP-kinases into particular micro-environments that are "wired" to perform particular physiological roles. The following sections give an overview of recently discovered factors controlling PIP-kinase functionality.

6.1 Interaction of PIP-Kinases with Lipids

An important aspect of PIP-kinase function regards the recognition of substrate lipids. PtdIns4P and other PtdIns-monophosphates may be represented in plant cells by molecular species differing in the associated fatty acids (König et al. 2007; Heilmann 2008), and plant PIP-kinases can likely discriminate for certain substrate-species. Preferences for substrate species containing certain fatty acids have been demonstrated for various PI-modifying enzymes from different source-organisms, including *Arabidopsis* PI-synthases (Löfke et al. 2008), invertebrate PIP-kinases and PLC (Carricaburu and Fournier 2001), and mammalian PI phosphatases (Schmid et al. 2004). Although no biochemical characterizations of substrate specificities with regard to fatty acid species has been reported for plant enzymes of PI metabolism other than PI-synthases, the available data suggest that different functional pools of PIs may be generated by enzymes that have preferences for precursor lipids based on the respective fatty acid patterns.

As PIP-kinases are soluble proteins (Rao et al. 1998), which nonetheless modify membrane-lipids, it raises the question how PIP-kinases are recruited to membranes to associate with their particular signaling environment. The possible mechanisms by which PIP-kinases may be guided into particular metabolic contexts include their interaction with membrane lipids, proteins, or both. Data supporting all the options have previously been reported, and it is important to note that a combination of lipid- and protein-interactions has previously been proposed as a requirement for the correct localization of enzymes related to phosphoinositide signaling in mammalian cells (Varnai et al. 2005). Using liposome-binding tests and on-blot-overlay assays, specific binding of the N-terminal domains of *Arabidopsis* type I PIP-kinase 1 to PtdIns4P and PtdIns(4,5)P_2 and phosphatidic acid has been demonstrated (Im et al. 2007). Interestingly, lipids binding to the PIP5K1 protein changed the kinetic parameters for catalysis and increased the specific activity of recombinant PIP5K1 in in vitro (Im et al. 2007), suggesting that PIP-kinases are affected in their catalytic activity by the immediate lipid-environment.

An interesting aspect of PIP kinase–lipid interactions is suggested by the proposed mechanism by which the yeast type I PIP-kinase, Mss4p, is recruited to the plasma membrane of yeast cells. The function of Mss4p is essential for yeast budding, and it has been proposed that PtdIns(4,5)P_2-mediated control of F-actin attachment to the plasma membrane is a prerequisite for the budding process (Desrivieres et al. 1998). Plasma membrane association of Mss4p has been found to depend on the capability of the yeast cells to form complex sphingolipids (Kobayashi et al. 2005), such as inositol-phosphorylceramide (IPC). In the absence of complex sphingolipids, a conditional Mss4p-mutant was not rescued by the expression of wild type Mss4p. This lack of complementation correlated positively with mistargeting of the wild type Mss4p-enzyme to the cytosol (Kobayashi et al. 2005). Based on genetic and pharmacological evidence, it was concluded that (1) Mss4P required complex sphingolipids for plasma membrane-association, and (2) correct targeting of Mss4p was required for its essential physiological function. According to a widely recognized model, sphingolipids in eukaryotic cells may be enriched in detergent-resistant membrane-areas ("lipid-rafts"), which possibly host enzymes and proteins involved in lipid-signaling processes (Laux et al. 2000; Golub et al. 2004; Mongrand et al. 2004; Borner et al. 2005). In this context, it is interesting to note that in mammalian cells, membrane recruitment of ceramide kinase depends on the presence of PtdIns(4,5)P_2 (Kim et al. 2006), suggesting another functional link between sphingolipid- and phosphoinositide-signals. So far no information is available about a function for sphingolipids in the recruitment of PIP-kinases to target membranes in plant cells, and it must be noted that the sphingolipid-composition of plants differs from those of yeast or mammalian cells. In light of a common evolutionary origin of the phosphoinositide system in all eukaryotes, it will be interesting to see whether sphingolipid-dependent PIP-kinase recruitment is shared by representatives of the organismic kingdoms of animals, fungi, and plants.

6.2 Interaction of PIP-Kinases with Proteins

Of particular interest with regard to the recruitment of soluble PIP-kinase proteins to membranes is the finding that type I PIP-kinase activity associates with a monomeric Rac-type GTPase found in growing tobacco pollen tubes (Kost et al. 1999). While the identity of the PIP-kinase isoform(s) representing this activity has not been revealed and it remains unclear whether the interaction was direct or indirect, it is tempting to speculate that plant PIP-kinases may be recruited by small GTPases, as is discussed for mammalian cells (Santarius et al. 2006). In addition to such PIP-kinase–GTPase interactions possibly guiding PIP-kinase localization, it has been demonstrated in *Arabidopsis* roots that the type I PIP-kinase 9 interacts with a cytosolic invertase, controlling cellular carbohydrate-balancing important for root growth (Lou et al. 2007). In addition to the interactions mentioned, it has been demonstrated that recombinant *Arabidopsis* PIP-kinases bind actin and, most interestingly, a number of other lipid-signaling proteins in vitro (Davis et al. 2007); however, the physiological relevance of these interactions is unknown. Still, it is clear that PIP-kinases interact with specific partner proteins, likely dictating the regulatory effects of the PtdIns(4,5)P_2 produced. More experimental work is needed to identify binding partners and to better define the effects of protein–protein interactions on PIP-kinase functionality in vivo.

7 Conclusions

PIP-kinases control the production of PtdInsP_2. Overall, several levels of regulation appear to be at work in the regulation of PtdInsP_2-formation, from organ-specific and inducible expression of PIP-kinases over their dynamic subcellular distribution and interaction with specific partner-proteins to the selectivity of substrate conversion possibly guided by the presence or absence of certain molecular species of substrate lipids. Clearly, the function of PIP-kinases presents the researcher with a multitude of unresolved questions, as is true for the entire phosphoinositide network itself. As research moves forward to ask ever more precise questions, and with the availability of ever better genetic tools and analytical techniques, these are exciting times for the investigation of PIP-kinases as key players in the function of eukaryotic cells.

Acknowledgments We thank Wendy Boss, Imara Perera, and Yang Ju Im (North Carolina State University) for helpful discussion. We also gratefully acknowledge the financial support through an Emmy Noether grant from the German Research Foundation (DFG, to I. H.).

References

Audhya A, Emr SD (2003) Regulation of PI4, 5P2 synthesis by nuclear-cytoplasmic shuttling of the Mss4 lipid kinase. EMBO J 22:4223–4236

Balla T (2005) Inositol-lipid binding motifs: signal integrators through protein-lipid and protein-protein interactions. J Cell Sci 118:2093–2104

Balla T (2006) Phosphoinositide-derived messengers in endocrine signaling. J Endocrinol 188: 135–153

Berridge MJ (1983) Rapid accumulation of inositol trisphosphate reveals that agonists hydrolyse polyphosphoinositides instead of phosphatidylinositol. Biochem J 212:849–858

Borner GH, Sherrier DJ, Weimar T, Michaelson LV, Hawkins ND, Macaskill A, Napier JA, Beale MH, Lilley KS, Dupree P (2005) Analysis of detergent-resistant membranes in Arabidopsis. Evidence for plasma membrane lipid rafts. Plant Physiol 137:104–116

Braun M, Baluska F, von Witsch M, Menzel D (1999) Redistribution of actin, profilin and phosphatidylinositol-4, 5-bisphosphate in growing and maturing root hairs. Planta 209: 435–443

Burden LM, Rao VD, Murray D, Ghirlando R, Doughman SD, Anderson RA, Hurley JH (1999) The flattened face of type II beta phosphatidylinositol phosphate kinase binds acidic phospholipid membranes. Biochemistry 38:15141–15149

Carricaburu V, Fournier B (2001) Phosphoinositide fatty acids regulate phosphatidylinositol 5-kinase, phospholipase C and protein kinase C activities. Eur J Biochem 268:1238–1249

Cho H, Lee D, Lee SH, Ho WK (2005) Receptor-induced depletion of phosphatidylinositol 4, 5-bisphosphate inhibits inwardly rectifying K + channels in a receptor-specific manner. Proc Natl Acad Sci USA 102:4643–4648

Ciruela A, Hinchliffe KA, Divecha N, Irvine RF (2000) Nuclear targeting of the beta isoform of type II phosphatidylinositol phosphate kinase (phosphatidylinositol 5-phosphate 4-kinase) by its alpha-helix 7. Biochem J 346(Pt 3):587–591

Cooke FT, Dove SK, McEwen RK, Painter G, Holmes AB, Hall MN, Michell RH, Parker PJ (1998) The stress-activated phosphatidylinositol 3-phosphate 5-kinase Fab1p is essential for vacuole function in S. cerevisiae. Curr Biol 8:1219–1222

Cote GG, Yueh YG, Crain RC (1996) Phosphoinositide turnover and its role in plant signal transduction. Subcell Biochem 26:317–343

Davis AJ, Im YJ, Dubin JS, Tomer KB, Boss WF (2007) Arabidopsis phosphatidylinositol phosphate kinase 1 binds F-actin and recruits phosphatidylinositol 4-kinase beta1 to the actin cytoskeleton. J Biol Chem 282:14121–14131

Desrivieres S, Cooke FT, Parker PJ, Hall MN (1998) MSS4, a phosphatidylinositol-4-phosphate 5-kinase required for organization of the actin cytoskeleton in Saccharomyces cerevisiae. J Biol Chem 273:15787–15793

Doughman RL, Firestone AJ, Anderson RA (2003) Phosphatidylinositol phosphate kinases put PI4, 5P(2) in its place. J Membr Biol 194:77–89

Dove SK, Cooke FT, Douglas MR, Sayers LG, Parker PJ, Michell RH (1997) Osmotic stress activates phosphatidylinositol-3, 5-bisphosphate synthesis. Nature 390:187–192

Dowd PE, Coursol S, Skirpan AL, Kao TH, Gilroy S (2006) Petunia phospholipase c1 is involved in pollen tube growth. Plant Cell 18:1438–1453

Drobak BK, Watkins PAC, Valenta R, Dove SK, Lloyd CW, Staiger CJ (1994) Inhibition of plant plasma membrane phosphoinositide phospholipase C by the actin-binding protein, profilin. Plant J 6:389–400

Elge S, Brearley C, Xia HJ, Kehr J, Xue HW, Mueller-Roeber B (2001) An Arabidopsis inositol phospholipid kinase strongly expressed in procambial cells: synthesis of PtdIns(4, 5)P2 and PtdIns(3, 4, 5)P3 in insect cells by 5-phosphorylation of precursors. Plant J 26:561–571

Fan LM, Zhao Z, Assmann SM (2004) Guard cells: a dynamic signaling model. Curr Opin Plant Biol 7:537–546

Gary JD, Wurmser AE, Bonangelino CJ, Weisman LS, Emr SD (1998) Fab1p is essential for PtdIns(3)P 5-kinase activity and the maintenance of vacuolar size and membrane homeostasis. J Cell Biol 143:65–79

Golub T, Caroni P (2005) PI(4, 5)P2-dependent microdomain assemblies capture microtubules to promote and control leading edge motility. J Cell Biol 169:151–165

Golub T, Wacha S, Caroni P (2004) Spatial and temporal control of signaling through lipid rafts. Curr Opin Neurobiol 14:542–550

Gubbels MJ, Vaishnava S, Boot N, Dubremetz JF, Striepen B (2006) A MORN-repeat protein is a dynamic component of the Toxoplasma gondii cell division apparatus. J Cell Sci 119: 2236–2245

Heilmann I (2008) Tails wagging the dogs – On phosphoinositides and their fatty acyl moieties. Plant Signal Behav 3:768–771

Heilmann I, Perera IY, Gross W, Boss WF (1999) Changes in phosphoinositide metabolism with days in culture affect signal transduction pathways in *Galdieria sulphuraria*. Plant Physiol 119:1331–1339

Heilmann I, Perera IY, Gross W, Boss WF (2001) Plasma membrane phosphatidylinositol 4, 5-bisphosphate levels decrease with time in culture. Plant Physiol 126:1507–1518

Helling D, Possart A, Cottier S, Klahre U, Kost B (2006) Pollen tube tip growth depends on plasma membrane polarization mediated by tobacco PLC3 activity and endocytic membrane recycling. Plant Cell 18:3519–3534

Hilgemann DW, Feng S, Nasuhoglu C (2001) The complex and intriguing lives of PIP2 with ion channels and transporters. Sci STKE 2001:RE19

Hinchliffe KA, Ciruela A, Irvine RF (1998) PIPkins1, their substrates and their products: new functions for old enzymes. Biochim Biophys Acta 1436:87–104

Im YJ, Davis AJ, Perera IY, Johannes E, Allen NS, Boss WF (2007) The N-terminal membrane occupation and recognition nexus domain of Arabidopsis phosphatidylinositol phosphate kinase 1 regulates enzyme activity. J Biol Chem 282:5443–5452

Kilian J, Whitehead D, Horak J, Wanke D, Weinl S, Batistic O, D'Angelo C, Bornberg-Bauer E, Kudla J, Harter K (2007) The AtGenExpress global stress expression data set: protocols, evaluation and model data analysis of UV-B light, drought and cold stress responses. Plant J 50:347–363

Kim TJ, Mitsutake S, Igarashi Y (2006) The interaction between the pleckstrin homology domain of ceramide kinase and phosphatidylinositol 4, 5-bisphosphate regulates the plasma membrane targeting and ceramide 1-phosphate levels. Biochem Biophys Res Commun 342:611–617

Kobayashi T, Takematsu H, Yamaji T, Hiramoto S, Kozutsumi Y (2005) Disturbance of sphingolipid biosynthesis abrogates the signaling of Mss4, phosphatidylinositol-4-phosphate 5-kinase, in yeast. J Biol Chem 280:18087–18094

König S, Ischebeck T, Lerche J, Stenzel I, Heilmann I (2008) Salt stress-induced association of phosphatidylinositol-4, 5-bisphosphate with clathrin-coated vesicles in plants. Biochem J 415:387–399

König S, Mosblech A, Heilmann I (2007) Stress-inducible and constitutive phosphoinositide pools have distinct fatty acid patterns in *Arabidopsis thaliana*. FASEB J 21:1958–1967

Kost B, Lemichez E, Spielhofer P, Hong Y, Tolias K, Carpenter C, Chua NH (1999) Rac homologues and compartmentalized phosphatidylinositol 4, 5-bisphosphate act in a common pathway to regulate polar pollen tube growth. J Cell Biol 145:317–330

Kunz J, Fuelling A, Kolbe L, Anderson RA (2002) Stereo-specific substrate recognition by phosphatidylinositol phosphate kinases is swapped by changing a single amino acid residue. J Biol Chem 277:5611–5619

Kunz J, Wilson MP, Kisseleva M, Hurley JH, Majerus PW, Anderson RA (2000) The activation loop of phosphatidylinositol phosphate kinases determines signaling specificity. Mol Cell 5:1–11

Kusano H, Testerink C, Vermeer JEM, Tsuge T, Shimada H, Oka A, Munnik T, Aoyama T (2008) The *Arabidopsis* phosphatidylinositol phosphate 5-kinase PIP5K3 is a key regulator of root hair tip growth. Plant Cell 20:367–380

Lacalle RA, Peregil RM, Albar JP, Merino E, Martinez AC, Merida I, Manes S (2007) Type I phosphatidylinositol 4-phosphate 5-kinase controls neutrophil polarity and directional movement. J Cell Biol 179(7):1539–1553

Laux T, Fukami K, Thelen M, Golub T, Frey D, Caroni P (2000) GAP43, MARCKS, and CAP23 modulate PI(4, 5)P(2) at plasmalemmal rafts, and regulate cell cortex actin dynamics through a common mechanism. J Cell Biol 149:1455–1472

Lee Y, Kim YW, Jeon BW, Park KY, Suh SJ, Seo J, Kwak JM, Martinoia E, Hwang I (2007) Phosphatidylinositol 4, 5-bisphosphate is important for stomatal opening. Plant J 52:803–816
Lemmon MA (2003) Phosphoinositide recognition domains. Traffic 4:201–213
Lemmon MA, Ferguson KM, Abrams CS (2002) Pleckstrin homology domains and the cytoskeleton. FEBS Lett 513:71–76
Löfke C, Ischebeck T, König S, Freitag S, Heilmann I (2008) Alternative metabolic fates of phosphatidylinositol produced by PI-synthase isoforms in Arabidopsis thaliana. Biochem J 413:115–124
Lou Y, Gou JY, Xue HW (2007) PIP5K9, an Arabidopsis phosphatidylinositol monophosphate kinase, interacts with a cytosolic invertase to negatively regulate sugar-mediated root growth. Plant Cell 19:163–181
Maple J, Vojta L, Soll J, Moller SG (2007) ARC3 is a stromal Z-ring accessory protein essential for plastid division. EMBO Rep 8:293–299
McEwen RK, Dove SK, Cooke FT, Painter GF, Holmes AB, Shisheva A, Ohya Y, Parker PJ, Michell RH (1999) Complementation analysis in PtdInsP kinase-deficient yeast mutants demonstrates that Schizosaccharomyces pombe and murine Fab1p homologues are phosphatidylinositol 3-phosphate 5-kinases. J Biol Chem 274:33905–33912
Meijer HJ, Berrie CP, Iurisci C, Divecha N, Musgrave A, Munnik T (2001) Identification of a new polyphosphoinositide in plants, phosphatidylinositol 5-monophosphate (PtdIns5P), and its accumulation upon osmotic stress. Biochem J 360:491–498
Meijer HJ, Divecha N, van den Ende H, Musgrave A, Munnik T (1999) Hyperosmotic stress induces rapid synthesis of phosphatidyl-D-inositol 3, 5-bisphosphate in plant cells. Planta 208:294–298
Meijer HJ, Munnik T (2003) Phospholipid-based signaling in plants. Annu Rev Plant Biol 54:265–306
Mikami K, Katagiri T, Iuchi S, Yamaguchi-Shinozaki K, Shinozaki K (1998) A gene encoding phosphatidylinositol-4-phosphate 5-kinase is induced by water stress and abscisic acid in Arabidopsis thaliana. Plant J 15:563–568
Mongrand S, Morel J, Laroche J, Claverol S, Carde JP, Hartmann MA, Bonneu M, Simon-Plas F, Lessire R, Bessoule JJ (2004) Lipid rafts in higher plant cells: purification and characterization of Triton X-100-insoluble microdomains from tobacco plasma membrane. J Biol Chem 279:36277–36286
Mueller-Roeber B, Pical C (2002) Inositol phospholipid metabolism in Arabidopsis. Characterized and putative isoforms of inositol phospholipid kinase and phosphoinositide-specific phospholipase C. Plant Physiol 130:22–46
Perera IY, Davis AJ, Galanopoulou D, Im YJ, Boss WF (2005) Characterization and comparative analysis of Arabidopsis phosphatidylinositol phosphate 5-kinase 10 reveals differences in Arabidopsis and human phosphatidylinositol phosphate kinases. FEBS Lett 579:3427–3432
Perera IY, Heilmann I, Boss WF (1999) Transient and sustained increases in inositol 1, 4, 5-trisphosphate precede the differential growth response in gravistimulated maize pulvini. Proc Natl Acad Sci USA 96:5838–5843
Perera IY, Heilmann I, Chang SC, Boss WF, Kaufman PB (2001) A role for inositol 1, 4, 5-trisphosphate in gravitropic signaling and the retention of cold-perceived gravistimulation of oat shoot pulvini. Plant Physiol 125:1499–1507
Preuss ML, Schmitz AJ, Thole JM, Bonner HK, Otegui MS, Nielsen E (2006) A role for the RabA4b effector protein PI-4Kbeta1 in polarized expansion of root hair cells in Arabidopsis thaliana. J Cell Biol 172:991–998
Rao VD, Misra S, Boronenkov IV, Anderson RA, Hurley JH (1998) Structure of type IIbeta phosphatidylinositol phosphate kinase: a protein kinase fold flattened for interfacial phosphorylation. Cell 94:829–839
Roth MG (2004) Phosphoinositides in constitutive membrane traffic. Physiol Rev 84:699–730
Santarius M, Lee CH, Anderson RA (2006) Supervised membrane swimming: small G-protein lifeguards regulate PIPK signalling and monitor intracellular PtdIns(4, 5)P2 pools. Biochem J 398:1–13

Schmid AC, Wise HM, Mitchell CA, Nussbaum R, Woscholski R (2004) Type II phosphoinositide 5-phosphatases have unique sensitivities towards fatty acid composition and head group phosphorylation. FEBS Lett 576:9–13

Shimada H, Koizumi M, Kuroki K, Mochizuki M, Fujimoto H, Ohta H, Masuda T, Takamiya K (2004) ARC3, a chloroplast division factor, is a chimera of prokaryotic FtsZ and part of eukaryotic phosphatidylinositol-4-phosphate 5-kinase. Plant Cell Physiol 45:960–967

Shyng SL, Cukras CA, Harwood J, Nichols CG (2000) Structural determinants of PIP(2) regulation of inward rectifier K(ATP) channels. J Gen Physiol 116:599–608

Stenzel I, Ischebeck T, König S, Holubowska A, Sporysz M, Hause B, Heilmann I (2008) The type B phosphatidylinositol-4-phosphate 5-kinase 3 is essential for root hair formation in Arabidopsis thaliana. Plant Cell 20:124–141

Stevenson JM, Perera IY, Heilmann I, Persson S, Boss WF (2000) Inositol signaling and plant growth. Trends Plant Sci 5:252–258

Suh BC, Hille B (2005) Regulation of ion channels by phosphatidylinositol 4, 5-bisphosphate. Curr Opin Neurobiol 15:370–378

Takeshima H, Komazaki S, Nishi M, Iino M, Kangawa K (2000) Junctophilins: a novel family of junctional membrane complex proteins. Mol Cell 6:11–22

Van Leeuwen W, Vermeer JEM, Gadella TWJ Jr, Munnik T (2007) Visualisation of phosphatidylinositol 4, 5-bisphosphate in the plasma membrane of suspension-cultured tobacco BY-2 cells and whole Arabidopsis seedlings. Plant J 52:1014–1026

Varnai P, Balla T (1998) Visualization of phosphoinositides that bind pleckstrin homology domains: calcium- and agonist-induced dynamic changes and relationship to myo-[3H]inositol-labeled phosphoinositide pools. J Cell Biol 143:501–510

Varnai P, Bondeva T, Tamas P, Toth B, Buday L, Hunyady L, Balla T (2005) Selective cellular effects of overexpressed pleckstrin-homology domains that recognize PtdIns(3, 4, 5)P3 suggest their interaction with protein binding partners. J Cell Sci 118:4879–4888

Wasteneys GO, Galway ME (2003) Remodeling the cytoskeleton for growth and form: an overview with some new views. Annu Rev Plant Biol 54:691–722

Wenk MR, De Camilli P (2004) Protein-lipid interactions and phosphoinositide metabolism in membrane traffic: insights from vesicle recycling in nerve terminals. Proc Natl Acad Sci USA 101:8262–8269

Westergren T, Dove SK, Sommarin M, Pical C (2001) AtPIP5K1, an Arabidopsis thaliana phosphatidylinositol phosphate kinase, synthesizes PtdIns(3, 4)P(2) and PtdIns(4, 5)P(2) in vitro and is inhibited by phosphorylation. Biochem J 359:583–589

Whatmore J, Morgan CP, Cunningham E, Collison KS, Willison KR, Cockcroft S (1996) ADP-ribosylation factor 1-regulated phospholipase D activity is localized at the plasma membrane and intracellular organelles in HL60 cells. Biochem J 320(Pt 3):785–794

Zimmermann P, Hirsch-Hoffmann M, Hennig L, Gruissem W (2004) GENEVESTIGATOR. Arabidopsis microarray database and analysis toolbox. Plant Physiol 136:2621–2632

Zonia LE, Munnik T (2004) Osmotically-induced cell swelling versus cell shrinking elicits specific changes in phospholipid signals in tobacco pollen tubes. Plant Physiol 134:813–823

Plant Phosphatidylinositol 3-Kinase

Yuree Lee, Teun Munnik, and Youngsook Lee

Abstract Phosphatidylinositol 3-kinase (PI3K) phosphorylates the D-3 position of phosphoinositides. In Arabidopsis, only one PI3K exists, which belongs to the class-III PI3K subfamily which makes phosphatidylinositol 3-phosphate (PtdIns3P). The single *AtPI3K* gene is essential for survival, since loss of its expression results in lethality. Although not much is known about the molecular mechanism of its function, recent studies show that plant PI3K is important for development and signaling, similar to yeast and animal systems. This includes involvement in endocytosis, reactive oxygen species (ROS) production, and transcriptional activity. Many more interesting stories about the role of this enzyme in the core of cellular activities of plants will be unfold as refined technologies are applied to study this important enzyme.

1 Introduction

The phosphatidylinositol 3-kinase (PI3K) family of enzyme is the central player in cell cycle regulation, signaling, and development in animal systems and thus has been studied extensively (reviewed in García et al. 2006). In plants, PI3K is also important for development and signaling (Welters et al. 1994; Jung et al. 2002; Park et al. 2003; Joo et al. 2005; Lee et al. 2008a, 2008b), though it has not been studied as much as in animals. This is mainly because plants without the enzyme cannot survive, and even reduction of expression of the enzyme results in severe retardation in growth and development (Welters et al. 1994). In addition, the low quantity of 3′-phosphorylated inositol lipids makes biochemical detection very difficult. Using diverse methods to overcome these problems, new aspects on the

Y. Lee (✉)
POSTECH-UZH Cooperative Laboratory, Division of Molecular Life Sciences, Pohang University of Science and Technology, Pohang 790-784, Korea
e-mail: ylee@postech.ac.kr

T. Munnik (ed.), *Lipid Signaling in Plants*, Plant Cell Monographs 16,
DOI 10.1007/978-3-642-03873-0_6,

role of PI3K have been found. Here, we briefly review studies on plant PI3K, with the emphasis on its role in signal transduction and vesicle trafficking.

2 Molecular Classification of PI3K

PI3K phosphorylates the D-3 position of inositol phospholipids. Three different classes can be distinguished, based on sequence homology and in vitro substrate specificity (Wymann and Pirola 1998). *Class-I PI3Ks* are heterodimers composed of a regulatory subunit and a PI3K catalytic subunit. They are involved in diverse cellular phenomena, such as control of growth (Leevers et al. 1996), regulation of cell cycle progression (Klippel et al. 1998; Gille and Downward 1999), DNA synthesis (Roche et al. 1994; Vanhaesebroeck et al. 1999), cell survival (Yao and Cooper 1995), actin rearrangements (Servant et al. 2000), and Ca^{2+} channel trafficking (Viard et al. 2004), by generating the phospholipid second messengers, phosphatidylinositol-3,4,5-trisphosphate [PtdIns(3,4,5)P_3], and PtdIns(3,4)P_2 in the plasma membrane of target cells.

Class-II PI3Ks are structurally distinct from the class I PI3Ks, and use only phosphatidylinositol and phosphatidylinositol-4-phosphate as substrates. They are constitutively associated with membrane structures (including plasma and intracellular membranes) and with nuclei. Several lines of evidence suggest a potential role for these enzymes in agonist-mediated signal transduction (Foster et al. 2003), migration of cancer cells (Maffucci et al. 2005), suppression of apoptotic cell death (Kang et al. 2005), exocytosis (Meunier et al. 2005), pattern formation (MacDougall et al. 2004), cytoskeletal organization (Katso et al. 2006), and insulin signaling (Falasca et al. 2007).

Class-III PI3Ks use only phosphatidylinositol as a substrate, producing PtdIns3P. The prototype for this enzyme, Vps34p, was first identified in *Saccharomyces cerevisiae*, where it is required for delivery of soluble proteins to the vacuole (Herman et al. 1992; Schu et al. 1993). Subsequently, a human homolog was identified, and meanwhile Vps34p-related PI3Ks are known to exist in a wide range of eukaryotes, including *Dictyostelium* (Zhou et al. 1995) and *Drosophila* (Linassier et al. 1997), and it is this isoform that is found in plants too (Hong and Verma 1994; Welters et al. 1994; Molendijk and Irvine 1998). Since plants lack the class-I and -II PI3Ks, differences between plant and animal PI3K signaling can be expected.

3 Processes in Plants that Require Normal PI3K Activity

Pharmacological studies using the PI3K inhibitors, Wortmannin (WM) or LY294002 (LY), have implicated a role for PI3K in various physiological events. These include auxin-induced gravitropism (Joo et al. 2005; Jaillais et al. 2006), the formation of infection threads in *Medicago truncatula* roots inoculated with *Sinorhizobium meliloti* (Peleg-Grossman et al. 2007), the salt-tolerance response in Arabidopsis roots (Leshem et al. 2007), ABA-induced stomatal closure (Jung et al. 2002; Park et al. 2003), actin reorganization (Choi et al. 2008), and tip growth in root hairs (Lee et al. 2008a). PI3K is thought to modulate these processes by regulating endocytosis and reactive oxygen species (ROS) production. Another potential mechanism of plant PI3K action is via modulation of transcriptional activity (Bunney et al. 2000).

Molecular genetic evidence suggests that PI3K is crucial for plant development, both in vegetative and reproductive organs. Using antisense to reduce *PI3K* expression was found to impair leaf and stem development (Welters et al. 1994), while a T-DNA insertion KO mutant is lethal and impaired in pollen development (Lee et al. 2008b).

3.1 *Roles of PI3K in Endocytosis and Protein Trafficking*

In vivo, PtdIns3P can be tracked using a genetically encoded biosensor, which is a fusion between GFP (or any other color) and two FYVE (from Fab1, YOTB, Vac1 and EEA1) domains in tandem which specifically bind PtdIns3P (Gillooly et al. 2000). Stably expressing lines of Arabidopsis plants and suspension-cultured tobacco BY2 cells revealed strong colocalization with the late endosomal/prevacuolar marker, AtRABF2b, and was found to partially colocalize with the endosomal tracer FM4-64 (Voigt et al. 2005; Vermeer et al. 2006; see chapter, "Imaging lipids in living plants").

PI3K seems to play a role at different stages of vesicular trafficking, depending on the cell type, as PI3K inhibitors have been found to suppress the initial uptake of FM4-64 in tobacco cells and Arabidopsis roots under salt stress (Emans et al. 2002; Leshem et al. 2007), the endocytic recycling of endosomes to the plasma membrane in tobacco pollen tubes (Helling et al. 2006), and the fusion of late endosomes with the tonoplast (Lee et al. 2008a).

PI3K-related endocytic routes have been suggested to deliver molecules important for plant signal transduction. For example, diacylglycerol (DAG), generated from PtdIns(4,5)P_2 (and/or PtdIns4P) via PI-PLC hydrolysis (see chapter, "The Emerging Roles of Phospholipase C in Plant Growth and Development") is delivered to a specific region of the plasma membrane in pollen tubes (Helling et al. 2006). Inhibition of PI3K disturbed the DAG localization pattern, as judged by the accumulation of a DAG biosensor into YFP_{FYVE}-labeled endocytic compartment, with no or only weak accumulation at the plasma membrane. Based on these results, Helling et al. (2006) suggested that DAG, generated at the flanks of the pollen-tube tip, is internalized and reinserted into the plasma membrane at the apex via PI3K-related endocytic routes. PI3K-related endocytic routes also deliver PIN (auxin efflux transporters) proteins to specific regions of the plasma membrane. Vesicular trafficking between the plasma membrane and endosomal compartments is necessary to maintain the polar distribution of PIN proteins (Geldner et al. 2001; 2003; Abas et al. 2006). This polar PIN localization is the primary factor determining the direction of auxin flow in roots during the gravity response (Wisniewska et al. 2006). Inhibition of PI3K by wortmannin leads to the relocalization of PIN2 into wortmannin-induced endosomal compartments, but did not affect PIN1 localization, suggesting a specific role of PI3K in PIN2 cycling (Jaillais et al. 2006).

PI3K is essential for normal trafficking of proteins to and from vacuoles. PI3K inhibitor interferes with targeting of vacuolar proteins in tobacco suspension cells (Matsuoka et al. 1995). Reduction of free PtdIns3P level by expression of PtdIns3P-binding protein interferes with vacuolar protein targeting in Arabidopsis protoplasts (Kim et al. 2001). Moreover, PI3K inhibitors cause swelling or vacuolation of the prevacuolar compartment (Tse et al. 2004) and block retrograde transport of vacuolar sorting receptors to the TGN (daSilva et al. 2005; Oliviusson et al. 2006). Vesicular trafficking mediated by PI3K may rely on dynamic changes in the actin cytoskeleton, since profilin, a regulator of actin dynamics, binds PI3K in phosphorylation-dependent manner in *Phaseolus vulgaris* (Aparicio-Fabre et al. 2006).

3.2 Roles of PI3K in ROS Generation and ROS-Mediated Signaling

ROS production is reduced by PI3K inhibitors in various cell types of plants including root hair, guard cell, and pollen tube (Foreman et al. 2003; Park et al. 2003; Kwak et al. 2003; Potocký et al. 2007). This effect of the inhibitors is likely due to the inhibition of PI3K-mediated activation/delivery of NADPH oxidase (NOX), a major source of ROS. Based on the function of PI3K in endosomal trafficking, there are three possible mechanisms by which PI3K could modulate ROS production (Fig. 1): (1) by affecting the activity or distribution of plasma-membrane localized NOX, (2) by transferring exogenously produced ROS into cytoplasm, (3) by regulating NOX activity in the endosomes. The first and the second hypotheses are based on NOX localization at the plasma membrane and ROS being produced at the apoplast, whereas the third one suggests that ROS is produced inside endosomes. Two recent papers suggest that PI3K-dependent plasma membrane internalization is linked to ROS production. Leshem et al. (2007) reported that salt stress triggers PI3K-dependent plasma membrane internalization and ROS production within endosomes of root cells. Intracellular ROS were encapsulated by endosomal membrane in root cells, and were interpreted as the product of NOX internalized from the plasma membrane in response to salt stress. In root hair cells, Lee et al. (2008a) also showed ROS inside endosomes and the level of ROS in these organelles was reduced after treatment with LY.

In animal cells, PtdIns3P stimulates endosomal ROS generation through binding the PX domain of $p40^{phox}$, a soluble factor of the NOX complex (Ellson et al. 2006). More and more endocytic organelles are considered as intracellular signaling

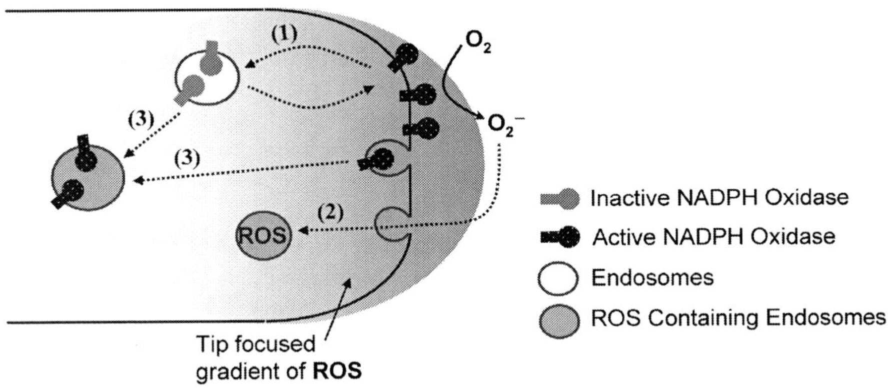

Fig. 1 Diagram depicting three possible mechanisms of modulation of ROS generation by PI3K in root hair system. First, PI3K-related endocytic recycling route can affect activity or distribution of plasma membrane localized NADPH Oxidase (NOX), which produce ROS outside of cells (1). Second, PI3K can affect import of exogenously produced ROS into cytosol. Diffusion of ROS across lipid layers is very low and endocytosis mediated by PtdIns3P may contribute to transfer of ROS into the cell (2). Finally, NOX can be recruited to the endosomes, where PtdIns3P is localized, and produce ROS inside endosomes (3)

stations, where downstream cascades are activated after receptor–ligand complexes are internalized into the endosomal compartment (Miaczynska et al. 2004), and endosomal ROS plays a key role in regulating their activity (Li et al. 2006). In plants, neither cytosolic factors of the NOX complex, nor intramembrane ROS generation mediated by NOX has been shown. However, activated forms of receptors, e.g., the LRR receptors, FLAGELLIN SENSITIVE2 (FLS2), and the steroid receptor kinase BRI1, have been observed to accumulate in endosomes (Robatzek et al. 2006; Geldner et al. 2007). Whether and how endosome-localized FLS2 and BRI1 activate downstream signaling cascades is unknown, but these results show the potential of PI3K and endosomes as plant signaling components.

3.3 *Roles of PI3K in Nucleus*

Involvement of PI3K in nuclear function is based on the observation that PI3Ks are associated with active nuclear transcription sites in plants (Bunney et al. 2000). A catalytically active PI3K was demonstrated in isolated, detergent-resistant plant nuclei and a monoclonal antibody raised against a truncated form of the soybean PI3K was located at, or near, active transcription sites, both in the nucleolus and in the nucleoplasm. The presence of PI3K and its product PtdIns3P in the nucleus is not unique to plants. Nuclear PtdIns3P has been reported in BHK cells, human fibroblasts, and HL-60 cells (Gillooly et al. 2000; Visnjic et al. 2003). In HL-60 cells, PtdIns3P level increases at G2/M phase of the cell cycle (Visnjic et al. 2003), suggesting a role of the lipid in cell cycle. Although there are no reports yet for a link of PI3K with transcriptional regulation in animal cells, class I PI3Ks of animals have been reported as important factors in various steps of cell division, such as control of cell cycle entry (Álvarez et al. 2003), regulation of cyclin/Cdk (Olson et al. 1995; Klippel et al. 1998), and progression of G2/M phases (Álvarez et al. 2001).

Distinct features of FYVE containing proteins of plants also provide some clues to the possible roles of PI3K in nucleus. Among the 16 proteins having FYVE domain in Arabidopsis, nine contain tandem repeats of regulator of chromosome condensation-1 (RCC1)-like domain (Van Leeuwen et al., 2004). RCC1 is a protein that contains seven tandem repeats of a domain of about 50–60 amino acids and functions as a nucleotide exchange factor for the nuclear Ran G-protein (Bischoff and Ponstingl 1991). It regulates diverse biological processes including G1/S phase transition (Matsumoto and Beach 1991), mating (Clark and Sprague 1989), the processing and export of mRNAs (Kadowaki et al. 1993), and chromatin condensation (Sazer and Nurse 1994) in various eukaryotes. Although RCC1 homologs have not been reported from plants, the RCC1 domain is found in many plant proteins. Some of these may function similarly as RCC1 in the nucleus as suggested from the in vitro assay using purified GST-RCC1 domain of PRAF1 in Arabidopsis, which demonstrated the guanine nucleotide exchange of a Rab small GTPase (Jensen et al. 2001). Further studies are required to understand whether plant FYVE proteins with a RCC1-like domain function similar to RCC1 proteins in animals.

3.4 Roles of PI3K in Growth and Development of Plants

The broad and significant role of PI3K in plant growth and development was first suggested by the results of Welters et al. (1994), who regenerated Arabidopsis plants from calli transformed with an antisense construct of *AtVPS34*. Regeneration of shoot and root was slow, flowers were formed, but the seed-set was poor. The next generation of plants could not survive in kanamycin-containing medium. Even in normal medium without antibiotics, leaves were abnormal in shape, and petiole elongation and stem formation were impaired. In soybean, a *PI3K* is induced during nodule development when membrane proliferation is required to establish the peribacteroid membrane (Hong and Verma 1994).

In addition to the role of PI3K in vegetative tissue development, the enzyme also plays a role in reproductive tissues (Lee et al. 2008b). When *VPS34/vps34* heterozygous plants, harboring a T-DNA insertion, were self-fertilized, a segregation ratio of 1:1:0 for wild type, heterozygous-, and homozygous mutant plants, respectively, was obtained, thus homozygous mutants without *PI3K* expression were lacking. These results suggested a gametophytic defect, which was further supported by reciprocal crosses between heterozygous and wild-type plants. There was no transmission of the T-DNA insertion allele through the male gametophyte, indicating an important role for PI3K during male gametophyte development. Male gametophytes of the heterozygous mutant plants showed reduced number of nuclei, enlarged vacuoles, and reduced germination rate more often than the wild type.

Is PI3K also required for female gametophyte development? Considering its basic functions in cellular trafficking and nuclear division, it seems likely that PI3K is also important for the development of female gametophyte, especially because it involves many rounds of cell division. Consistent with this explanation, plants expressing the *AtVPS34*-antisense construct were severely reduced in seed-set (Welters et al. 1994), which suggested a role of the PI3K in development and/or function of female reproductive organ as well. But how can we explain then the results from the reciprocal crosses, which suggested that female allele of the *pi3k* knockout is transmitted normally? A potential explanation for this discrepancy is that a sufficient quantity of sporophytic gene product persists to complete megagametogenesis. The PI3K enzyme from previous generation may provide the lipid during development of female gametophyte; female gametophyte inherits more PI3K from the cytosol of previous generation than male gametophyte, is able to complete its development normally. Such an explanation is consistent with the observation that only the later steps in the mutant male gamete development were defective, while the early steps of the process was normal.

4 Signal Transduction Pathway Activated at Downstream of PI3K

The remarkably diverse and potent effect of PI3K-mediated signal transduction in animal cells depends on the interaction of the lipid products of the kinases with multiple protein partners. Signaling molecules related to class I PI3K of animal cells have been identified and include phosphatidylinositol 3-phosphatase (PTEN), 3-phosphoinositide-dependent protein kinase-1 (PDK1), and protein kinase B (PKB)/c-Akt. PTEN is a lipid phosphatase which hydrolyzes the phosphate from D3-position of inositol phospholipids, thus attenuating PI3K-mediated signaling. AKT1 is recruited to the plasma membrane by binding PtdIns(3,4,5)P_3 which is produced by activated class I PI3K and is phosphorylated by PDK1. Phosphorylated AKT1, in turn, phosphorylates numerous target proteins and thereby induces multifaceted effects of PI3K. Identification of plant homologs of mammalian downstream molecules of PI3K can be one way to obtain further clues about plant PI3K signaling. Indeed, homologues of PTEN and PDK1 have been identified in plants (Deak et al. 1999; Gupta et al. 2002). AtPTEN1 was shown to have phosphatase activity against PtdIns(3,4,5)P_3 and to play an important role in pollen maturation after mitosis (Gupta et al. 2002). AtPDK1 has been shown to complement a yeast mutant lacking PDK1, to activate mammalian PKB in vitro, and to bind a broad range of lipids, including PA, PtdIns3P, PtdIns(3,4)P_2, PtdIns(4,5)P_2, and PtdIns(3,4,5)P_3 (Deak et al. 1999). Further analysis of AtPDK1 revealed its substrates, AGC2-1 kinase (OXI1), PINOID, and S6 kinase (Anthony et al. 2004; Otterhag et al. 2006; Zegzouti et al. 2006) and its capacity to be regulated by PA and PtdIns(4,5)P_2 (Anthony et al. 2004).

Although the conservation of AtPTEN1 and AtPDK1 indicate that PI3K-related signaling is well conserved, plants lack the class-I PI3K and its product, PtdIns (3,4,5)P_3 (Meijer and Munnik 2003; Munnik and Testerink 2009). Thus, plant cells may differ from animal cells in the downstream pathways. But if AtPTEN1 is indeed the plant ortholog of animal PTEN, then what is its substrate(s)? Plants do contain PtdIns3P and PtdIns(3,5)P_2 as 3-phosphorylated phosphoinositides (Meijer et al., 1999; Munnik and Testerink 2009), which may function as substrates of PTEN. Similarly, if the AtPDK1 is, indeed, the plant ortholog of animal PDK1, the immediate question to be resolved is whether any of the 3-phosphorylated phosphoinositides provides a specific site for recruitment of AtPDK1.

Signaling targets related to class-III PI3K of animal and yeast cells are proteins that contain FYVE-, PH-, or PX-domains. Plants contain several of such proteins (Van Leeuwen et al. 2004) and some of them have even been shown to bind PtdIns3P (Deak et al. 1999; Jensen et al. 2001; Heras and Drøbak 2002; Vermeer et al. 2006), although none have been characterized functionally in depth. Interestingly, Arabidopsis contains three proteins with a putative PX domain which are members of the sorting nexin-like (SNX) proteins which are involved in endosomal trafficking in yeast and animals. Recently, AtSNX1 has been shown to play a role in auxin-carrier trafficking which is sensitive to WM and was proposed to define a sorting endosome (Jaillais et al. 2006, 2008). Clearly, more of these studies are required to reveal the roles of proteins functioning downstream of PtdIns3P in vesicular trafficking and protein targeting in plants.

5 Conclusion and Prospects

PI3K is emerging as important enzyme in plant signal transduction, regulating ROS production and modulating the recycling of plasma membrane proteins and lipids. It is also likely to be important for cell cycle regulation, via its role in nuclear division and transcriptional control. To better understand the whole picture of PI3K-mediated pathways, downstream effector molecules have to be identified. In addition, improved genetic analyses are required, using conditional mutations driven by specific promoters. The cell biology of PtdIns3P dynamic and its targets may also provide further information on the role of PI3K in signaling and protein trafficking. With all these new tools, exciting times are ahead of us.

References

Abas L, Benjamins R, Malenica N, Paciorek T, Wiśniewska J, Moulinier-Anzola JC, Sieberer T, Friml J, Luschnig C (2006) Intracellular trafficking and proteolysis of the Arabidopsis auxin-efflux facilitator PIN2 are involved in root gravitropism. Nat Cell Biol 8:249–256

Álvarez B, Martínez AC, Burgering BM, Carrera AC (2001) Forkhead transcription factors contribute to execution of the mitotic programme in mammals. Nature 413:744–747

Álvarez B, Garrido E, Garcia-Sanz JA, Carrera AC (2003) PI3K activation regulates cell division time by coordinated control of cell mass and cell cycle progression rate. J Biol Chem 278:26466–26473

Anthony RG, Henriques R, Helfer A, Mészáros T, Rios G, Testerink C, Munnik T, Deák M, Koncz C, Bögre L (2004) A protein kinase target of a PDK1 signalling pathway is involved in root hair growth in Arabidopsis. EMBO J 23:572–581

Aparicio-Fabre R, Guillén G, Estrada G, Olivares-Grajales J, Gurrola G, Sánchez F (2006) Profilin tyrosine phosphorylation in poly-L-proline-binding regions inhibits binding to phosphoinositide 3-kinase in *Phaseolus vulgaris*. Plant J 47:491–500

Bischoff FR, Ponstingl H (1991) Catalysis of guanine nucleotide exchange on Ran by the mitotic regulator RCC1. Nature 354:80–82

Bunney TD, Watkins PA, Beven AF, Shaw PJ, Hernandez LE, Lomonossoff GP, Shanks M, Peart J, Drøbak BK (2000) Association of phosphatidylinositol 3-kinase with nuclear transcription sites in higher plants. Plant Cell 12:1679–1688

Choi Y, Lee Y, Jeon BW, Staiger CJ, Lee Y (2008) Phosphatidylinositol 3- and 4-phosphate modulate actin reorganization in dayflower guard cells. Plant Cell Environ 31:366–377

Clark KL, Sprague GFJ (1989) Yeast pheromone response pathway: characterization of a supressor that restores mating to receptorless mutants. Mol Cell Biol 9:2682–2694

daSilva LLP, Taylor JP, Hadlington JL, Hanton SL, Snowden CJ, Fox SJ, Foresti O, Brandizzi F, Denecke J (2005) Receptor salvage from the prevacuolar compartment is essential for efficient vacuolar protein targeting. Plant Cell 17:132–148

Deak M, Casamayor A, Currie RA, Downes CP, Alessi DR (1999) Characterisation of a plant 3-phosphoinositide-dependent protein kinase-1 homologue which contains a pleckstrin homology domain. FEBS Lett 451:220–226

Ellson CD, Davidson K, Anderson K, Stephens LR, Hawkins PT (2006) PtdIns3P binding to the PX domain of p40phox is a physiological signal in NADPH oxidase activation. EMBO J 25: 4468–4478

Emans N, Zimmermann S, Fischer R (2002) Uptake of a fluorescent marker in plant cells is sensitive to brefeldin A and wortmannin. Plant Cell 14:71–86

Falasca M, Hughes WE, Dominguez V, Sala G, Fostira F, Fang MQ, Cazzolli R, Shepherd PR, James DE, Maffucci T (2007) The role of phosphoinositide 3-kinase C2alpha in insulin signaling. J Biol Chem 282:28226–28236

Foreman J, Demidchik V, Bothwell JHF (2003) Reactive oxygen species produced by NADPH oxidase regulate plant cell growth. Nature 422:442–446

Foster FM, Traer CJ, Abraham SM, Fry MJ (2003) The phosphoinositide (PI) 3-kinase family. J Cell Sci 116:3037–3040

García Z, Kumar A, Marqués M, Cortés I, Carrera AC (2006) Phosphoinositide 3-kinase controls early and late events in mammalian cell division. EMBO J 25:655–661

Geldner N, Friml J, Stierhof YD, Jürgens G, Palme K (2001) Auxin transport inhibitors block PIN1 cycling and vesicle trafficking. Nature 413:425–428

Geldner N, Anders N, Wolters H, Keicher J, Kornberger W, Muller P, Delbarre A, Ueda T, Nakano A, Jürgens G (2003) The Arabidopsis GNOM ARF-GEF mediates endosomal recycling, auxin transport, and auxin-dependent plant growth. Cell 112:219–230

Geldner N, Hyman DL, Wang X, Schumacher K, Chory J (2007) Endosomal signaling of plant steroid receptor kinase BRI1. Genes Dev 21:1598–1602

Gille H, Downward J (1999) Multiple Ras effector pathways contribute to G_1 cell cycle progression. J Biol Chem 274:22033–22040

Gillooly DJ, Morrow IC, Lindsay M, Gould R, Bryant NJ, Gaullier JM, Parton RG, Stenmark H (2000) Localization of phosphatidylinositol 3-phosphate in yeast and mammalian cells. EMBO J 19:4577–4588

Gupta R, Ting JT, Sokolov LN, Johnson SA, Luan S (2002) A tumor suppressor homolog, AtPTEN1, is essential for pollen development in Arabidopsis. Plant Cell 14:2495–2507

Helling D, Possart A, Cottier S, Klahre U, Kost B (2006) Pollen tube tip growth depends on plasma membrane polarization mediated by tobacco PLC3 activity and endocytic membrane recycling. Plant Cell 18:3519–3534

Heras B, Drøbak BK (2002) PARF-1: an Arabidopsis thaliana FYVE-domain protein displaying a novel eukaryotic domain structure and phosphoinositide affinity. J Exp Bot 53:565–567

Herman PK, Stack JH, Emr SD (1992) An essential role for a protein and lipid kinase complex in secretory protein sorting. Trends Cell Biol 2:363–368

Hong Z, Verma DPS (1994) A PtdIns 3-kinase is induced during soybean nodule organogenesis and is associated with membrane proliferation. Proc Natl Acad Sci USA 91:9617–9621

Jaillais Y, Fobis-Loisy I, Miège C, Rollin C, Gaude T (2006) AtSNX1 defines an endosome for auxin-carrier trafficking in Arabidopsis. Nature 443:106–109

Jaillais Y, Fobis-Loisy I, Miège C, Gaude T (2008) Evidence for a sorting endosome in Arabidopsis root cells. Plant J. 53:237–247

Jensen RB, La Cour T, Albrethsen J, Nielsen M, Skriver K (2001) FYVE zinc-finger proteins in the plant model Arabidopsis thaliana: identification of PtdIns3P-binding residues by comparison of classic and variant FYVE domains. Biochem J 359:165–173

Joo JH, Yoo HJ, Hwang I, Lee JS, Nam KH, Bae YS (2005) Auxin-induced reactive oxygen species production requires the activation of phosphatidylinositol 3-kinase. FEBS Lett 14:1243–1248

Jung JY, Kim YW, Kwak JM, Hwang JU, Young J, Schroeder JI, Hwang I, Lee Y (2002) Phosphatidylinositol 3-and 4-phosphate are required for normal stomatal movements. Plant Cell 14:2399–2412

Kadowaki T, Goldfarb D, Spitz LM, Tartakoff AM, Ohno M (1993) Regulation of RNA processing and transport by a nuclear guanine nucleotide release protein and members of the Ras superfamily. EMBO J 12:2929–2937

Kang S, Song J, Kang J, Kang H, Lee D, Lee Y, Park D (2005) Suppression of the alpha-isoform of class II phosphoinositide 3-kinase gene expression leads to apoptotic cell death. Biochem Biophys Res Commun 329:6–10

Katso RM, Pardo OE, Palamidessi A, Franz CM, Marinov M, De Laurentiis A, Downward J, Scita G, Ridley AJ, Waterfield MD, Arcaro A (2006) Phosphoinositide 3-Kinase C2beta regulates cytoskeletal organization and cell migration via Rac-dependent mechanisms. Mol Biol Cell 17:3729–3744

Kim DH, Eu YJ, Yoo CM, Kim YW, Pih KT, Jin JB, Kim SJ, Stenmark H, Hwang I (2001) Trafficking of phosphatidylinositol 3-phosphate from the trans-Golgi network to the lumen of the central vacuole in plant cells. Plant Cell 13:287–301

Klippel A, Escobedo MA, Wachowicz MS, Apell G, Brown TW, Giedlin MA, Kavanaugh WM, Williams LT (1998) Activation of phosphatidylinositol 3-kinase is sufficient for cell cycle entry and promotes cellular changes characteristic of oncogenic transformation. Mol Cell Biol 18:5699–5711

Kwak JM, Mori IC, Pei ZM, Leonhardt N, Torres MA, Dangl JL, Bloom RE, Bodde S, Jones JD, Schroeder JI (2003) NADPH oxidase AtrbohD and AtrbohF genes function in ROS-dependent ABA signaling in Arabidopsis. EMBO J 22:2623–2633

Lee Y, Bak G, Choi Y, Chuang W-I, Cho H-T, Lee Y (2008a) Roles of Phosphatidylinositol 3-kinase in root hair growth. Plant Physiol 147:624–635

Lee Y, Kim E-S, Choi Y, Hwang I, Staiger CJ, Chung Y-Y, Lee Y (2008b) The Arabidopsis phosphatidylinositol-3-kinase is essential for pollen development. Plant Physiol 147:1886–1897

Leevers SJ, Weinkove D, MacDougall LK, Hafen E, Waterfield MD (1996) The Drosophila phosphoinositide 3-kinase Dp110 promotes cell growth. EMBO J 15:6584–6594

Leshem Y, Seri L, Levine A (2007) Induction of phosphatidylinositol 3-kinase-mediated endocytosis by salt stress leads to intracellular production of reactive oxygen species and salt tolerance. Plant J 51:185–197

Li Q, Harraz MM, Zhou W, Zhang LN, Ding W, Zhang Y, Eggleston T, Yeaman C, Banfi B, Engelhardt JF (2006) Nox2 and Rac1 regulate H2O2-dependent recruitment of TRAF6 to endosomal interleukin-1 receptor complexes. Mol Cell Biol 26:140–154

Linassier C, MacDougall LK, Domin J, Waterfield MD (1997) Molecular cloning and biochemical characterization of a *Drosophila* PtdIns-specific phosphoinositide 3-kinase. Biochem J 321: 849–856

MacDougall LK, Gagou ME, Leevers SJ, Hafen E, Waterfield MD (2004) Targeted expression of the class II phosphoinositide 3-kinase in Drosophila melanogaster reveals lipid kinase-dependent effects on patterning and interactions with receptor signaling pathways. Mol Cell Biol 24: 796–808

Maffucci T, Cooke FT, Foster FM, Traer CJ, Fry MJ, Falasca M (2005) Class II phosphoinositide 3-kinase defines a novel signaling pathway in cell migration. J Cell Biol 169:789–799

Matsumoto T, Beach D (1991) Premature initiation of mitosis in yeast lacking RCC1 or an interacting GTPase. Cell 66:347–360

Matsuoka K, Bassham DC, Raikhel NV, Nakamura K (1995) Different sensitivity to wortmannin of two vacuolar sorting signals indicates the presence of distinct sorting machineries in tobacco cells. J Cell Biol 130:1307–1318

Meijer HJ, Munnik T (2003) Phospholipid-based signaling in plants. Annu Rev Plant Biol 54:265–306

Meunier FA, Osborne SL, Hammond GR, Cooke FT, Parker PJ, Domin J, Schiavo G (2005) PI3-Kinase C2{alpha} is essential for ATP-dependent priming of neurosecretory granule exocytosis. Mol Biol Cell 16:4841–4851

Miaczynska M, Pelkmans L, Zerial M (2004) Not just a sink: endosomes in control of signal transduction. Curr Opin Cell Biol 16:400–406

Molendijk AJ, Irvine RF (1998) Inositide signalling in Chlamydomonas: characterization of a phosphatidylinositol 3-kinase gene. Plant Mol Biol 37:53–66

Munnik T, Testerink C (2009) Plant phospholipid signaling: "in a nutshell". J Lipid Res 50: S260–S265

Oliviusson P, Heinzerling O, Hillmer S, Hinz G, Tse YC, Jiang L, Robinson DG (2006) Plant retromer, localized to the prevacuolar compartment and microvesicles in *Arabidopsis*, may interact with vacuolar sorting receptors. Plant Cell 18:1239–1252

Olson MF, Ashworth A, Hall A (1995) An essential role for Rho, Rac and Cdc42 GTPases in cell cycle progression through G1. Science 269:1270–1272

Otterhag L, Gustavsson N, Alsterfjord M, Pical C, Lehrach H, Gobom J, Sommarin M (2006) Arabidopsis PDK1: identification of sites important for activity and downstream phosphorylation of S6 kinase. Biochimie 88:11–21

Park KY, Jung JY, Park J, Hwang JU, Kim YW, Hwang I, Lee Y (2003) A role for phosphatidylinositol 3-phosphate in abscisic acid-induced reactive oxygen species generation in guard cells. Plant Physiol 132:92–98

Peleg-Grossman S, Volpin H, Levine A (2007) Root hair curling and Rhizobium infection in Medicago truncatula are mediated by phosphatidylinositide-regulated endocytosis and reactive oxygen species. J Exp Bot 58:1637–1649

Potocký M, Jones MA, Bezvoda R, Smirnoff N, Zárský V (2007) Reactive oxygen species produced by NADPH oxidase are involved in pollen tube growth. New Phytol 174:742–751

Robatzek S, Chinchilla D, Boller T (2006) Ligand-induced endocytosis of the pattern recognition receptor FLS2 in Arabidopsis. Genes Dev 20:537–542

Roche S, Koegl M, Courtneidge SA (1994) The phosphatidylinositol 3-kinase alpha is required for DNA synthesis induced by some, but not all, growth factors. Proc Natl Acad Sci USA 91:9185–9189

Sazer S, Nurse P (1994) A fission yeast RCC1-related protein is required for the mitosis to interphase transition. EMBO J 13:606–615

Schu PV, Takegawa K, Fry MJ, Stack JH, Waterfield MD, Emr SD (1993) Phosphatidylinositol 3-kinase encoded by yeast VPS34 gene essential for protein sorting. Science 260:88–91

Servant G, Weiner OD, Herzmark P, Balla T, Sedat JW, Bourne HR (2000) Polarization of Chemoattractant Receptor Signaling During Neutrophil Chemotaxis. Science 287:1037–1040

Tse YC, Mo B, Hillmer S, Zhao M, Lo SW, Robinson DG, Jiang L (2004) Identification of multivesicular bodies as prevacuolar compartments in *Nicotiana tabacum* BY-2 cells. Plant Cell 16:672–693

Vanhaesebroeck B, Jones GE, Allen WE, Zicha D, Hooshmand-Rad R (1999) Distinct PI(3)Ks mediate mitogenic signalling and cell migration in macrophages. Nat Cell Biol 1:69–71

van Leeuwen W, Okrész L, Bögre L, Munnik T (2004) Learning the lipid language of plant signalling. Trends Plant Sci 9:378–384

Vermeer JEM, van Leeuwen W, Tobeña-Santamaria R, Laxalt AM, Jones DR, Divecha N, Gadella TWJ Jr, Munnik T (2006) Visualization of PtdIns3P dynamics in living plant cells. Plant J 47:687–700

Viard P, Butcher AJ, Halet G, Davies A, Nurnberg B, Heblich F, Dolphin AC (2004) PI3K promotes voltage-dependent calcium channel trafficking to the plasma membrane. Nat Neurosci 7:939–946

Visnjic D, Curic J, Crljen V, Batinic D, Volinia S, Banfic H (2003) Nuclear phosphoinositide 3-kinase C2beta activation during G2/M phase of the cell cycle in HL-60 cells. Biochim Biophys Acta 1631:61–71

Voigt B, Timmers AC, Samaj J, Hlavacka A, Ueda T, Preuss M, Nielsen E, Mathur J, Emans N, Stenmark H, Nakano A, Baluska F, Menzel D (2005) Actin-based motility of endosomes is linked to the polar tip growth of root hairs. Eur J Cell Biol 84:609–621

Welters P, Takegawa K, Emr SD, Chrispeels MJ (1994) ATVPS34, a PtdIns 3-kinase of *Arabidopsis thaliana* is an essential protein with homology to a calcium-dependent lipid-binding domain. Proc Natl Acad Sci USA 91:11398–11402

Wisniewska J, Xu J, Seifertová D, Brewer PB, Ruzicka K, Blilou I, Rouquié D, Benková E, Scheres B, Friml J (2006) Polar PIN localization directs auxin flow in plants. Science 312:883

Wymann MP, Pirola L (1998) Structure and function of phosphoinositide 3-kinases. Biochim Biophys Acta 8:127–150

Yao R, Cooper GM (1995) Requirement for phosphatidylinositol-3 kinase in the prevention of apoptosis by nerve growth factor. Science 267:2003–2006

Zegzouti H, Anthony RG, Jahchan N, Bögre L, Christensen SK (2006) Phosphorylation and activation of PINOID by the phospholipid signaling kinase 3-phosphoinositide-dependent protein kinase 1 (PDK1) in Arabidopsis. Proc Natl Acad Sci USA 103:6404–6409

Zhou K, Takegawa K, Emr SD, Firtel RA (1995) A PtdIns (PI) kinase gene family in *Dictyostelium discoideum*: Biological roles of putative mammalian p110 and yeast Vps34p PI3-kinase homologues during growth and development. Mol Cell Biol 15:5645–5651

Diacylglycerol Kinase

Steven A. Arisz and Teun Munnik

Abstract Diacylglycerol kinase (DGK, EC 2.7.1.107) is a lipid kinase that phosphorylates diacylglycerol (DAG) to generate phosphatidic acid (PA). DGK belongs to a well-conserved family of proteins found in diverse species, such as *Saccharomyces cerevisiae*, *Drosophila melanogaster*, *Caenorhabditis elegans*, mammals, plants, and bacteria (Merida et al. Biochem J 409:1–18, 2008). In eukaryotes, DGK can be activated in response to various stimuli, and its product PA is emerging as a novel lipid second messenger (Munnik, Trends Plant Sci 6:227–233, 2001; Testerink and Munnik, Trends Plant Sci 10:368–375, 2005). Despite similarities to the mammalian system, plant DGKs display several distinctive features.

1 Regulatory Domains and Classification

Eukaryotic DGKs typically contain a kinase domain consisting of a conserved catalytic domain with a presumed ATP-binding site, and an accessory domain that is thought to interact with the catalytic domain. The ten mammalian DGK isozymes are characterized by the presence of two C1-type cysteine-rich domains, C1a and C1b, together with a great variety of additional domains that are involved in their regulation and form the basis of their classification into five different types (Fig. 1a; Merida et al. 2008). In plants, less structural diversity has evolved. Basically, their DGKs can be subdivided into three phylogenetic clusters (Fig. 1b, c; Gomez-Merino et al. 2004). The *Arabidopsis thaliana* genome encodes seven DGK isozymes. Cluster I, epitomized by *At*DGK1 and *At*DGK2, resembles the mammalian DGKε type, containing the kinase domain, the C1a and C1b, and a transmembrane helix.

T. Munnik (✉)
Section Plant Physiology, Swammerdam Institute for Life Sciences, University of Amsterdam, Science Park 904, NL-1098 XH Amsterdam, The Netherlands
e-mail: t.munnik@uva.nl

T. Munnik (ed.), *Lipid Signaling in Plants*, Plant Cell Monographs 16,
DOI 10.1007/978-3-642-03873-0_7, © Springer-Verlag Berlin Heidelberg 2010

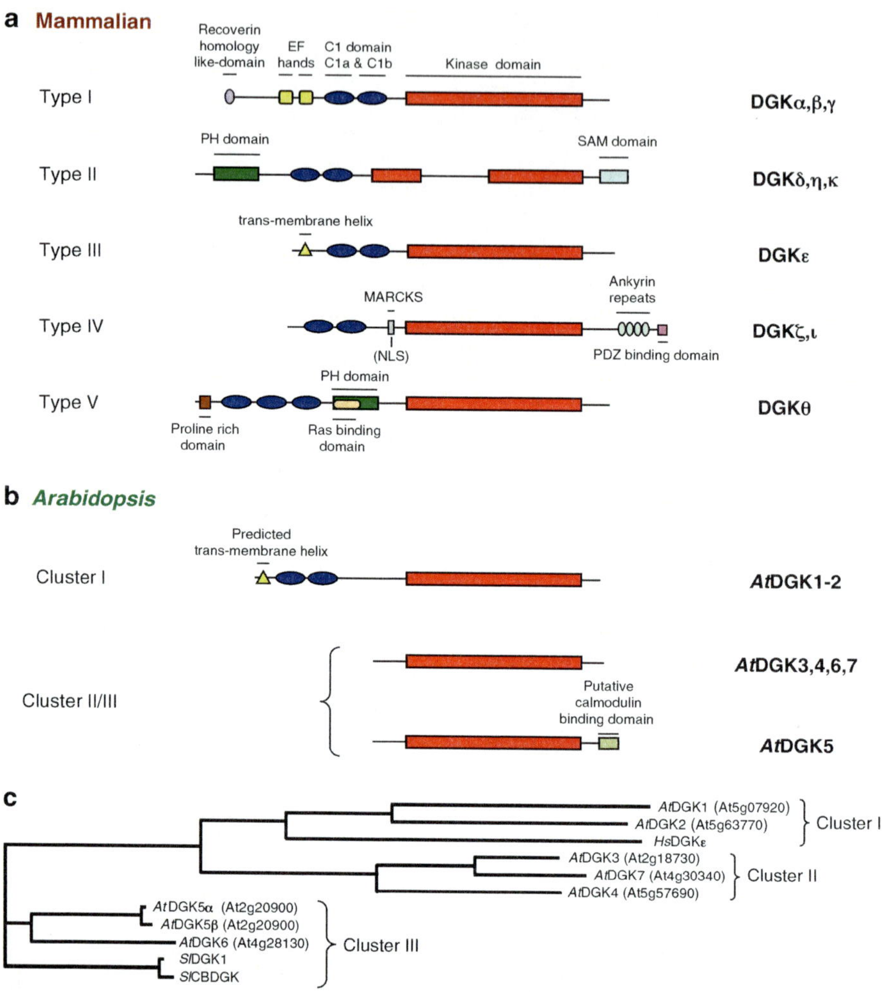

Fig. 1 Protein structures of members of the DGK family of mammals (**a**) and *Arabidopsis* (**b**), and phylogenetic clustering of *Arabidopsis* and tomato DGKs (**c**). (**a**) Characteristic regulatory domains of typical mammalian DGK isoforms are indicated. *Abbreviations*: MARCKS, myristoylated alanine-rich C-kinase substrate, NLS, nuclear localization sequence, PH, pleckstrin homology, SAM, sterile α-motif. (**b**) The three basic structures of plant DGKs present in *Arabidopsis*. The putative carboxyterminal calmodulin-binding domain is in a splice variant of *At*DGK5, and is similar to the structure of *Sl*CBDGK in tomato. In addition to the domains indicated, there are two putative Ca^{2+} binding EF hand domains in *At*DGK1 (Katagiri et al. 1996). *At*DGK7 has an incomplete kinase-accessory domain (Gomez-Merino et al. 2005). (**c**) Phylogram (Clustalw2) based on amino acid sequences of the *Arabidopsis* DGKs, containing tomato DGK and human *Hs*DGKε for comparison. All known plant DGKs fall into these phylogenetic clusters. (Adapted from Arisz et al. 2009)

The other DGKs, *At*DGK3–7, share the conserved kinase domain but lack any other motifs. Their molecular mass of ~55 kDa is also considerably lower than that of *At*DGK1 and *At*DGK2 (~80 kDa). Importantly, in cluster III, *At*DGK5 has two alternative splice variants, *At*DGK5α and *At*DGK5β. The latter has a C-terminal extension of around 25 amino acids, which has been speculated to represent a calmodulin-binding domain (CBD), reminiscent of *Sl*CBDGK, and a CBD-containing splice variant of *Sl*DGK1 of tomato (Snedden and Blumwald 2000). The *Sl*CBDGK protein binds membranes in vitro, in a Ca^{2+}- and calmodulin-dependent manner. Although such translocations are an important mode of activation for animal DGKs, a calmodulin-dependent mechanism would be unique to plants.

The absence of C1 domains from clusters II and III DGK is not prohibitive for enzyme activity as heterologously expressed *At*DGK7 is biochemically active in vitro (Gomez-Merino et al. 2005). Although the function of the prototypical C1 domain of mammalian PKC is to bind DAG and translocate it to the membrane, it is unclear whether the cysteine-rich region in DGK has a similar function.

2 Localization of the Activity and Gene Expression

DGK activity has been detected in a wide variety of plant systems and tissues (Arisz et al. 2009). Membrane fractionation experiments showed that the activity is primarily associated with the plasma membrane, and, less abundantly, with the cytoskeleton, the chloroplast, and nuclear membranes (Munnik et al. 1998). In *Arabidopsis* suspension-cultured cells, activity was associated with fractions enriched in plasma membrane, Golgi, and nuclear membranes, but the highest specific activity was found in the ER fraction (Vaultier et al. 2008).

According to Genevestigator and Northern analyses, DGKs are widely and differentially expressed (Arisz et al. 2009). A promoter–GUS fusion of *AtDGK2* revealed a shift from a uniform expression pattern during the first week after germination towards a more selective expression near the vascular bundles (Gomez-Merino et al. 2004). Also, in the first few weeks, expression altered from the root–shoot junction to the root tip; in later stages exhibiting a marked preference for vascular tissue again.

3 DGK Signaling

The rapid formation of PA has been established as a general element in the acute response to several abiotic and biotic environmental stress conditions (Testerink and Munnik 2005). Two pathways are predominantly held responsible for this. Phospholipase D (PLD), which directly generates PA through hydrolysis of a structural phospholipid, such as phosphatidylcholine (PC) and phosphatidylethanolamine (PE), and the combined activities of phospholipase C (PLC) and DGK, where PI-PLC hydrolyzes PIP and/or PIP_2, and DGK phosphorylates the resulting DAG to produce PA (see chapter, “Phosphatidylinositol 4-Phosphate is Required for Tip Growth in Arabidopsis thaliana”). As PLC also produces water-soluble inositol polyphosphates, which may act on releasing cytoplasmic Ca^{2+} or may have other signaling functions (Munnik and Testerink 2009), the result of a PLD- or PLC/DGK-generated PA will not be the same. Moreover, PIP and PIP_2 are emerging as lipid-second messengers themselves, which would be attenuated by PLC.

In order to experimentally distinguish between both pathways, a differential ^{32}P-labeling protocol can be used (Arisz et al. 2009, Munnik 2001, Munnik et al. 1998). When plants, plant parts, or cells, are only briefly (1–5 min) incubated with a solution containing ^{32}P-orthophosphate, the radiolabel is rapidly taken up and incorporated into organic molecules, ATP being among the first. When subsequently a stress treatment is given, which triggers the PLC/DGK pathway, the resultant PA is labeled as soon as ATP. In contrast, after longer prelabeling times (hours–days), ^{32}P-orthophosphate is also incorporated into the phosphodiesters of structural phospholipids such as PE and PC. Only then, PLD activity would be able to generate a radiolabeled PA. Using this technique, DGK has been implicated in various abiotic- and biotic-stress responses.

3.1 Abiotic Stresses

Following the seminal work on the PLC/DGK- and PLD activation during salt and osmotic stresses in green algae, similar responses were found in higher plants (Arisz et al. 2009). In general, exposure to adverse environmental factors, such as cold, drought, osmotic stress, and mechanical wounding, have been found to trigger a rapid and transient accumulation of PA. Cold shock treatment at 0°C of *Arabidopsis* suspension-cultured cells triggered a PA increase which was in the first 10 min predominantly caused by PLC/DGK activity, albeit, with different kinetics, with a minor contribution of PLD (Ruelland et al. 2002). Gene expression studies in *Arabidopsis* demonstrated the transcriptional upregulation of cluster I DGKs. Although this cannot account for the rapid activation of the pathway, it may suggest the involvement of *At*DGK1 and *At*DGK2 in the cold response. Using pharmacological inhibitors, a set of cold-regulated genes has been implicated downstream of the PLC/DGK- or PLD pathway, or both (Vergnolle et al. 2005). In *Arabidopsis* seedlings, PLDδ has been suggested to be instrumental in cold acclimation, which confers freezing resistance in plants previously exposed to chilling temperatures (Li et al. 2004). During episodes of freezing and thawing, which induce a dramatic drop in extracellular water potential, PLD rather than PLC/DGK prevails; PLD activity being unleashed as membranes get severely disorganized. Nevertheless, in rosette leaves of adult *Arabidopsis* plants, the fatty acid composition of a minor fraction of the freezing-induced PA pools suggested that it is derived from monogalactosyl diacylglycerol (MGDG) (Welti et al. 2002). This implies that, upon freezing, MGDG is hydrolyzed to provide a DAG as substrate for DGK. A galactolipid:galactolipid galactosyltransferase (GGGT) could account for the generation of DAG from MGDG (Benning and Ohta 2005).

3.2 *Biotic Stresses*

A number of studies have demonstrated the formation of PA in plant–pathogen interactions. Accumulation of PA from the PLC/DGK pathway seems to be one of the earliest host responses as evidenced upon treatment with various pathogen-derived elicitors, virulent or avirulent, e.g., xylanase, chitotetraose, flagellin, and Avr4. A late phase of PA formation was found to be dominated by PLD activity (Andersson et al. 2006). Reactive oxygen species (ROS) accumulate with similar kinetics, reflecting a hypersensitive response, which involves not only programmed cell death of infected tissues, but also expression of pathogenesis-related (PR) genes. Evidence suggests that ROS accumulation in some cases is a downstream response to PA. A potential PA target in oxidative stress signaling is the protein kinase OXI1 (Anthony et al. 2004).

The transcription of a DGK gene of rice, *OsBIDK1*, was shown to be induced by treatment with benzothiadiazole, a structural analog of salicylic acid (SA), which is involved in the induced resistance response during infection by virulent pathogens, such as the blast fungus *Magnaporthe grisea* (Zhang et al. 2008). Transgenic tobacco plants, overexpressing *OsBIDK1* exhibited enhanced resistance against tobacco mosaic virus and *Phytophtora parasitica*. Also, the *AtDGK5* gene of *Arabidopsis* is suggested to be involved in the response to virulent *Pseudomonas* bacteria, since a *Atdgk5* knock-out mutant failed to induce *PR1* gene expression during either pathogen infection or treatment with SA (S. Van Wees, B. Van Schooten, T. Munnik, pers. com.). Moreover, the basal resistance against the bacteria was decreased. Both *OsBIDK1* and *AtDGK5* are members of cluster III. As a GFP-fusion of the latter localizes to the plasma membrane, it will be interesting to learn whether this is due to the CBD domain (Fig. 1b) or via another mechanism.

4 Aspects of DAG, PA and the Regulation of DGK Activity

PA is also a central intermediate in the de novo synthesis of glycerolipids, being primarily formed in the plastid and the ER by two successive acylations of glycerol-3-phosphate. PA is either converted to CDP-DAG for the synthesis of PG, PI, and PS, or hydrolyzed to DAG for the synthesis of PE, PC, and galactolipids. As PA and DAG are the precursors for different lipid classes, their interconversion by PA phosphatase (PAP; see chapter, "Phosphatidic Acid Phosphatases in Seed Plants") and DGK may function in regulating membrane compositions. In nonplants, DGK activity has been proposed as an attenuator of DAG signaling, a salvage pathway for PA formation, and a step in the resynthesis of PI via CDP-DAG, following PLC-catalyzed hydrolysis of PIP_2 in signaling. The latter function has been assigned to mammalian DGKs of the ε type, which structurally resemble *At*DGK1 and *At*DGK2. In plants, this may be different, though, as signaling PA can also be metabolized to diacylglycerol pyrophosphate (Van Schooten et al. 2006; see chapter, "Diacylglycerol Pyrophosphate, a Novel Plant Signaling Lipid"). The enzyme responsible, PA kinase (PAK), is lacking from mammalian cells and the encoding gene is still to be identified in plants (Van Schooten et al. 2006).

Since PA is emerging as a second messenger in plants, DGK activity has mainly been associated with the regulation of responses to environmental stress conditions. In recent years, PA's mode of action has become more defined as PA is capable of binding various target proteins (Arisz et al. 2009; Testerink et al. 2004). Such binding can provide a mechanism of membrane recruitment and/or activation or inhibition of protein kinases/phosphatases (see chapter, "Phosphatidic acid – an Electrostatic/Hydrogen- Bond Switch?").

Although plant DGKs lack the multitude of structural regulatory domains of mammalian DGKs, several potential mechanisms of activity modulation are emerging.

1. The C1a and C1b domains of cluster I DGKs could interact with DAG, or alternatively, with proteins that modulate enzyme activity, or that are themselves targets of the resulting lipid signal (Merida et al. 2008).
2. *At*DGK1 has two putative EF hands which implies regulation through Ca^{2+}-binding (Katagiri et al. 1996).
3. Some DGKs of cluster III have a CBD which suggests regulation by calmodulin and intracellular Ca^{2+} increases. The CBD in these DGKs is exclusively encoded by alternatively spliced gene transcripts, which suggests additional regulation on the level of mRNA splicing.
4. Bacterial DGKB (e.g., of *Staphylococcus aureus*) lacks C1 domains, similar to *Arabidopsis* DGK3–7, and requires dimerization for activity (Jerga et al. 2009). Even though the amino acids responsible for dimerization in bacteria are lacking, similar interactions of *Arabidopsis* proteins cannot be excluded.
5. All DGK activities assayed in vitro were highly dependent on the lipid environment, generally showing dependence of anionic lipids such as polyphosphoinositides and PG. Mg^{2+} is essential for DGK′s interaction with the lipid bilayer, and the three aspartate residues responsible for Mg^{2+} coordination in DGKB are conserved in mammalian and *Arabidopsis* DGKs. Binding of the membrane triggers a conformational change, which increases the affinity for ATP, and is required for enzyme activity.

Although the targeting of DGK to specific subcellular locations and its integration into signaling pathways may largely depend on auxillary domains, the kinase domains are critical to the interaction with phospholipids and the change to an active conformation, providing a clue to its regulation. Specificity of DGK function in plants may then be achieved through the availability and access to substrate DAG and cofactors, in combination with posttranscriptional/posttranslational modifications and interactions with protein partners. Although in stress signaling DGK is thought to act in tandem with PLC, DAG substrate is also generated potentially via the activities of inositolphosphorylceramide synthase (IPCS; Wang et al. 2008), GGGT (Benning and Ohta 2005), or nonspecific PLC (NPC; Gaude et al. 2008; see chapter, "The Emerging Roles of Phospholipase C in Plant Growth and Development"), which have been implicated in pathogen resistance signaling (IPCS), and the response to phosphate limitation (GGGT, NPC).

In the future, the functions of DGK will be further studied using (multiple-) knock-out and overexpression mutants, combined with fluorescent probes to monitor phospholipid signaling in vivo. Understanding the functional coupling of DGK and other enzymes in PA metabolism to specific stress responses may help to pave the way for generating more tolerant crops.

References

Andersson MX, Kourtchenko O, Dangl JL, Mackey D, Ellerstrom M (2006) Phospholipase-dependent signalling during the AvrRpm1- and AvrRpt2-induced disease resistance responses in Arabidopsis thaliana. Plant J 47:947–959

Anthony RG, Henriques R, Helfer A, Meszaros T, Rios G, Testerink C, Munnik T, Deak M, Koncz C, Bogre L (2004) A protein kinase target of a PDK1 signalling pathway is involved in root hair growth in Arabidopsis. EMBO J 23:572–581

Arisz SA, Testerink C, Munnik T (2009) Plant PA signaling via diacylglycerol kinase. Biochim Biophys Acta 1791:869–875

Benning C, Ohta H (2005) Three enzyme systems for galactoglycerolipid biosynthesis are coordinately regulated in plants. J Biol Chem 280:2397–2400

Gaude N, Nakamura Y, Scheible WR, Ohta H, Dormann P (2008) Phospholipase C5 (NPC5) is involved in galactolipid accumulation during phosphate limitation in leaves of Arabidopsis. Plant J 56:28–39

Gomez-Merino FC, Brearley CA, Ornatowska M, Abdel-Haliem ME, Zanor MI, Mueller-Roeber B (2004) AtDGK2, a novel diacylglycerol kinase from Arabidopsis thaliana, phosphorylates 1-stearoyl-2-arachidonoyl-sn-glycerol and 1, 2-dioleoyl-sn-glycerol and exhibits cold-inducible gene expression. J Biol Chem 279:8230–8241

Gomez-Merino FC, Arana-Ceballos FA, Trejo-Tellez LI, Skirycz A, Brearley CA, Dormann P, Mueller-Roeber B (2005) Arabidopsis AtDGK7, the smallest member of plant diacylglycerol kinases (DGKs), displays unique biochemical features and saturates at low substrate concentration: the DGK inhibitor R59022 differentially affects AtDGK2 and AtDGK7 activity in vitro and alters plant growth and development. J Biol Chem 280:34888–34899

Jerga A, Miller DJ, White SW, Rock CO (2009) Molecular determinants for interfacial binding and conformational change in a soluble diacylglycerol kinase. J Biol Chem 284:7246–7254

Katagiri T, Mizoguchi T, Shinozaki K (1996) Molecular cloning of a cDNA encoding diacylglycerol kinase (DGK) in *Arabidopsis thaliana*. Plant Mol Biol 30:647–653

Li W, Li M, Zhang W, Welti R, Wang X (2004) The plasma membrane-bound phospholipase Dδ enhances freezing tolerance in *Arabidopsis thaliana*. Nat Biotechnol 22:427–433

Merida I, Avila-Flores A, Merino E (2008) Diacylglycerol kinases: at the hub of cell signalling. Biochem J 409:1–18

Munnik T (2001) Phosphatidic acid: an emerging plant lipid second messenger. Trends Plant Sci 6:227–233

Munnik T, Testerink C (2009) Plant phospholipid signaling – ‘in a nutshell’. J Lipid Res 50: S260–S265

Munnik T, Irvine RF, Musgrave A (1998) Phospholipid signalling in plants. Biochim Biophys Acta 1389:222–272

Ruelland E, Cantrel C, Gawer M, Kader JC, Zachowski A (2002) Activation of phospholipases C and D is an early response to a cold exposure in Arabidopsis suspension cells. Plant Physiol 130:999–1007

Snedden WA, Blumwald E (2000) Alternative splicing of a novel diacylglycerol kinase in tomato leads to a calmodulin-binding isoform. Plant J 24:317–326

Testerink C, Munnik T (2005) Phosphatidic acid: a multifunctional stress signaling lipid in plants. Trends Plant Sci 10:368–375

Testerink C, Dekker HL, Lim ZY, Johns MK, Holmes AB, Koster CG, Ktistakis NT, Munnik T (2004) Isolation and identification of phosphatidic acid targets from plants. Plant J 39:527–536

Van Schooten B, Testerink C, Munnik T (2006) Signalling diacylglycerol pyrophosphate, a new phosphatidic acid metabolite. Biochim Biophys Acta 1761:151–159

Vaultier MN, Cantrel C, Guerbette F, Boutte Y, Vergnolle C, Cicek D, Bolte S, Zachowski A, Ruelland E (2008) The hydrophobic segment of Arabidopsis thaliana cluster I diacylglycerol kinases is sufficient to target the proteins to cell membranes. FEBS Lett 582:1743–1748

Vergnolle C, Vaultier MN, Taconnat L, Renou JP, Kader JC, Zachowski A, Ruelland E (2005) The cold-induced early activation of phospholipase C and D pathways determines the response of two distinct clusters of genes in Arabidopsis cell suspensions. Plant Physiol 139:1217–1233

Wang W, Yang X, Tangchaiburana S, Ndeh R, Markham JE, Tsegaye Y, Dunn TM, Wang GL, Bellizzi M, Parsons JF, Morrissey D, Bravo JE, Lynch DV, Xiao S (2008) An inositolphosphorylceramide synthase is involved in regulation of plant programmed cell death associated with defense in Arabidopsis. Plant Cell 20:3163–3179

Welti R, Li W, Li M, Sang Y, Biesiada H, Zhou HE, Rajashekar CB, Williams TD, Wang X (2002) Profiling membrane lipids in plant stress responses. Role of phospholipase D α in freezing-induced lipid changes in Arabidopsis. J Biol Chem 277:31994–32002

Zhang W, Chen J, Zhang H, Song F (2008) Overexpression of a rice diacylglycerol kinase gene OsBIDK1 enhances disease resistance in transgenic tobacco. Mol Cells 26:258–264

Part III
Phosphatases

Signaling and the Polyphosphoinositide Phosphatases from Plants

Glenda E. Gillaspy

Abstract Polyphosphoinositides (PPIs) are important signaling molecules involved in membrane and vesicular trafficking events, regulation of the cytoskeleton, and response to stress in plants. PPI phosphatases dephosphorylate the D-3, -4, and -5 positions of the inositol ring within these molecules, and as such, they have the ability to regulate the PPIs present in the cell. There are three categories of PPI phosphatases, and they are classified as to which position phosphate is removed during catalysis. Within each category, there are subgroups of enzymes with unique protein domains, substrate preferences, and expression patterns. Recent genetic analyses of knock-out mutants highlight the fact that PPI phosphatases regulate crucial events during the growth and development of plants.

1 Introduction

All organisms require the ability to respond to their environment in order to adapt and survive. In response to extracellular signals, many organisms utilize polyphosphoinositides (PPIs) , a group of signaling molecules that play a role in a variety of critical eukaryotic cellular processes (Di Paolo and De Camilli 2006; Meijer and Munnik 2003). These molecules are unique in that their soluble head group contains the polyol *myo*-inositol (inositol) linked via the D-1 position of the inositol ring to phosphate, glycerol, and fatty acids. The soluble inositol headgroup is known to be phosphorylated at different positions, and the resulting PPIs constitute a type of cellular language in which specific phosphates convey different information within the cell. As such, the kinases and phosphatases that modify PtdIns and PPIs have the

G.E. Gillaspy
Department of Biochemistry, Virginia Tech, Blacksburg, Virginia 24061, USA
e-mail: gillaspy@vt.edu

T. Munnik (ed.), *Lipid Signaling in Plants*, Plant Cell Monographs 16,
DOI 10.1007/978-3-642-03873-0_8,

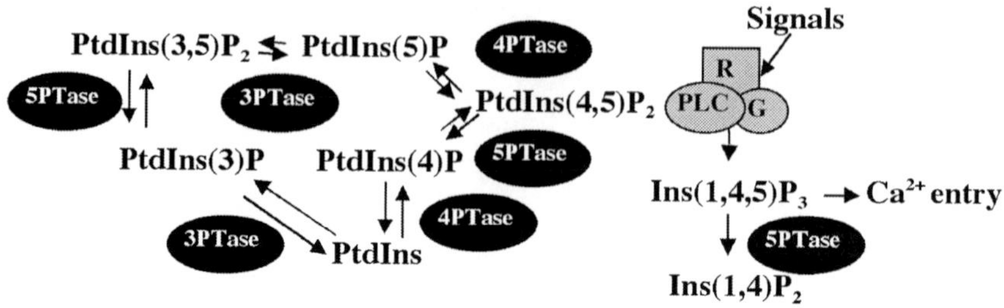

Fig. 1 Phosphatase action on polyphosphoinositides. PPI phosphatases are indicated by black ovals. Each *double set of arrows* indicates a point where a kinase can act to catalyze the opposing dephosphorylation reaction. PtdIns(4,5)P_2 can serve as a substrate for phospholipase C (PLC), which could act in conjunction with a G protein (G) and a receptor (R) to convey external signals to the interior of the cell via production of second messenger Ins(1,4,5)P_3

potential to regulate signaling within higher eukaryotes, including plants (Fig. 1) (Blero et al. 2007). This chapter focuses on the PPI phosphatases and will provide an overview of the three categories of PPI phosphatases that are known to impact plant biochemistry, signal transduction, growth, and development.

PPI phosphatases differ in their substrate preferences and appear to discriminate primarily based on the position and number of phosphates present in the inositol ring (Blero et al. 2007). For example, PPIs containing a 3-phosphate can serve as substrates for the 3-phosphatases (3PTases), while PPIs containing a 4-or 5-phosphate can serve as substrates for the 4-phosphatases (4PTases) and 5-phosphatases (5PTases), respectively. Within each category of PPI phosphatase, there are additional substrate preference requirements, with some PPI phosphatases acting to hydrolyze mono-, bis- or tris-phosphorylated substrates, while some are more specific with regard to the number of phosphate groups present in the substrate. The reader should keep in mind that for each phosphatase that removes a phosphate from a PPI, there is almost always a corresponding kinase that catalyzes the addition of a phosphate at the same position (Clarke et al. 2007) (see chapters, "Phosphatidylinositol 4-Phosphate is Required for Tip Growth in Arabidopsis thaliana,"."PIP-Kinases as Key Regulators of plant Function," "Plant Phosphatidylinositol 3-Kinase").

PPIs can also serve as substrates for phospholipases, and this has an impact on signal transduction as well (Bargmann and Munnik 2006; Wang 2004) (Fig. 1). For example, PtdIns(4,5)P_2 is hydrolyzed by phospholipase C (PLC) in plants in response to drought, salt, cold, gravity, etc., and production of second messenger inositol 1,4,5-trisphosphate [Ins(1,4,5)P_3] occurs (Stevenson et al. 2000). Since Ins(1,4,5)P_3 is water soluble, it can act within the interior of the cell. In animal cells, Ins(1,4,5)P_3 is known to stimulate intracellular calcium release (Berridge et al. 1999; Berridge 1993, 2005), and data from plants support this same action within plant cells although no Ins(1,4,5)P_3 receptor has yet been identified (Krinke et al. 2007).

2 The PPI 3-Phosphatases

The PPI 3-phosphatases remove a phosphate from the D-3 position of various PPI substrates and are of two general types, the phosphatase and tensin homologues (PTENs) (Harris et al. 2008; Tamguney and Stokoe 2007) and the myotubularins (MTMs) (Bolis et al. 2007). Both types of phosphatases also dephosphorylate tyrosine residues on substrate proteins and are thus classified as dual-specificity protein/lipid phosphatases (Taylor and Dixon 2003).

2.1 *PTEN*

PTEN (E.C. 3.1.3.67) was first identified in humans as a gene locus mutated in several types of tumors, including Cowden and Bannayan-Zonana syndromes, in which patients often suffer from increased risk of breast and thyroid cancers (Maehama and Dixon 1998). In vitro, human PTEN can hydrolyze PtdIns3P, PtdIns(3,4)P_2, and PtdIns(3,4,5)P_3. In vivo, however, PtdIns(3,4,5)P_3 is probably the most important substrate, such that PTEN activity results in an increase in PtdIns(4,5)P_2 (Leslie and Downes 2004). Lack of PtdIns(3,4,5)P_3 hydrolysis in humans is thought to increase activation of the AKT protein kinase signaling pathway, which contributes to the aberrant, increased growth in tumors (Li and Ross 2007). Some additional functions of PTEN, such as hyperpolarization of pancreatic cells, requires both the lipid and protein phosphatase activities (Ning et al. 2006).

The Arabidopsis genome contains one gene shown to encode a functional PTEN protein (AtPEN1; At5G39400), and two other related genes (At3G19420 and At3G50110) (Gupta et al. 2002) (Table 1). All the three AtPENs contain a conserved dual specificity phosphatase domain, like the animal PTEN proteins. AtPTEN1 recombinant protein was shown to dephosphorylate a ^{32}P-labeled synthetic peptide and to release free phosphate from PtdIns(3,4,5)P_3. Gupta et al., also found AtPTEN1 expression exclusively in pollen grains during later stages of development. RNAi suppression of AtPTEN1 resulted in pollen cell death after mitosis. Thus, this gene plays a critical role in pollen development. These results verify the expected activity of AtPTEN1 in vitro, and its functional significance in development, but they do not address the intriguing and somewhat perplexing issue of the in vivo substrate(s) for this enzyme. Many different studies performed in plants have failed to detect PtdIns(3,4,5)P_3, and the general consensus is that PtdIns (3,4,5)P_3 either does not exist in plants, or is only present at very low levels under specific conditions not currently identified. Thus, it seems likely that removal of a D-3 phosphate from a PPI substrate is required for pollen development, but which PPI molecule provides the substrate for AtPTEN1 is not known. It is also possible that AtPTEN1 acts only on an as yet, unidentified protein substrate required for pollen development.

Table 1 PPI phosphatases in plants

3PTases		#	Predicted or known substrates	Domains	Mutants altered in:
	PTEN	3	**PtdIns(3,4,5)$_3$** *PtdIns(3,4)P$_2$* PtdIns3P	Dual specificity phosphatase	Pollen development
	MTM	2	*PtdIns(3,5)P$_2$* *PtdIns3P*	Myotubularin-related, GRAM, MBOAT, VPS51,67	n.d.
4PTases					
	RHD4	1	**PtdIns4P**	Sac I homology	Root hair growth
	SAC6, 8	2	*PtdIns4P*	SacI homology	n.d.
5PTases					
	5PTase 1-10 (*cvp2*, *rhd5*)	10	**Ins(1,4,5)P$_3$** **Ins(1,3,4,5)P$_4$** **PtdIns(4,5)P$_2$** **PtdIns(3,4,5)$_3$**	Exo_endo_phos	*5ptase1,2:* Seedling development, *cvp2*: cotyledon vascular development, *rhd5*: root hair development
	5PTase11	1	**PtdIns(3,4,5)$_3$** **PtdIns(4,5)P$_2$** **PtdIns(3,5)P$_2$**	Exo_endo_phos	Seedling development
	FRA3	1	**Ins(1,4,5)P$_3$** **PtdIns(4,5)P$_2$** **PtdIns(3,4,5)$_3$**	Exo_endo_phos	Fiber cell development
	5PTase 12-14	4	**Ins(1,4,5)P$_3$ and/or PtdIns(4,5)P$_2$**	Exo_endo_phos, WD40	*5ptase13*: response to stress, sugars, auxin and blue light
	SAC9	1	*PtdIns(4,5)P$_2$*	SacI homology, WW	Stress responses
	SAC 1	1	**PtdIns(3,5)P$_2$**	SacI homology	Fiber cell, vascular development
	SAC2-5	4	*PtdIns(3,5)P$_2$*	SacI homology	n.d.

The name of each PPI phosphatase and number of *Arabidopsis* genes encoding these enzymes are listed. Predicted substrates are *italicized*, while *bold-faced substrates* have been addressed experimentally. Domains were predicted and annotated as found in Pfam (http://pfam.janelia.org/), and are described in the text. Inclusion required an *e* value of less than 0.01. Phenotypes of mutants are described in the text *n.d.* not determined

2.2 *Myotubularins*

The MTMs (E.C.3.1.3.48) are dual specificity lipid/protein phosphatases that act on PtdIns3P and PtdIns(3,5)P$_2$ in animals leading to the production of PtdIns and PtdIns5P (Taylor and Dixon 2003; Laporte et al. 2002). The MTM gene family in humans has 14 members (Clague and Lorenzo 2005), with three associated with genetic diseases including X-linked myotubular myopathy (Laporte et al. 1996). Interestingly, there are many MTMs either predicted or shown to be catalytically inactive phosphatases, and these inactive MTMs form heterodimers with the active MTMs. Heterodimerization results in enhancement of the active MTM subunit and/or alterations in the subcellular location of the heterodimer (Robinson and Dixon 2005). Removal of the D-3 phosphate from PtdIns3P and PtdIns(3,5)P_2 in animals is speculated to function in negative regulation of endosomal–lysosomal membrane trafficking, although the mechanistic details are unknown (Xue et al. 2003; Parrish et al. 2004). Animal MTM proteins have a potential link to calcium signaling in that one isoform (MTM6) associates with the Ca^{2+}-activated K^+ channel via a special domain called the GRAM domain (Srivastava et al. 2005).

Arabidopsis contains two potential MTM genes (At5g04540, At3g10550) (Table 1). Both are predicted to encode catalytically active MTM enzymes, which may mean that plant genomes are unique in their lack of catalytically inactive MTM isoforms. Both contain the conserved MTM-related domain, and both also contain the GRAM, MBOAT and VPS51,67 domains. The GRAM domain is found in the animal MTMs, glucosyltransferases, and other putative membrane-associated proteins (Doerks et al. 2000). As mentioned previously, the GRAM domain is required for the animal MTM6 protein to bind to a K^+ channel in the plasma membrane (Srivastava et al. 2005). The MBOAT (membrane bound *O*-acyl transferase) and VPS51,67 domains are unique to the plant MTMs. The VPS51,67 domain is found in a number of components of vesicular transport in eukaryotes (Reggiori et al. 2003), suggesting that plant MTMs may also function in trafficking. No biochemical or genetic studies on the plant MTMs have been reported, although it is clear that rice contains several putative MTM genes, and potential Arabidopsis T-DNA knock-out mutants are available within the Salk mutant collection.

3 The PPI 4-phosphatases

Less is known about the 4-phosphatases (E.C. 3.1.3.66), and only recent data from both animals and plants has highlighted the functions of this class of enzymes. The first characterized 4-phosphatase was the invasion plasmid D gene (IpgD) from the pathogen *Shigella flexneri* (Niebuhr et al. 2000). IpgD is homologous to the animal inositol 4-phosphatases which hydrolyze specific soluble inositol phosphates that contain a 4-phosphate. However, IpgD was shown instead to remove a 4-phosphate from PtdIns(4,5)P_2 resulting in conversion to PtdIns5P. This action of the pathogen is thought to manipulate host cell PPI metabolism, leading to alterations in membrane and the actin cytoskeleton (Pendaries et al. 2006), indicating the importance of PtdIns(4,5)P_2 for the animal cytoskeleton. Subsequently, other pathogens were found to encode PPI-phosphatases (such as SopB, etc.) (Marcus et al. 2001). Thus, it appears that manipulation of host PPI levels may be a common strategy for some animal pathogens. This possibility offers potential new drug targets for treatments (Kuijl et al. 2007).

Two types of mammalian PPI 4-phosphatases were identified that contain very little similarity at the amino acid level outside of the conserved CX_5R phosphatase motif (Ungewickell et al. 2005). Both mammalian enzymes localize to the endosomal/lysosomal membranes. The Type I PPI 4-phosphatase enzyme has been found to translocate to the nucleus in response to apoptotic stimuli, and to increase the levels of cellular and nuclear PtdIns5P as measured by ^{3}H-inositol labeling (Ungewickell et al. 2005; Zou et al. 2007).

There are no obvious orthologues of the PPI 4-phosphatases in plants. Such enzymes could be important regulators of PtdIns(4,5)P_2 levels which has been correlated with membrane and vesicular trafficking and root hair development in plants (Kost 2008). However, recent data using forward genetics has identified an enzyme that functions as a PPI 4-phosphatase. Root hair defective-4 (RHD4) was identified as the affected locus in *Arabidopsis thaliana* plants containing root hairs with altered regulation of polarized membrane trafficking, which results in wider, sometimes, bulging root hairs (Thole et al. 2008) (Table 1). The authors of this work showed that recombinant RHD4 protein hydrolyzed PtdIns4P preferentially over other mono- or bis-phosphorylated PtdInsPs, and RHD4 binds to PtdIns3P and PtdIns4P on lipid blots. Most importantly, *rhd4* mutants contain elevated PtdIns4P levels as seen by ^{3}H-inositol labeling. To determine how PtdIns4P contributes to spatial control of root hair development, an enhanced yellow fluorescent protein–RHD4 construct was used to localize the RHD4 protein and a green fluorescent protein biosensor was used to determine the distribution of PtdIns4P. The results indicate that RHD4 contributes to regulation of root hair growth through its localization within the post-Golgi secretory compartment in growing root hair tips. This compartment also contains the known regulator of root hair growth, RabA4b, and as well, PtdIns 4-kinase β1, which catalyzes synthesis of PtdIns4P (Preuss et al. 2004, 2006). RHD4 could act to regulate levels of newly synthesized PtdIns4P in vesicles, resulting in low PtdIns4P levels in vesicles and higher levels in the plasma membrane. In *rhd4* mutants, loss of the PPI 4-phosphatase activity could cause an increased accumulation of PtdIns4P in vesicles, resulting in wider root hairs (Thole et al. 2008).

One surprise in the identification of the RHD4 PPI 4-phosphatase is that the *RHD4* gene has been previously known as a plant suppressor of actin (SAC) orthologue. SAC genes are found in yeast, animal, and plants, and all of the SAC proteins are known or predicted to function as PtdInsP-phosphatases (Zhong and Ye 2003). RHD4 (AtSAC7), thus, does not contain homology to the mammalian or bacterial PtdInsP-4-phosphatases, but presumably uses its SacI homology domain to catalyze removal of a 4-phosphate from PtdIns4P. The SAC gene family in Arabidopsis has nine members and can be divided into three subgroups (Zhong and Ye 2003). RHD4 (AtSAC7), AtSAC6, and AtSAC8 form one subgroup and these proteins are most similar in amino acid identity to the yeast Sac1p protein (Thole et al. 2008). AtSAC6–8 cDNAs can rescue the yeast Sac1p mutation (Despres et al. 2003), suggesting that they function in a similar manner as Sac1p. Disruption of Sac1p in yeast results in a 2-fold elevation in PtdIns3P levels and a moderate decrease in PtdIns(4,5)P_2; however, PtdIns4P is probably the major substrate as it becomes elevated 10-fold in *sac1p* mutants (Nemoto et al. 2000). This along with the elevated PtdIns4P levels found in *rhd4* mutants suggests a likely role for AtSAC6–8 enzymes as PPI 4-phosphatases.

AtSAC6–8 differ from the other plant SAC genes in that they encode a shorter C-terminus and contain two putative transmembrane helices (Zhong and Ye 2003). Additionally, both promoter-β-glucuronidase studies and expression studies demonstrate that AtSAC7 and AtSAC8 are expressed broadly, while AtSAC6 is expressed only in pollen (Zhong and Ye 2003; Despres et al. 2003). It is important to note that these studies also show that AtSAC7 (RHD4) is the only member of this subgroup expressed in root hairs. Thus, this may explain why the *rhd4* mutant is affected in root hair development.

The other six AtSAC genes are present in the two other SAC subgroups, and are either known or hypothesized to function as PPI 5-phosphatases; thus, these gene products will discussed in the next section.

4 The PPI 5-Phosphatases

To remove a phosphate from the D-5 position of the inositol ring, plants and animals utilize two classes of enzymes, the Sac1 homology-containing enzymes introduced previously or the inositol polyphosphate 5-phosphatases (5PTases). The 5PTases are a large group of enzymes that have the ability to hydrolyze 5-phosphates from a variety of PPI substrates (Astle et al. 2006). Studies on human 5PTases have classified the 5PTases into four groups according to the substrates they hydrolyze in vitro. The Group 1 or Type I enzymes have a substrate preference for the soluble PPIs (3.1.3.56), while the Group 2 or Type II enzymes (E.C. 3.1.3.36) hydrolyze 5-phosphates from both soluble- and lipidated PPIs (Erneux et al. 1998). All 5PTases characterized to date contain a conserved catalytic domain referred to now in databases as the Exo_end_phos domain, which is also present in DNA exo- and endonucleases (Whisstock et al. 2002). In some plant, animal, and yeast 5PTases, the Exo_end_phos domain is found in conjunction with other domains that allow for either modification or association with specific signal transduction components (Blero et al. 2007; Astle et al. 2006). Plant genomes contain a large group of smaller (~36–75 kDa) 5PTase enzymes (Group A), and a smaller group of large (~110–150 kDa) 5PTases (Group B). Arabidopsis contains 11 Group A enzymes (At5PTase 1–11) and 4 Group B enzymes (FRA3 and At5PTase 12–14) (Berdy et al. 2001). In contrast to the yeast and animal 5PTases, only one identifiable domain is present in the plant 5PTases besides the catalytic domain, and only Group B enzymes contain this difference. The Group B 5PTases contain 5–7 WD40 repeat regions in their N-termini (Zhong and Ye 2004). WD40 repeats are found in many proteins, including the *cop1* repressor of light development (Deng et al. 1992), and usually function to facilitate protein:protein interactions (Smith et al. 1999). The presence of WD40 repeats in the plant and certain filamentous fungal 5PTases is unique, as no other genomes contain these. This most likely reflects the addition of a WD40 region before the diversification of plants and fungi around 100 million years ago. Since signaling and metabolic events are facilitated by such complexes, identifying 5PTase-interacting proteins will lead to a new understanding of 5PTase function.

Forward and reverse genetics, along with biochemical approaches have been used to understand the action of the plant 5PTase enzymes. 5PTase enzymes in both Group A and B appear to have different substrate preferences, and also impact different developmental and/or signaling events (Table 1) that will be described here.

4.1 *Group A 5PTases*

The At5PTase1 (At1g34120) and At5PTase2 (At4g18010) gene products were the first plant 5PTases characterized and both were originally shown to hydrolyze the 5-phosphate from Ins(1,4,5)P_3 and Ins(1,3,4,5)P_4 (Burnette et al. 2003; Sanchez and Chua 2001). In these studies, there was no evidence that these proteins could remove a 5-phosphate from radiolabeled PtdIns(4,5)P_2 as measured by TLC separation of products; thus they were described as similar to the Type I or Group 1 animal 5PTases. Since this time, activity assays utilizing fluorescent PPI substrates have been developed (Maehama et al. 2000) that allow for more sensitivity in detecting reaction products (Taylor and Dixon 2001). Using this assay, it has been shown that At5PTase1 and At5PTase2 can hydrolyze fluorescent PtdIns(4,5)P_2 in contrast to previous reports (Ercetin et al. 2008). This discrepancy is most likely due to the increased sensitivity of the newer assay, and indicates that At5PTase1 and At5PTase2 do not have a preference for soluble PPI substrates and should be considered Type II or Group 2-like enzymes.

Ectopic expression of either At5PTase1 or At5PTase2 alters abscisic acid (ABA) signaling by reducing both the basal and ABA-stimulated level of Ins(1,4,5)P_3 (Burnette et al. 2003; Sanchez and Chua 2001). In contrast, a genetic loss-of-function in either gene results in faster seed germination and small increases in growth of seedlings, as well as increased sensitivity to ABA (Gunesekera et al. 2007). These growth and ABA signaling changes were accompanied by increased levels of Ins (1,4,5)P_3 as seen by mass assays and by ^{3}H-inositol labeling experiments. It is important to note that although PtdIns(4,5)P_2 was shown to be an in vitro substrate for both At5PTase and At5PTase2, neither mutant contained alterations in PtdIns (4,5)P_2. Thus, At5PTase1 and At5PTase2 appear to act primarily as modulators of Ins (1,4,5)P_3 levels which impacts seedling growth control and ABA signaling.

Two other Group A 5PTases that function in plant development have been identified through genetic or microarray screens. The *cvp2* (At5PTase6; At1g05470), and *mrh3* (At5PTase5; At5g65090) mutants are altered in cotyledon vascular patterning (Carland and Nelson 2004), and root hair initiation (Jones et al. 2006), respectively. In WT plants, CVP2 expression is restricted to developing vascular elements within seedlings (Carland and Nelson 2004), while expression of MRH3 is greatest in the growing tip of root hairs (Jones et al. 2006). No biochemistry has been done on either enzyme; however, *cvp2* mutants have a 3-fold increase in Ins(1,4,5) P_3 levels (Carland and Nelson 2004). This suggests that CVP2 hydrolyzes Ins(1,4,5)

P_3 primarily in the developing vascular system of seedlings. Phylogenetic analysis of Group A 5PTases suggest that both CVP2 and MRH3 may be similar in substrate preference to At5PTase1 and At5PTase2 (unpublished data, G. Gillaspy).

The final member of the Group A 5PTases characterized is At5PTase11 (At1g47510), which is one of the smallest predicted 5PTases found in any organism. At5PTase11 has been shown to hydrolyze 5-phosphates from PtdIns(4,5)P_2, PtdIns(3,5)P_2, and PtdIns(3,4,5)P_3, but not InsP substrates; thus At5PTase11 has a more restricted substrate preference than either At5PTase1 or At5PTase2 (Ercetin and Gillaspy 2004). Localization of a 5PTase11:green fluorescent fusion protein is consistent with a cell surface or plasma membrane location, thus At5PTase11 appears to be located near its PPI substrates within the plant cell (Ercetin et al. 2008). It is also interesting to note that At5PTase11 does not hydrolyze monophosphorylated PtdIns5P, which has been found to be upregulated in response to osmotic stress in plants (Meijer et al. 2001). The At5PTase11 gene is regulated by abscisic acid (ABA), jasmonic acid (JA), and auxin, suggesting a role for PPI action in these signal transduction pathways (Ercetin and Gillaspy 2004).

Two T-DNA knock-out mutants in the At5PTase11 gene have been characterized and were found to have a decrease in dark-grown seedling hypocotyl growth (Ercetin et al. 2008). The opposite nature of the *5ptase11* phenotype as compared to *5ptase 1* and *5ptase2* phenotypes (increased growth) may be explained by differences in InsP and PPI levels in these mutants. Whereas *5ptase1* and *5ptase2* mutants only contain elevated Ins(1,4,5)P_3, *5ptase11* mutants labeled with ^{3}H-inositol reveals decreased hydrolysis of PtdIns(4,5)P_2 and PtdIns3P, along with increases in Ins(1,4,5)P_3 and Ins(4,5)P_2 (Ercetin et al. 2008; Gunesekera et al. 2007). The decrease in PtdIns3P is consistent with the ability of At5PTase11 to hydrolyze PtdIns(3,5)P_2 in vitro (Ercetin and Gillaspy 2004). Although PtdIns (3,5)P_2has not been reported in Arabidopsis yet, it is detectable and salt-induced in alfalfa, pea, tomato, and in the green alga *Chlamydomonas* (Meijer et al. 1999).

The increase in Ins(1,4,5)P_3 and Ins(4,5)P_2 found in *5ptase11* mutants is also perplexing, as neither of these molecules is a substrate for recombinant At5PTase11 in vitro (Ercetin et al. 2008). Thus, the decreased growth phenotype and PPI and InsP alterations found in *5ptase11* mutants most likely results from the elevation of PPIs followed by the action of phospholipases, as opposed to a direct loss of Ins (1,4,5)P_3 or Ins(4,5)P_2 hydrolysis. Together these data indicate that At5PTase11 plays a unique role in the early stages of seedling development and growth.

4.2 *Group B 5PTases*

The presence of 5–7 WD40 domains within the N-termini of the Group B 5PTases indicates that these proteins may form specific signaling protein complexes. Fragile Fiber 3 (FRA3; At1g65580) is a Group B 5PTase identified in a genetic screen for genes involved in fiber cell and vascular development in Arabidopsis (Zhong et al. 2004) (Table 1; see also the chapter, "Phosphoinositides and Plant Cell Wall

Synthesis" by Ye in this volume). Mutations in FRA3 result in reduction in secondary wall thickness, altered actin organization, and a decrease in stem strength. FRA3 expression is restricted to fiber cells and developing vasculature within the stem (Zhong et al. 2004). FRA3 recombinant protein hydrolyzes Ins (1,4,5)P_3, PtdIns(4,5)P_2, and PtdIns(3,4,5)P_3 in vitro, and interestingly *fra3* mutant stems, but not leaves, contain a 1.8-fold and 1.7-fold increase in Ins(1,4,5,)P_3 and PtdIns(4,5)P_2, respectively (Zhong et al. 2004). Together, this indicates that FRA3 regulates Ins(1,4,5,)P_3 and PtdIns(4,5)P_2 levels in specific tissues. Probably because *fra3* seedlings do not have large PPI and InsP alterations, they are not ABA hypersensitive in contrast to *5ptase1* and *5ptase2* mutant seedlings (Gunesekera et al. 2007; Zhong et al. 2004).

A second Group B 5PTase, the At5PTase13 gene (At1g05630) has also been examined. The At5PTase13 gene has a very low level of transcription, but can be induced by wounding and ABA (Zhong and Ye 2004), and repressed by blue light (Chen et al. 2008). Studies on recombinant At5PTase13 have conflicting results with one study finding that only Ins(1,4,5)P_3 is a substrate (Zhong and Ye 2004), while another reported that overnight incubation with Ins(1,4,5)P_3 or Ins(1,3,4,5)P_4 resulted in catalysis as measured by HPLC separation of products (Chen et al. 2008). The function of this WD40-containing 5PTase has been addressed with T-DNA knockout mutants which have alterations in auxin levels (Lin et al. 2005), stress and nutrient responses (Ananieva et al. 2009), and blue-light signaling (Chen et al. 2008). The blue-light study, in particular, made a major breakthrough in connecting 5PTase alterations with calcium signaling. These authors showed that blue light represses transcription of At5PTase13 via action of Phototropin 1. *5ptase13* mutants grown in blue light have deregulated blue light responses resulting in increased Ins(1,4,5)P_3 and calcium levels, and suppression of hypocotyl growth inhibition (Chen et al. 2008). This indicates that one function of At5PTase13 is to dampen or terminate Ins(1,4,5)P_3 and calcium-mediated hypocotyl growth suppression in the dark.

The intriguing assortment of phenotypes found in *5ptase13* mutants may indicate a complex role for At5PTase13 and/or involvement with multiple signaling complexes. Of major interest is whether the WD40 regions of the four different Group B 5PTases form protein complexes with other proteins, and whether these complexes are unique. Recent work utilizing the yeast 2-hybrid system has identified the SNF1-like kinase (SnRK) from Arabidopsis as an interactor of At5PTase13 (Ananieva et al. 2009). Together with the decreased response of this mutant to ABA and sugars, this indicates that this 5PTase may participate in nutrient sensing in plants. Further characterization of Group B 5PTase protein complexes may expand our understanding of PPI and InsP signaling in plants.

4.3 SAC Proteins that Function as PPI 5-Phosphatases

As previously discussed, some enzymes containing a SacI homology domain function to remove a 5-phosphate from PPI substrates. Of the nine Arabidopsis SAC genes, a related group of five genes (AtSAC1–5; Table 1) contain a larger C terminus and are more closely related to the yeast Fig4p enzyme, which preferentially hydrolyzes PtdIns(3,5)P_2 (Zhong and Ye 2003). Only one of these genes has been characterized to date and it was identified in the genetic screen for fiber cell alterations (see chapter, "Phosphoinositides and Plant Cell Wall Synthesis"). Fragile Fiber 7 (FRA7) or AtSAC1 enzyme hydrolyzes PtdIns(3,5)P_2 in vitro, but not other mono-, bis- or trisphosphorylated PPIs (Zhong et al. 2005). GFP-tagged FRA7 was found to colocalize with the Golgi, and *fra7* mutants have a dramatic decrease in wall thickness of fiber cells and vessel elements, and altered actin cables, which leads to a weak stem phenotype. This is similar to the phenotype of *fra3* mutants; however, the FRA3 enzyme has a substrate preference for PtdIns(4,5) P_2 (Zhong et al. 2004). Together this suggests that multiple PPIs are involved in regulation of fiber cell and vascular development. Changes in PPIs and InsPs have not been measured in *fra7* mutants. This is of interest as PtdIns(3,5)P_2 has been reported (Meijer et al. 1999) and several recombinant PPI phosphatases have been shown to hydrolyze this substrate in vitro. Perhaps the future analysis of the *fra7* mutant will shed light on whether PtdIns(3,5)P_2 is a bonafide signaling molecule in higher plants.

The last subgroup of AtSAC enzymes has a single member, AtSAC9 (At3g59770; Table 1). This gene was identified in a genetic screen for stress mutants, which revealed that *sac9* mutants have reduced growth, are hyponastic and have deeply purple leaves, indicative of a constitutive stress response (Williams et al. 2005). The AtSAC9 protein differs from other SAC proteins in that it has a very long C-terminus and is missing a short stretch of conserved residues in the SacI homology domain. The missing SacI homology domain residues are replaced by a WW domain, a 38 amino acid residue unit that forms a binding surface for the Pro-rich proteins, or may bind to phosphoserine or phosphothreonine residues (Ilsley et al. 2002). Williams et al. addressed the relationship of PPIs and InsPs in the constitutive stress response in *sac9* mutants by reciprocal grafting experiments and ^{3}H-inositol labeling of both wildtype and *sac9* roots and shoots. The results revealed that the *sac9* mutant has elevated Ins(1,4,5)P_3 and PtdIns(4,5)P_2 levels only in the root (Williams et al. 2005). Thus, SAC9 may function to regulate root levels of PtdIns(4,5)P_2 and Ins(1,4,5)P_3 during stress responses.

5 Conclusions

PPIs can communicate different information to the cell depending on the number and position of phosphates attached. Besides signaling, these molecules also bind to proteins and affect their localization within discrete subcellular regions (Sasaki et al. 2007; Krauss and Haucke 2007). The enzymes that phosphorylate and dephosphorylate these molecules thus have the potential to regulate many physiological events. This chapter seeks to bring together data from biochemical, genetic, and physiological studies, so that a better perspective of PPI phosphatases can be gained. One obvious conclusion made from the examination of Table 1 is that there is tremendous diversity in the gene family members and enzyme types that act as PPI 5-phosphatases. This diversity may reflect specialization of a limited number of 5PTases in a simple eukaryotic ancestor, to a larger number of tissue-specific and/or developmentally-regulated genes that might be required in plants. An alternative but not mutually exclusive source of diversity is that each PPI phosphatase functions solely to hydrolyze a specific substrate or a specific set of substrates. A second conclusion can be drawn about recent data reported for 4-phosphatases. There appears to be a lack of conservation of these enzymes at the amino acid level, and thus, we are just beginning to understand how removal of 4-phosphates may impact plant physiology. Examination of Fig. 1 indicates that other potential 4-phosphatases could hydrolyze PtdIns(4,5)P_2, producing PtdIns5P, an interesting and sometimes nuclear signaling molecule in animals (Meijer et al. 2001; Ye and Ahn 2008). A last point that should be considered is how enzymes, such as the PTENs, MTMs, and FRA7(AtSAC1) impact biology as their substrates have yet to be quantified in plants. Either these enzymes have the ability to hydrolyze PtdIns (3,4,5)P_3, and/or PtdIns(3,5)P_2 and our techniques have not been sensitive enough to measure these molecules, or hydrolysis of these substrates in vitro is a remnant of evolutionary history and we have not identified the correct in vivo substrates yet. Both possibilities, along with new technological approaches and more definitive biochemical and genetic work on the PPI phosphatases, will provide an exciting area for future research.

References

Ananieva EA, Gillaspy GE, Ely A, Burnette RN, Erickson FL (2008) Plant Physiol 148:1868–1882
Astle MV, Seaton G, Davies EM, Fedele CG, Rahman P, Arsala L, Mitchell CA (2006) IUBMB Life 58:451–456
Bargmann BO, Munnik T (2006) Curr Opin Plant Biol 9:515–522
Berdy S, Kudla J, Gruissem W, Gillaspy G (2001) Plant Physiol 126:801–810
Berridge MJ (1993) Nature 361:315–325
Berridge MJ (2005) Annu Rev Physiol 67:1–21
Berridge M, Lipp P, Bootman M (1999) Curr Biol 9:157–159
Blero D, Payrastre B, Schurmans S, Erneux C (2007) Pflugers Arch 455:31–44
Bolis A, Zordan P, Coviello S, Bolino A (2007) Mol Neurobiol 35:308–316
Burnette RN, Gunesekera BM, Gillaspy GE (2003) Plant Physiol 132:1011–1019
Carland FM, Nelson T (2004) Plant Cell 16:1263–1275
Chen X, Lin WH, Wang Y, Luan S, Xue HW (2008) Plant Cell 20:353–366
Clague MJ, Lorenzo O (2005) Traffic 6:1063–1069
Clarke JH, Richardson JP, Hinchliffe KA, Irvine RF (2007) Biochem Soc Symp:149–159
Deng XW, Matsui M, Wei N, Wagner D, Chu AM, Feldmann KA, Quail PH (1992) Cell 71:791–801
Despres B, Bouissonnie F, Wu HJ, Gomord V, Guilleminot J, Grellet F, Berger F, Delseny M, Devic M (2003) Plant J 34:293–306
Di Paolo G, De Camilli P (2006) Nature 443:651–657

Doerks T, Strauss M, Brendel M, Bork P (2000) Trends Biochem Sci 25:483–485
Ercetin ME, Gillaspy GE (2004) Plant Physiol 135:938–946
Ercetin ME, Ananieva EA, Safaee NM, Torabinejad J, Robinson JY, Gillaspy GE (2008) Plant Mol Biol 67:375–388
Erneux C, Govaerts C, Communi D, Pesesse X (1998) Biochim. Biophys. Acta 1436:185–199
Gunesekera B, Torabinejad J, Robinson J, Gillaspy GE (2007) Plant Physiol 143:1408–1417
Gupta R, Ting JT, Sokolov LN, Johnson SA, Luan S (2002) Plant Cell 14:2495–2507
Harris SJ, Parry RV, Westwick J, Ward SG (2008) J Biol Chem 283:2465–2469
Ilsley JL, Sudol M, Winder SJ (2002) Cell Signal 14:183–189
Jones MA, Raymond MJ, Smirnoff N (2006) Plant J 45:83–100
Kost B (2008) Trends Cell Biol 18:119–127
Krauss M, Haucke V (2007) FEBS Lett 581:2105–2111
Krinke O, Novotna Z, Valentova O, Martinec J (2007) J Exp Bot 58:361–376
Kuijl C, Savage ND, Marsman M, Tuin AW, Janssen L, Egan DA, Ketema M, van den Nieuwendijk R, van den Eeden SJ, Geluk A, Poot A, van der Marel G, Beijersbergen RL, Overkleeft H, Ottenhoff TH, Neefjes J (2007) Nature 450:725–730
Laporte J, Hu LJ, Kretz C, Mandel JL, Kioschis P, Coy JF, Klauck SM, Poustka A, Dahl N (1996) Nat Genet 13:175–182
Laporte J, Liaubet L, Blondeau F, Tronchere H, Mandel JL, Payrastre B (2002) Biochem Biophys Res Commun 291:305–312
Leslie NR, Downes CP (2004) Biochem J 382:1–11
Li L, Ross AH (2007) J Cell Biochem 102:1368–1374
Lin WH, Wang Y, Mueller-Roeber B, Brearley CA, Xu ZH, Xue HW (2005) Plant Physiol 139:1677–1691
Maehama T, Dixon JE (1998) J Biol Chem 273:13375–13378
Maehama T, Taylor GS, Slama JT, Dixon JE (2000) Anal Biochem 279:248–250
Marcus SL, Wenk MR, Steele-Mortimer O, Finlay BB (2001) FEBS Lett 494:201–207
Meijer HJ, Munnik TM (2003) Annu Rev Plant Biol 54:265–306
Meijer H, Divecha N, van den Ende H, Musgrave A, Munnik T (1999) Planta 208:294–298
Meijer HJ, Berrie CP, Iurisci C, Divecha N, Musgrave A, Munnik T (2001) Biochem J 360:491–498
Nemoto Y, Kearns BG, Wenk MR, Chen H, Mori K, Alb JG Jr, De Camilli P, Bankaitis VA (2000) J Biol Chem 275:34293–34305
Niebuhr K, Jouihri N, Allaoui A, Gounon P, Sansonetti PJ, Parsot C (2000) Mol Microbiol 38:8–19
Ning K, Miller LC, Laidlaw HA, Burgess LA, Perera NM, Downes CP, Leslie NR, Ashford ML (2006) EMBO J 25:2377–2387
Parrish WR, Stefan CJ, Emr SD (2004) Mol Biol Cell 15:3567–3579
Pendaries C, Tronchere H, Arbibe L, Mounier J, Gozani O, Cantley L, Fry MJ, Gaits-Iacovoni F, Sansonetti PJ, Payrastre B (2006) EMBO J 25:1024–1034
Preuss ML, Serna J, Falbel TG, Bednarek SY, Nielsen E (2004) Plant Cell 16:1589–1603
Preuss ML, Schmitz AJ, Thole JM, Bonner HK, Otegui MS, Nielsen E (2006) J Cell Biol 172:991–998
Reggiori F, Wang CW, Stromhaug PE, Shintani T, Klionsky DJ (2003) J Biol Chem 278:5009–5020
Robinson FL, Dixon JE (2005) J Biol Chem 280:31699–31707
Sanchez JP, Chua NH (2001) Plant Cell 13:1143–1154
Sasaki T, Sasaki J, Sakai T, Takasuga S, Suzuki A (2007) Biol Pharm Bull 30:1599–1604
Smith TF, Gaitatzes C, Saxena K, Neer EJ (1999) Trends Biochem Sci 24:181–185
Srivastava S, Li Z, Lin L, Liu G, Ko K, Coetzee WA, Skolnik EY (2005) Mol Cell Biol 25:3630–3638
Stevenson JM, Perera IY, Heilmann II, Persson S, Boss WF (2000) Trends Plant Sci 5:357
Tamguney T, Stokoe D (2007) J Cell Sci 120:4071–4079
Taylor GS, Dixon JE (2001) Anal Biochem 295:122–126

Taylor GS, Dixon JE (2003) Methods Enzymol 366:43–56
Thole JM, Vermeer JE, Zhang Y, Gadella TW Jr, Nielsen E (2008) Plant Cell 20:381–395
Ungewickell A, Hugge C, Kisseleva M, Chang SC, Zou J, Feng Y, Galyov EE, Wilson M, Majerus PW (2005) Proc Natl Acad Sci USA 102:18854–18859
Wang X (2004) Curr Opin Plant Biol 7:329–336
Whisstock JC, Wiradjaja F, Waters JE, Gurung R (2002) IUBMB Life 53:15–23
Williams ME, Torabinejad J, Cohick E, Parker K, Drake EJ, Thompson JE, Hortter M, Dewald DB (2005) Plant Physiol 138:686–700
Xue Y, Fares H, Grant B, Li Z, Rose AM, Clark SG, Skolnik EY (2003) J Biol Chem 278:34380–34386
Ye K, Ahn JY (2008) Front Biosci 13:540–548
Zhong R, Ye ZH (2003) Plant Physiol 132:544–555
Zhong R, Ye ZH (2004) Plant Cell Physiol 45:1720–1728
Zhong R, Burk DH, Morrison WH 3 rd, Ye ZH (2004) Plant Cell 16:3242–3259
Zhong R, Burk DH, Nairn CJ, Wood-Jones A, Morrison WH 3 rd, Ye ZH (2005) Plant Cell 17:1449–1466
Zou J, Marjanovic J, Kisseleva MV, Wilson M, Majerus PW (2007) Proc Natl Acad Sci USA 104:16834–16839

Phosphatidic Acid Phosphatases in Seed Plants

Yuki Nakamura and Hiroyuki Ohta

Abstract Phosphatidic acid phosphatase (PAP) dephosphorylates phosphatidic acid (PA) to yield diacylglycerol (DAG) and inorganic phosphate (Pi). In seed plants, PAP functions both as a crucial enzyme for membrane lipid biosynthesis and an attenuator for PA signal. Despite its importance, molecular cloning and functional analysis on plant PAP have only recently started to be reported. In this chapter, we initially overview earlier biochemical analyses on plant native PAP, followed by summarizing recent advances of molecular biological studies in *Arabidopsis*.

1 Introduction

Phosphatidic acid phosphatase (PAP; EC 3.1.3.4) dephosphorylates phosphatidic acid (PA) to yield diacylglycerol (DAG) and inorganic phosphate (Pi). In seed plants, PA functions both as an intermediate to membrane lipid biosynthesis and a signaling molecule. Glycerolipids, primary components of plant cellular membrane, are synthesized from glycerol-3-phosphate (G3P) by the so-called Kennedy pathway (Kennedy 1957). In this pathway, G3P is acylated by acyltransferases to yield PA and further dephosphorylated by PAP to produce DAG. Indeed, DAG can be utilized for biosynthesis of three classes of glycerolipids, namely phospholipids, glycolipids, and triacylglycerol. Since Kennedy pathway is localized both at chloroplasts and ER, PAP involved in this pathway also localizes in both organelles. As for the signaling, PA is widely recognized as an important signaling molecule that is involved in various physiological functions, such as biotic/abiotic stresses, phytohormone responses, etc (Wang 2004, 2005). As a signaling molecule, PA is

H. Ohta (✉)
Temasek Life Sciences Laboratory, Research Link, National University of Singapore, Singapore 117604
e-mail: yuki@tll.org.sg

T. Munnik (ed.), *Lipid Signaling in Plants*, Plant Cell Monographs 16,
DOI 10.1007/978-3-642-03873-0_9, © Springer-Verlag Berlin Heidelberg 2010

synthesized mainly from membrane phospholipids by phospholipase D (Munnik 2001; Testerink and Munnik 2005; Wang 2005) or from phosphatidylinositides by two-step reaction of phosphatidylinositol-specific phospholipase C (PI-PLC) and diacylglycerol kinase (DGK) (Munnik 2001; Testerink and Munnik 2005). In addition, recent isolation of nonspecific phospholipase (NPC) in Arabidopsis (Nakamura et al. 2005) implies that PA can be produced also by NPC + DGK from membrane phospholipids. PA can be dephosphorylated to DAG by PAP, phosphorylated by PA kinase (PAK) to diacylglycerol pyrophosphate (DGPP) or deacylated to lyso PA (LPA) by phospholipase A. Which PA is involved in lipid metabolism and which in signaling, as well as which pathway attenuates PA are not always clear.

PAP activity is categorized into either PAP1 or PAP2 by their enzymatic properties (Carman and Han 2006). PAP1 is primarily a soluble enzyme which occasionally translocates to membranes such as ER (Gomez-Munoz et al. 1992). The activity requires Mg^{2+} and is considered to play roles in the Kennedy pathway. The protein has recently been purified from the yeast *Saccharomyces cerevisiae* and was named phosphatidate phosphohydrolase (PAH) (Han et al. 2006). Interestingly, the purified protein was identical to the yeast homolog of Lipin, which has been long known for the key regulator in lipid metabolism (described more later). However, PAP2 is an integral membrane protein that dephosphorylates not only PA but also other lipid phosphates, such as LPA, DGPP, sphingosine-1-phosphate (S1P), and ceramide-1-phosphate in vitro (C1P) (Brindley et al. 2002). Proteins possessing PAP2 activity are therefore named lipid phosphate phosphatase (LPP), and the PAP2 activity is independent of Mg^{2+}. LPP is considered to be mainly involved in signaling function.

Due to the pivotal function in lipid metabolism, studies on plant PAP has been reported since 1950s (Kates 1955). However, it was until recently when the functional importance of PAP is started to be elucidated at the molecular level. In this chapter, we initially overview early biochemical analyses on plant PAP, followed by recent advances on molecular studies in Arabidopsis.

2 Biochemical Features of Plant PAP

Early studies on PAP were focused on the biochemical characterization of the activity in a wide variety of plant species. Here, we briefly summarize the results with a few representative examples. For further details, readers are referred to review Kocsis and Weselake (1996).

2.1 Leaf PAP Activity

The majority of the studies on leaf PAP activity was carried out with the chloroplast-localized one (Kocsis and Weselake 1996). By using spinach chloroplasts, Joyard and Douce first reported that chloroplast envelopes are capable of synthesizing DAG from PA (Joyard and Douce 1977). Then, this activity was further characterized with intact chloroplasts (Joyard and Douce 1979; Block et al. 1983). Unlike the extrachloroplastic PAP activity, that shows features similar to animal- or yeast-PAP (Stymne and Stobart 1987), chloroplastic PAP has unique enzymatic features in that it shows a pH optimum of around pH 9 (Joyard and Douce 1979). In addition, the activity is inhibited by Mg^{2+}, but activated by EDTA (Joyard and Douce 1979; Block et al. 1983). In spinach, chloroplastic PAP activity is localized exclusively to the inner envelope and tightly associated with the membranes (Block et al. 1983). This localization was also confirmed in Pea chloroplasts (Andrews et al. 1985).

The envelope PAP activity is regulated in vivo by the level of reaction product DAG (Malherbe et al. 1992). When DAG is low, PAP activity reaches its maximum. As the DAG level increases, the activity is decreased (Malherbe et al. 1995). Thus, the DAG/PA molar ratio may be monitored in vivo by the steady state activity of PAP so that the substrate PA can be utilized also for PG biosynthesis.

In seed plants, there are two pathways for membrane lipid biosynthesis, prokaryotic and eukaryotic pathways (Joyard et al. 1998). In the prokaryotic pathway, all reaction steps take place in plastids, whereas in eukaryotic pathway, fatty acids are exported from plastids to ER, where most of phospholipids are assembled. Furthermore, some of those ER-localized membrane lipids return to plastids to serve as a substrate for galactolipid biosynthesis. Therefore, galactolipids, which are primary components of photosynthetic membranes, are synthesized by both pathways. However, not all plants possess both pathways. As such, seed plants can be classified into two types: 18:3 and 16:3 plants, according to their fatty acid composition at the *sn*-2 position of MGDG.

MGDG can be synthesized both by prokaryotic and eukaryotic pathways. However, DAG (a substrate for MGDG synthesis) derived from prokaryotic pathway exclusively contains 16:0 fatty acid. This 16:0 acyl moiety is subsequently unsaturated into 16:3, which is not found in the eukaryotic pathway-derived lipids. Therefore, the existence of 16:3 moiety in MGDG indicates that a prokaryotic pathway is active (16:3 plants). By contrast, the lack of 16:3 in MGDG means that the prokaryotic pathway does not serve for galactolipid biosynthesis (hence 18:3 plants).

There is a clear difference in envelope PAP activity between 16:3 and 18:3 plants (Frentzen et al. 1983; Gardiner and Roughan 1983; Heinz and Roughan 1983). In contrast to 16:3 plants, in which the envelope-localized Kornberg–Pricer pathway contributes significantly to the galactolipid biosynthesis, the chloroplasts of 18:3 plants shows rather low PAP activity (Heinz and Roughan 1983). This activity was estimated to be too low for efficient galactolipid yield that requires sufficient supply of DAG as a substrate (Heinz and Roughan 1983). Therefore, it is considered that such difference might be the reason why 18:3 plants contain small amount of galactolipids (and a sulfolipid) with C16 fatty acid in the *sn*-2 position.

An attempt to purify the envelope PAP was made by Malherbe and coworkers (Malherbe et al. 1995). They solubilized the plastid PAP with 3-[(3-cholamido-propy)-dimethylammonio]propanesulfonate (CHAPS) and partially purified it to 4–5-fold from the solubilized fraction. However, no report succeeded in higher level of purification for amino acid sequencing since then. Marcel and coworkers cloned two cDNAs homologous to animal LPP from cowpea (*Vigna unguiculata* L.) leaves, named *VuPAP-α* and *VuPAP-β* (Marcel et al. 2000). They showed that *VuPAP-α*, but not *VuPAP-β*, has an *N*-terminal transit peptide and is targeted to chloroplasts by in vitro import assay. The expression of *VuPAP-α* is induced by rehydration whereas that of *VuPAP-β* is stimulated by air-dessication. Judging from the amino acid sequence, these proteins are membrane integral PAP. However, no in vitro activity assay was reported for these proteins.

2.2 PAP Activity in Seeds or Seedlings

In contrast to leaf PAP, most of the studies in seeds or seedlings have focused on extraplastidic activity. The cotyledons of germinating mung beans (*Vigna radiate*) was shown to contain at least two PAP activities, located at the ER and protein bodies, with a pH optimum of 7.5 and 5.0, respectively (Herman and Chrispeels 1980). The latter acidic activity was easily released into the soluble fraction, and its molecular mass was estimated to be ~37 kDa. Another case is that of castor bean endosperm in which the primary activity is localized to microsome fraction with the minor one in the soluble fraction (Moore et al. 1973). Interestingly, in more acidic condition, the majority of PAP activity was found in cytosolic fraction. Since the microsomal PAP was reported to be translocated from the cytosol to the ER to become functional (Gomez-Munoz et al. 1992), the PAP in mung bean may also translocate depending on pH condition.

In oilseeds, PAP produces DAG as a precursor for the biosynthesis of TAG as well as membrane lipids (Harwood and Price-Jones 1988). In developing seeds of groundnuts (*A. hypogaea*), PAP activity is detectable both in microsomal and mitochondrial fractions (Sukumar and Sastry 1987). Unlikely to the chloroplastic PAP, these activities are dependent on the addition of Mg^{2+}. No significant activity is detected in the soluble fraction. Another example is the microsomal PAP in developing safflower seeds, showing absolute Mg^{2+} requirement (Griffiths et al. 1985; Ichihara et al. 1989). Although no soluble form of activity was measured in safflower seeds, it was suggested to be a functionally inactive cytosolic reservoir of the microsomal enzyme (Ichihara et al. 1990). This again implies translocation of activity between microsomes and cytosol. The first succesful purification of plant PAP to homogeneity was achieved by Pearce and Slabas using avocado (*Persea Americana*) fruits (Pearce and Slabas 1998). Starting from total microsomes, they purified the activity to 7,000-fold and determined the molecular mass to be 49 kDa as a monomeric protein. The purified enzyme activity was tested in vitro using PA, LPA, *sn-2*-LPA, and ceramide-1-phosphate (C1P). Only PA and LPA were hydrolyzed, with a slightly higher preference to LPA. The optimal pH as PAP activity was shown to be 6 and low concentration of Triton X-100 stimulates PAP activity, as reported in yeast LPP (Lin and Carman 1989).

3 Arabidopsis LPP as a Stress-Responsive PAP

As mentioned above, two types of PAP are known in plants, referred as PAP1 (soluble PAP activity) or PAP2 (membrane-associated PAP activity). The first molecular cloning of PAP gene in Arabidopsis was carried out by Pierrugues and coworkers (Pierrugues et al. 2001). They identified an mRNA that rapidly increased in response to severe radiation stress and found it to be a PAP2-type, homologous to the mammalian PAP. Since earlier studies in mammalian or yeast PAP designated PAP2-type PAP genes as lipid phosphate phosphatase (LPP), it was named AtLPP1. They also found two more homologs, LPP2 and LPP3, and later on, owing to the accomplishment of Arabidopsis genome project, Katagiri and coauthors added LPP4 (Katagiri et al. 2005). Our group recently identified additional LPP isoforms that are not homologous to LPP1–4 or any other known LPP in yeast or mammals (Nakamura et al. 2007). These isoforms, designated *LPPβ, −γ, −δ, −ε1* and *−ε2*, contain three possible catalytic sites widely known in LPPs. Since these isoforms were identified from cyanobacterial orthologs, they are referred as prokaryotic LPPs, in comparison to LPP1–4, the eukaryotic LPPs. We actually proposed renaming LPP1–4 into LPPα1–α4 for comprehensiveness (Nakamura et al. 2007) (Table 1). Here, in the next section, we will summarize the properties and functions of each LPP isoform in Arabidopsis.

3.1 LPPα1 as a Stress and Elicitor-Inducible Isoform

Since *LPPα1* was isolated as a radiation stress-inducible isoform, Pierrugues and coworkers tested additional stress responses (Pierrugues et al. 2001). Result showed that expression of *LPPα1*, but not *LPPα2,* was induced transiently by radiating gamma rays or UV-B. Elicitation with mastoparan or harpin also induced a transient increase in *LPPα1* expression. *LPPα1* is predominantly expressed in leaves and roots, but subcellular localization of the LPPα1 protein has not been reported yet. In vitro, LPPα1 is capable of hydrolyzing both PA and DGPP, with significant preference to DGPP (Pierrugues et al. 2001). The PAP activity of LPPα1 was inhibited by *N*-ethylmaleimide (NEM), but was not affected by Mg^{2+}. Since characterization of *lppα1* mutant has not been reported yet, in vivo function of LPPα1 regarding the stress response is still an open question.

Table 1 List of PAP isoformsin Arabidopsis

Name	AGI code	Molecular mass (kDa)	Gene structure
Eukaryotic LPP			
LPPα1	At2g01180	36.6	
LPPα2	At1g15080	32.7	
LPPα3	At3g02600	40.8	
LPPα4	At3g18220	35.1	
Prokaryotic LPP			
LPPβ	At4g22550	23.4	
LPPγ	At3g03080	25.7	
LPPδ	At3g58490	46.2	
LPPε1	At3g50920	30.6	
LPPε2	At5g66450	31.5	

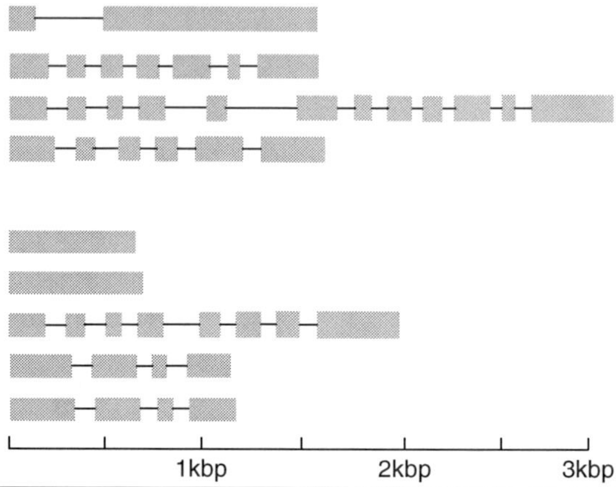

3.2 *LPPα2 as a Negative Regulator of ABI4 in ABA Signaling*

LPPα2 dephosphorylates both PA and DGPP with no preference (Pierrugues et al. 2001). The activity was independent of NEM treatment and induced by Mg^{2+}, with a maximal activity at 5 mM. However, in contrast to *LPPα1, LPPα2* showed no induction upon stress or elicitation. Later, Katagiri and coworkers reported that LPPα2 plays an important role in abscisic acid (ABA) signaling during seed germination (Katagiri et al. 2005). They found that PA and DGPP transiently increased during germination, in an ABA-dependent manner. During this stage, *LPPα2* and *LPPα3* were found to be expressed. The expression of *LPPα2* by GUS analysis was somewhat ubiquitous in seedlings, and staining was observed in vascular cells and filaments in adult plants. A T-DNA insertional *lpp2-1* mutant showed higher level of endogenous PA. Treatment with either ABA, uniconazole (UCZ; a glbberellic acid biosynthesis inhibitor), mannitol, or NaCl, revealed hypersensitive responses to the *lpp2-1* mutant, which resulted in decreased germination rates. Creating double mutants of *lpp2-1* with two known ABA-insensitive mutants, i.e., *abi3-6* and *abi4-1*, revealed that the *lpp2-1abi4-1* mutant exhibited an ABA insensitivity that was similar to the *abi4-1* single mutant, but that the *lpp2-1abi3-6* displayed a lower germination rate (less ABA-insensitive) than the *abi3-6* single mutant. Therefore, the authors suggested that PA may act upstream of ABI4, in parallel with the ABI3-mediated pathway. Additional mutation of *lpp2-1* suppressed ABA-insensitive phenotype in *abi3-6,* probably because LPPα2 represses ABI4-mediated signaling by its PA/DGPP-dephosphorylating activity. Since the pattern of DGPP production was similar to that of PA, DGPP may be also involved in ABA signaling during germination.

3.3 LPPα3 and LPPα4

Pierrugues and coworkers could not detect in vitro activity of LPPα3 (Pierrugues et al. 2001). After the Arabidopsis genome became available, additional coding region can be identified. This might be the reason why they failed to detect activity although the sequence they cloned contained all three conserved regions required for activity. Alternatively, LPPα3 might act specific to a lipid phosphate other than PA or DGPP, which was not tested. Further studies are anticipated for this isoform.

LPPα4 was discovered by Katagiri and coworkers using a BLAST database search (Katagiri et al. 2005). The reason why Perrugues and coworkers did not detect this isoform is probably because *LPPα4* was not expressed in the tissue they used to create cDNA library. Since neither characterization nor molecular cloning has been reported, the function of LPPα4 remains an open question.

4 Plastidic PAP as Prokaryotic LPP Subgroup

Efforts to unravel plastidic PAP gene in a model plant Arabidopsis has been made for long time because of its importance in plastid-localized membrane lipid metabolism. An early study identified a mutant, designated *gly-1*, in which MGDG contains reduced level of 16:3 compared to wild type (Miquel et al. 1998). Since 16:3 fatty acid in MGDG is indicative of a DAG backbone derived from plastid-localized metabolism, it was assumed that the plastidic PAP gene was affected in the mutant (Miquel et al. 1998). However, *GLY1* mapped to be a dihydroxy acetone phosphate dehydrogenase (Kachroo et al. 2004), leaving the plastidic PAP gene an open question. Fortunately, our group recently identified the plastidic PAP isoforms in a distinct way (Nakamura et al. 2007), as described below.

4.1 LPPγ, LPPε1 and LPPε2 as Plastidic PAP Isoforms

As mentioned earlier in this chapter, the biochemical studies on PAP were carried out with regard to the function of plastidic PAP in leaves. The molecular cloning of the plastidic PAP, however, remained obscure, mainly because all known PAP isoforms (*LPPα1–LPPα4*) lacked apparent transit peptides to localize it to the plastid. Our group recently isolated a distinct subgroup of LPP, designated prokaryotic LPPs, containing three isoforms, LPPγ, LPPε1 and LPPε2 that are localized to chloroplast (Nakamura et al. 2007). Because cyanobacteria, a possible ancestor of chloroplasts, have no LPP homologous to yeast/mammalian LPP, yet exhibit significant PAP activity, it was hypothesized that cyanobacteria possess a distinct type of LPP that might have been inherited to plastids of seed plants. To identify the cyanobacterial LPP, we found a possible ancestral LPP in a more primitive

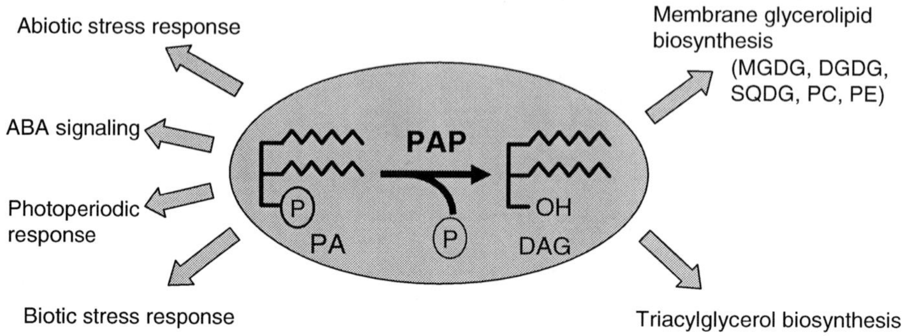

Fig. 1 Suggested physiological function of PAP

organism, *Chlorobium tepidum*. Using this sequence, one homolog (*sll0545*) was found in a representative cyanobacterium species, *Synechocystis sp.* PCC6803. Indeed, Arabidopsis contained five homologs of *sll0545,* which were completely different than *LPPα1–LPPα4*, but contained the three consensus sequences that are conserved within the LPP family (Nakamura et al. 2007). Specific antibodies confirmed that three of the five homologs localized to the chloroplasts (Nakamura et al. 2007). These isoforms, as well as *sll0545* in Synechosystis, complimented the yeast triple mutant *Δlpp1Δdpp1Δpah1* in which PAP activity is significantly decreased and causes a temperature-sensitive phenotype (Carman and Han 2006; Han et al. 2006). In vitro PAP activity assays for these isoforms showed that LPPγ most resembles the native chloroplast activity with respect to pH optimum and Mg^{2+} inhibition (Nakamura et al. 2007). T-DNA tagged mutants for *LPPε1* and *LPPε2* were isolated but they did not show any change in lipid metabolism profile, nor in the double mutant *lppε1/lppε2*. However, homozygous *lppγ-1* mutant was found to be lethal, possibly due to the aborted pollen tube elongation (Nakamura et al. 2007). Thus, although no direct evidence was obtained to show that LPPγ is a major plastidic PAP involved in lipid metabolism, to our knowledge, this is the most possible candidate for plastidic PAP. The physiological function of LPPγ, LPPε1, and LPPε2 awaits further investigation (Fig. 1).

4.2 *LPPδ/SPP1 as Sphingosine-1-Phosphate Phosphatase*

Although *LPPδ* is homologous to other prokaryotic LPPs, this isoform is structurally distinct in that it encodes larger protein (46 kDa compared to 23 kDa–30 kDa for the other isoforms) and contains much more exon/introns (Nakamura et al. 2007). Indeed, LPPδ is the only homolog of mammalian sphingosine-1-phosphate (S1P) phosphatase (SPP). Imai and coworkers recently found that LPPδ encodes a functional SPP protein and named it SPP1 (Imai et al., personal communication). The expression level of *LPPδ/SPP1* was decreased by drought stress. A T-DNA knock-out mutant of *LPPδ/SPP1* showed reduced transpiration rate. Since S1P has been implicated in ABA-mediated stomatal closure (Ng et al. 2001; Coursol et al. 2003), these results suggest that *SPP1* expression may modulate intracellular S1P level and subsequent stomatal apperture.

4.3 LPPβ as a Photoperiodically Expressed Protein Homolog

Among the five prokaryotic LPPs, LPPβ displays the highest homology to cyanobacterial LPPs (Nakamura et al. 2007). Although LPPβ showed PAP activity in vitro (Nakamura and Ohta, unpublished observation), no further data is available so far. However, a homolog of *LPPβ*, designated *PnFL-1*, was isolated from flower-induced cotyledons of a short-day plant morning glory (*Pharbitis nil*) by mRNA differential display (Kim et al. 2003). *PnFL-1* is expressed and gradually increases during the inductive dark period. However, the expression is extinguished by a night break treatment (a brief exposure to light during the dark period). Therefore, it is suggested that *Pn*FL-1's function may be associated with photoperiodic events. By analogy, Arabidopsis LPPβ might have similar function. Further studies are expected on this issue.

5 Future Perspective

As described so far, significant progress has been made in recent years on PAP research. However, we are aware of several important issues yet to be clarified. Firstly, although LPPγ is an indispensable plastidic PAP, we still have no direct evidence that this protein indeed affects prokaryotic lipid metabolism. Since a transporter unit, named TGD, imports eukaryotic PA into chloroplasts (Xu et al. 2005; Awai et al. 2006; Lu et al. 2007), plastidic PAP may also be involved in eukaryotic lipid metabolism. Secondly, ER-localized PAP is still unknown. In seed plants, it was suggested that PAP1 activity plays important roles in phospholipid metabolism and TAG biosynthesis. Until recently, molecular cloning of PAP1 has not been reported, also not in other organisms despite its suggested importance in lipid metabolism. In 2006, Han and coworkers purified native PAP1 in the yeast *S. cerevisiae* and sequenced the corresponding protein to be a homolog of Lipin in animals (Han et al. 2006). Lipin is an old protein, the defect in which causes abnormality in lipid metabolism, the so-called lipodistrophy in human (Reue and Zhang 2008). Indeed, human Lipin-1 possesses PAP activity in vitro and the yeast mutant of lipin, ΔSmp2/Δpah1, shows altered lipid metabolism (Santos-Rosa et al. 2005; Han et al. 2006). These evidences prompted us to be confident that PA metabolism by PAH plays central roles in lipid metabolism. Intriguingly, Arabidopsis has two Lipin homologs, which we designated AtPAH1 and AtPAH2 (Nakamura et al., submitted). A double knock out of *pah1pah2* showed PA accumulation and altered membrane lipid metabolism that is characteristic to eukaryotic pathway, suggesting that PAH also plays important roles in seed plants (Nakamura et al., submitted). Further investigation is ongoing in our group to answer this question. Lastly, physiological function of PAP as an attenuator of PA signal needs to be uncovered. Since the recent report by Katagiri and coworkers suggests AtLPPα2 as an attenuator of PA induced by ABA signal (Katagiri et al. 2005), it is likely that other PAP enzymes are involved in PA signaling. Future research progress is anticipated to unravel these and other important issues.

Acknowledgments The authors are grateful to Dr. Hiroyuki Imai, Konan University, Japan, for personal communication regarding unpublished results. The authors also thank Dr. Teun Munnik, Swammerdam Institute for Life Science, University of Amsterdam, The Netherlands, and Dr. Toshiro Ito, Temasek Life Sciences Laboratory, Singapore, for critical reading of the manuscript.

References

Andrews J, Ohlrogge JB, Keegstra K (1985) Plant Physiol 78:459–465
Awai K, Xu C, Tamot B, Benning C (2006) Proc Natl Acad Sci USA 103(28):10817–10822
Block MA, Dorne A-J, Joyard J, Douce R (1983) FEBS Lett 164:111–115
Brindley DN, English D, Pilquil C, Buri K, Ling Z-C (2002) Biochim Biophys Acta 1582:33–44
Carman GM, Han G-S (2006) Trends Biochem Sci 31:694–699
Coursol S, Fan LM, Le SH, Spiegel S, Gilroy S, Assmann SM (2003) Nature 423:651–654
Frentzen M, Heinz E, McKeon TA, Stumpf PK (1983) Eur J Biochem 129:629–636
Gardiner SE, Roughan PG (1983) Biochem J 210(3):949–952
Gomez-Munoz A, Hatch GM, Martin A, Jamal Z, Vance DE, Brindley DN (1992) FEBS Lett 301:103–106
Griffiths G, Stobart AK, Stymne S (1985) Biochem J 230:379–388
Han G-S, Wu W-I, Carman GM (2006) J. Biol Chem 281:9210–9218
Harwood JL, Price-Jones MJ (1988) In Brindley DN (ed) Phosphatidate phosphohydrolase, vol 2. CRC, Boca Ranton, Florida, pp 1–37
Heinz E, Roughan PG (1983) Plant Physiol 72:273–279
Herman EM, Chrispeels MJ (1980) Plant Physiol 66:1001–1007
Ichihara K, Norikura S, Fujii S (1989) Plant Physiol 90:413–419
Ichihara K, Murota N, Fujii S (1990) Biochim Biophys Acta 1043:227–234
Joyard J, Douce R (1977) Biochim Biophys Acta 486(2):273–285
Joyard J, Douce R (1979) FEBS Lett 102(1):147–150
Joyard J, Marecial E, Miege C, Block M, Dorne A-J, Douce R (1998) In Siegenthaler P-A, Murata N (eds) Lipids in photosynthesis: structure, function and genetics. Kluwer Academic, The Netherland, pp 21–52
Kachroo A, Venugopal SC, Lapchyk L, Falcone D, Hildebrand D, Kachroo P (2004) Proc Natl Acad Sci USA 101(14):5152–5157
Katagiri T, Ishiyama K, Kato T, Tabata S, Kobayashi M, Shinozaki K (2005) Plant J. 43:107–117
Kates M (1955) Can J Biochem Physiol 33:575–589
Kennedy EP (1957) Fed Proc Am Soc Exp Biol 16:853–855
Kim KC, Hur Y, Maeng J (2003) Mol Cells 16(1):54–59
Kocsis MG, Weselake RJ (1996) Lipids 31(8):785–802

Lin Y-P, Carman GM (1989) J Biol Chem 264(15):8641–8645
Lu B, Xu C, Awai K, Jones AD, Benning C (2007) J Biol Chem 282(49):35945–35953
Malherbe A, Block MA, Joyard J, Douce R (1992) J Biol Chem 267(33):23546–23553
Malherbe A, Block MA, Douce R, Joyard J (1995) Plant Physiol Biochem 33:149–161
Marcel GC, Matos A, d'arcy-Lameta A, Kader JC, Zuily-Fodil Y, Pham-Thi A (2000) Biochem Soc Trans 28(6):915–917
Miquel M, Cassagne C, Browse J (1998) Plant Physiol 117(3):923–930
Moore TS Jr, Lord JM, Kagawa T, Beevers H (1973) Plant Physiol 52:50–53
Munnik T (2001) Phosphatidic acid – an emerging plant lipid second messenger. Trends Plant Sci 6:227–233
Nakamura Y, Awai K, Masuda T, Yoshioka Y, Takamiya K, Ohta H (2005) J Biol Chem 280:7469–7476
Nakamura Y, Tsuchiya M, Ohta H (2007) J Biol Chem 282(39):29013–29021
Ng CK-Y, Carr K, McAinsh MR, Powell B, Hetherington AM (2001) Nature 410:596–599
Pearce ML, Slabas AR (1998) Plant J 14(5):555–564
Pierrugues O, Brutesco C, Oshiro J, Gouy M, Deveaux Y, Carman GM, Thuriaux P, Kazmaier M (2001) J Biol Chem 276:20300–20308
Reue K, Zhang P (2008) FEBS Lett 582:90–96
Santos-Rosa H, Leung J, Grimsey N, Peak-Chew S, Siniossoglou S (2005) EMBO J 24 (11):1931–1941
Stymne S, Stobart AK (1987) Lipids: structure and function. In Stumpf PK, Conn EE (eds) The biochemistry of plants, vol 9. Academic, New York, pp 175–214
Sukumar V, Sastry PS (1987) Biochem Int 14:1153–1158
Testerink C, Munnik T (2005) Phosphatidic acid – a multifunctional stress-signalling lipid in plants. Trends Plant Sci 10:368–375
Wang X (2004) Curr Opin Plant Biol 7:329–336
Wang X (2005) Plant Physiol 139(2):566–573
Xu C, Fan J, Froehlich JE, Awai K, Benning C (2005) Plant Cell 17:3094–3110

Part IV
PPI Metabolism

$InsP_3$ in Plant Cells

Yang Ju Im, Brian Q. Phillippy, and Imara Y. Perera

Abstract D-*myo*-Inositol 1,4,5-trisphosphate ($InsP_3$) is an important second messenger in eukaryotic cells. Although the phosphoinositide (PI) pathway has been well studied in plants, there is much that is not understood about PI-mediated signaling and there are fundamental differences between the plant and animal models. Many researchers have shown that plants produce $InsP_3$ in response to multiple stimuli and that $InsP_3$-mediated Ca^{2+} release is a component of plant signaling, although the candidate intracellular target of $InsP_3$ in plants remains elusive. As plants are sessile organisms with multiple back-up systems, the $InsP_3$-mediated signaling pathway may be one of the many signaling pathways in plants and its role may be more significant in specialized cells. This chapter provides an overview of $InsP_3$ metabolism in plants, the current methods of analysis, and a review of the role of $InsP_3$ in plants gathered from recent studies using mutants or transgenic plants with altered PI metabolism.

1 Introduction

Membrane-associated phospholipids and soluble inositol phosphates are ubiquitous in eukaryotic cells. The key steps in the phosphoinositide (PI) pathway and $InsP_3$ metabolism are outlined in Fig. 1. Phosphatidylinositol (PtdIns) is sequentially phosphorylated by specific lipid kinases to form phosphatidylinositol 4-phosphate (PtdInsP) and phosphatidylinositol 4,5-bisphosphate ($PtdInsP_2$). In response to a stimulus, $PtdInsP_2$ is hydrolyzed by phospholipase C (PLC) to produce the second messengers inositol 1,4,5-trisphosphate ($InsP_3$) and diacylglycerol (DAG). $InsP_3$

I.Y. Perera (✉)
Department of Plant Biology, North Carolina State University, Raleigh, NC 27695, USA
e-mail: imara_perera@ncsu.edu

T. Munnik (ed.), *Lipid Signaling in Plants*, Plant Cell Monographs 16,
DOI 10.1007/978-3-642-03873-0_10, © Springer-Verlag Berlin Heidelberg 2010

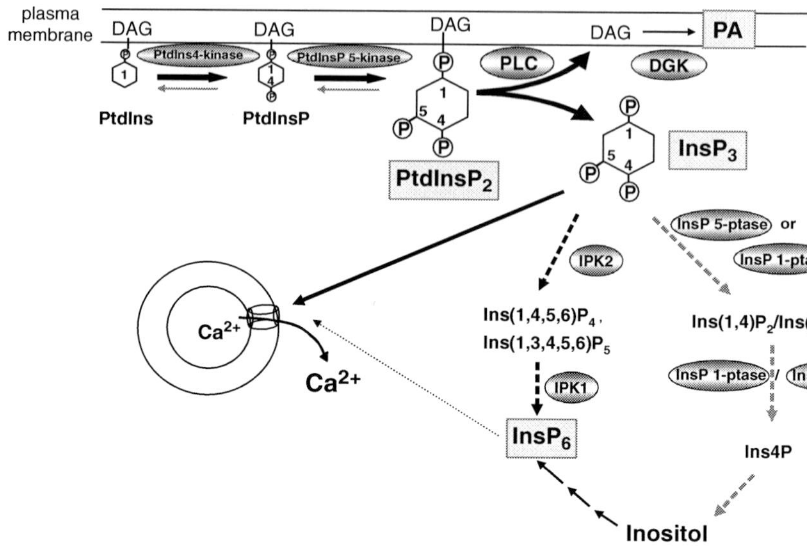

Fig. 1 Model of the phosphoinositide pathway and $InsP_3$ metabolism. *Solid arrows* denote major steps in the PI pathway and *dashed arrows* indicate routes of $InsP_3$ turnover (kinases, *black* and phosphatases, *gray*). *Shaded ovals* denote enzymes in the pathway and *gray boxes* denote key intermediates that have been implicated in signaling in plants

triggers the release of Ca^{2+} from intracellular stores activating a signaling cascade (Berridge 1993).

The major steps of the PI pathway are well conserved and many of the lipid intermediates and the enzymes responsible for their generation are present in plants (Mueller-Roeber and Pical 2002). However, there are fundamental differences in the plant and animal models. Some of the differences in plants may be due to their sessile life style and their multiple back-up systems. One critical difference is the relatively low steady-state level of $PtdInsP_2$ in plants compared with animals (Drøbak et al. 1998; Meijer and Munnik 2003). This may be partly explained by the fact that the plant PtdInsP kinases are not as active as their animal counterparts and seem to constitute a flux-limiting step in the pathway (Perera et al. 2005; Im et al. 2007; chapter "PIP-kinases as key regulators of plant function"). In addition, the DAG-activated protein kinase C (PKC) is not found in plants, although DAG is readily converted to phosphatidic acid (PA), which is emerging as an important signaling molecule in plants (Wang 2004; Testerink and Munnik 2005; chapter "Phosphatidic acid – an Electrostatic/Hydrogen- Bond Switch?"). Another interesting difference is that animal cells have several types of phospholipase C (PLC) enzymes which have distinct modes of regulation (Rebecchi and Pentyala 2000; Cockcroft 2006), while in plants, all PLCs appear to be most related to the mammalian PI-PLC ζ type and are regulated by Ca^{2+} (Hunt et al. 2004; Tasma et al. 2008). Additionally, the basic building block of the PI pathway, *myo*-inositol, is central to many diverse plant-specific metabolic processes, including cell wall synthesis, auxin physiology, seed storage, and responses to salt and drought stress, and its synthesis is under complex regulation (Loewus and Murthy 2000).

2 $InsP_3$ Metabolism in Plants

2.1 Synthesis

It is generally believed that $InsP_3$ involved in signaling is generated via the hydrolysis of $PtdInsP_2$ by PLC. In addition to the PLC-mediated route, theoretically there are two alternate ways of Ins(1,4,5)P_3 biosynthesis. It can be produced either from $InsP_2$ isomers such as Ins(1,4)P_2 and Ins(4,5)P_2 or through the degradation of $InsP_4$ isomers such as Ins(1,4,5,6)P_4 and Ins(1,3,4,5)P_4. Synthesis of Ins(1,4,5)P_3 from $InsP_2$ has been demonstrated in both plants and animals by an inositol trisphosphate kinase from barley (Josefsen et al. 2007) and by a multikinase from rat (Saiardi et al. 2001). Formation of Ins(1,4,5)P_3 from $InsP_4$ has been reported in *Dictyostelium* (Dijken et al. 1997; Williams et al. 1999) by a multiple inositol polyphosphate phosphatase (MIPP) as well as in animal cells (Yu et al. 2003); however, it is unclear if this represents a significant route of Ins(1,4,5)P_3 synthesis in plants.

2.2 Degradation

In order to affect spatially and temporally discrete responses, the duration of the $InsP_3$ signal must be tightly regulated and in mammalian cells there are two distinct routes for $InsP_3$ removal. $InsP_3$ may be phosphorylated on the C3 position of the inositol ring to form Ins(1,3,4,5)P_4, by a specific Ins(1,4,5) trisphosphate 3-kinase in mammalian cells (Takazawa et al. 1991) or by a multikinase in yeast (Shears 2004). The second method of signal termination is by dephosphorylation of the C5 position of the inositol ring by the mammalian type I inositol polyphosphate 5-phosphatase (InsP 5-ptase) to form $InsP_2$ (Laxminarayan et al. 1993, 1994; Majerus et al. 1999; see also chapter "Signaling and the Polyphosphoinositide Phosphatases from Plants").

As shown in Fig. 1, $InsP_3$ can be metabolized by phosphorylation or dephosphorylation in plants. Although plants do not appear to have a homolog of the mammalian Ins(1,4,5) trisphosphate 3-kinase, they do possess other inositol phosphate kinases that are capable of phosphorylating $InsP_3$. In *Arabidopsis*, Ins(1,4,5)P_3 is phosphorylated to Ins(1,4,5,6)P_4 and Ins(1,3,4,5,6)P_5 by the multikinase IPK2 (Stevenson-Paulik et al. 2002), which was first observed in pea roots (Chattaway et al. 1992). The barley inositol 1,3,4-trisphosphate 5/6-kinase, ITPK, is also able to convert, albeit weakly, Ins(1,4,5)P_3 to Ins(1,3,4,5)P_4 (Josefsen et al. 2007), similar to the enzyme from *Entamoeba* (Field et al. 2000). $InsP_5$ can be further phosphorylated to $InsP_6$ by IPK1 (an inositol polyphosphate 2-kinase). Together with PLC, IPK2 and IPK1 constitute a lipid-dependent pathway for generating $InsP_6$ from $PtdInsP_2$ (Stevenson-Paulik et al. 2002, 2005; Raboy 2003).

Dephosphorylation of Ins(1,4,5)P_3 in plants may occur via a 1-phosphatase to produce Ins(4,5)P_2 or by a 5-phosphatase to form Ins(1,4)P_2. Early biochemical studies (Joseph et al. 1989; Drobak et al. 1991; Martinoia et al. 1993) showed that both Ins(1,4)P_2 and Ins(4,5)P_2 were produced in plant extracts. The preferred path of $InsP_3$ hydrolysis may vary in different tissues. In tobacco cells, similar levels of 1-phosphatase and 5-phosphatase activity were present (Joseph et al. 1989), while in pea roots, the 1-phosphatase activity was predominant (Drobak et al. 1991). Brearley et al. (1997) used an in vivo labeling strategy and showed that the major product formed in permeabilized plant protoplasts was Ins(4,5)P_2, indicating a role for a 1-phosphatase. However, at the molecular level, not much is known about the inositol polyphosphate 1-phosphatases. The FRY1/SAL1 family of multifunctional enzymes possess both $3'(2')$, $5'$-bisphosphate nucleotidase activity and 1-phosphatase activity. However, the 1-phosphatase activity is higher with Ins(1,4)P_2 as a substrate rather than with Ins(1,4,5)P_3 (Quintero et al. 1996; Gil-Mascarell et al. 1999; Xiong et al. 2001), suggesting that this enzyme may work at the second step of $InsP_3$ hydrolysis, in the dephosphorylation of Ins(1,4)P_2 to InsP. Additionally, although plants do not seem to have the canonical type I inositol 5-phosphastase, *Arabidopsis* has a large family of inositol 5-phosphatases (Berdy et al. 2001; Burnette et al. 2003 and chapter “Signaling and the Polyphosphoinositide Phosphatases from Plants”). Several members of this family (including AtPTase 1, 2, 13, and 14) have been shown to hydrolyze $InsP_3$ (Berdy et al. 2001; Sanchez and Chua 2001; Zhong and Ye 2004). Furthermore, over expression and null mutants of these genes show reduced and increased levels of $InsP_3$, respectively (Burnette et al. 2003; Carland and Nelson 2004).

3 Analysis of $InsP_3$ in Plant Tissues

Analytical methods for $InsP_3$ differ greatly in their abilities to deal with the challenge of specificity. There are 20 possible stereoisomers of *myo*-inositol trisphosphate, and although relatively few have been identified in eukaryotic cells, less abundant ones are surely present due to the promiscuity of the inositol phosphate kinases and phosphatases. Plants are known to contain at least six IP_3 isomers (Brearley and Hanke 2000). Many of these are metabolic intermediates and therefore would not be expected to change rapidly in response to a stimulus. A variety of procedures have been used to measure inositol phosphates (Kukis 2003; Berrie et al. 2007). NMR and enzymatic methods are among the more specific methods but they are also the most complex and not widely used. The methods most commonly used by plant biologists are HPLC and the Ins(1,4,5)P_3 binding assay.

Numerous HPLC protocols capable of measuring $InsP_3$ have been published but their success in distinguishing among the stereoisomers is frequently questioned. The original methods developed in the 1980s for measuring $InsP_3$ in animal cells were concerned primarily with separating Ins(1,4,5)P_3 from Ins(1,3,4)P_3 and Ins (1,3,4,5)P_4. Subsequently, the presence of other isomers such as Ins(1,2,3)P_3 and Ins(1,5,6)P_3 in high nanomolar or low micromolar concentrations (Barker et al. 1995, 2004) made it clear that more robust separation methods are needed. Although resolution of the 12 trisphosphate peaks theoretically separable by HPLC has not been achieved (enantiomers always coelute), the optimized procedures by Carlsson et al. (2001) and Chen and Li (2003) come closest, with complete or partial resolution of up to nine $InsP_3$ peaks.

The other method frequently used to measure $InsP_3$ is the Amersham D-*myo*-inositol 1,4,5-trisphosphate (IP_3) [^{3}H] Biotrak assay system sold by GE Healthcare (Buckinghamshire, UK). The assay depends on the specific binding of Ins(1,4,5)P_3 to a bovine adrenal-binding protein, as first described by Challiss et al. (1988). The assay is specific for Ins(1,4,5)P_3 over other inositol phosphates, including Ins(1,3,4) P_3, Ins(2,4,5)P_3, Ins(1,3,4,5)P_4, Ins(1,3,4,5,6)P_5, and $InsP_6$. Pretreatment of plant extracts with the mammalian type I inositol 5-phosphatase, an enzyme that specifically hydrolyzes $InsP_3$ and $InsP_4$ (Laxminarayan et al. 1993, 1994), removed >90% of the signal measured by this assay, indicating that it is a valid assay for plants (Perera et al. 2001).

The limitation of this method is that the Ins(1,4,5)P_3 levels in plant cells may be small compared with the total inositol phosphate pool (Brearley and Hanke 2000), and the sum of their interferences could impact the results obtained from unpurified extracts. Therefore, it may not be valid to compare the absolute values obtained from such assays between different plant tissues and species. Nevertheless, the assay is useful to compare relative levels of $InsP_3$ (mutants to wild type or control to treated, etc.). Furthermore, this is a relatively quick procedure for measuring large numbers of samples (as in a stimulation time course) that would be extremely time-consuming to analyze by HPLC. Validation of the receptor-binding assay using HPLC or other methods would be helpful, but only to the extent of the specificity of the validating procedure. Similar results of $InsP_3$ changes in pathogen-challenged soybean cells have been reported using either the receptor-binding assay or HPLC analysis (Shigaki and Bhattacharyya 2000).

Fluorescent biosensors have been developed to detect the presence of $InsP_3$ in vivo in animal cells using confocal microscopy. The pleckstrin homology (PH) domain of PLCδ1 fused to enhanced green fluorescent protein was used to monitor both $PtdInsP_2$ and $InsP_3$ in animal cells (Varnai and Balla 1998; Hirose et al. 1999). This strategy has been successfully applied to plants to monitor either $PtdInsP_2$ (Dowd et al. 2006; Helling et al. 2006; Lee et al. 2007; van Leeuwen et al. 2007; Stenzel et al. 2008) or $InsP_3$ (Tang et al. 2007). What should be noted is that this biosensor reports changes in PtdIns(4,5)P_2 concentration rather than $InsP_3$ (van der Wal et al. 2001). More recently, a Förster resonance energy transfer (FRET)-based $InsP_3$ biosensor has been developed, with the $InsP_3$-binding domain of the mammalian $InsP_3$ receptor attached between cyan and yellow fluorescent proteins

(Tanimura et al. 2004; Remus et al. 2006). Although the FRET-based $InsP_3$ biosensor has been used effectively to monitor $InsP_3$ in a variety of different mammalian cell systems, there have been no reports of its use in plants. If this technique could be adapted for plant cells, it would provide a valuable noninvasive tool to monitor $InsP_3$ in vivo.

4 $InsP_3$ as a Signaling Molecule in Plants

Over 20 years ago, the pioneering work of Ruth Satter's group showed that light stimulates the release of $InsP_3$ from the motor pulvinus of *Samanea saman* (Morse et al. 1987a, 1987b). Since then, there have been numerous examples of rapid changes in $InsP_3$ in different plant species in response to many abiotic and biotic stimuli (Stevenson et al. 2000; Meijer and Munnik 2003; Krinke et al. 2007). These include abiotic stresses such as cold (SmolenskaSym and Kacperska 1996), heat (Liu et al. 2006), salt (Takahashi et al. 2001), and hyperosmotic stress (DeWald et al. 2001; Heilmann et al. 2001), gravistimulation (Perera et al. 1999, 2006), plant growth regulators such as ABA (Lee et al. 1996), external Ca^{2+} (Tang et al. 2007), and numerous fungal and bacterial elicitors (Aducci and Marra 1990; Legendre et al. 1993; Cho et al. 1995; Franklin-Tong et al. 1996). In most examples, $InsP_3$ levels have been shown to increase rapidly and transiently with stimulation, although in some instances, $InsP_3$ levels have been found to decrease (for more details see Table 1 in Krinke et al. 2007). In addition, sustained increases in $InsP_3$ have been documented preceding growth in graviresponsive maize and oat pulvini (Perera et al. 1999, 2001) and during the early stages of tracheary element differentiation (Zhang et al. 2002), suggesting a role for $InsP_3$ in the regulation of growth and development.

Table 1 Altered PI signaling affects inositol phosphate levels

Cell type	[^{3}H] Inositol-labeled inositol phosphates (% of total [^{3}H]InsPx recovered from the water soluble fraction)			
	$InsP_2$	$InsP_3$	$InsP_5$	$InsP_6$
WT	7.50	0.73	0.75	6.43
GFP	7.06	0.59	0.89	6.47
InsP 5-ptase	7.45	**0.17**↓	0.33↓	**3.90**↓
PIP5K Iα	6.70	**1.98**↑	1.08↑	**9.44**↑

Three-day-old tobacco cells were labeled in vivo with [^{3}H] inositol for 18 h. Cells were harvested by filtration, ground in liquid N_2, and treated with 5% PCA to precipitate proteins and lipids. The soluble fraction was analyzed for inositol phosphates by HPLC. Cell lines are as follows, wild-type cells (WT), tobacco cells expressing either GFP only (GFP), or the mammalian type I inositol polyphosphate 5-phosphatase (InsP 5-ptase), or GFP-tagged mammalian PIP 5-kinase Iα, (PIP5K Iα). *Arrows* indicate differences in inositol phosphate levels between wild type and transgenic lines

4.1 Mechanism of Action

In the classic signaling model, $InsP_3$ acts as a second messenger to release Ca^{2+} from intracellular stores. In mammalian systems, the $InsP_3$-responsive Ca^{2+} channel or $InsP_3$ receptor ($InsP_3R$) is primarily located on ER membranes and has been extensively studied (for recent reviews see Foskett et al. 2007; Mikoshiba 2007). In plants, Ca^{2+} release mediated by exogenously added $InsP_3$ is well documented. $InsP_3$ was first shown to trigger the release of Ca^{2+} from plant microsomal membranes (Drobak and Ferguson 1985). Subsequently, the plant vacuole was shown to be the primary $InsP_3$-sensitive Ca^{2+} store (Schumaker and Sze 1987; Alexandre et al. 1990; Canut et al. 1993), although it may not be the only $InsP_3$-sensitive Ca^{2+} store (Muir and Sanders 1997; Sanders et al. 1999, 2002). There are also reports of endogenous increases in $InsP_3$ accompanied by changes in cystosolic Ca^{2+} in response to a stimulus (Franklin-Tong et al. 1996; DeWald et al. 2001). $InsP_3$ is implicated in the propagation of Ca^{2+} oscillations in guard cells and root hair cells (Staxen et al. 1999; Engstrom et al. 2002), in cell-to-cell communication (Tucker and Boss 1996) and in regulating diurnal Ca^{2+} oscillations (Tang et al. 2007).

One critical component that is still missing or unknown in the plant model is the plant $InsP_3$ receptor. To date, a gene homologous to the animal Ins(1,4,5)P_3 receptor has not been found in either yeast or plants. This implies that there is low homology between plant and animal Ins(1,4,5)P_3 receptors due to evolutionary divergence. Alternatively, the plant receptor may be a heteromeric subunit complex (Krinke et al. 2007). The closest identification of the putative plant receptor was its purification to apparent homogeneity from *Vigna radiate* (Biswas et al. 1995). The protein purified from mung-bean was significantly smaller (~110 kDa) compared with the animal $InsP_3Rs$ (which are 260–310 kDa in size) and no sequence information was reported. It is, perhaps, not surprising that plants do not have a gene for the canonical $InsP_3R$; the animal receptor has an extensive coupling domain (~60% of the primary sequence) that contains sites for interaction with numerous regulatory factors, many of these represent mechanisms of regulation not commonly found in plants. It is very likely that plant and animal receptors share little homology beyond the $InsP_3$-binding domain and possibly the transmembrane domains that form the Ca^{2+} channel. The physiological characteristics of the putative plant $InsP_3$ receptors and how they contrast with their animal counterparts has been the focus of a recent review (Krinke et al. 2007).

It is important to recognize that while $InsP_3$-mediated Ca^{2+} release is a major signaling pathway in certain mammalian tissues (Berridge 1993), it may be one of many mechanisms to regulate Ca^{2+} homeostasis in plant cells. Calcium signaling in plants is additionally regulated by voltage-gated (slow vacuolar) channels and cyclic ADP ribose-gated channels (Pottosin and Schonknecht 2007). The different types of calcium channels can operate separately or coordinately to fine-tune cellular functions. The contribution of PI-mediated signaling may be significant in tip-growing cells like pollen tubes and root hairs or cells exhibiting rapid and reversible responses such as guard cells and motor pulvini cells. For instance,

$PtdInsP_2$ synthesis is highly regulated in *Arabidopsis* root hairs (Kusano et al. 2008; Stenzel et al. 2008), and PLC-mediated $PtdInsP_2$ hydrolysis is critical for pollen tube growth (Dowd et al. 2006; Helling et al. 2006; Kost 2008).

4.2 PI Turnover in Guard Cell Regulation

Probably the best-characterized system where many components of the PI pathway have been implicated in signaling is guard cells (Li et al. 2006). One early demonstration of $InsP_3$ involvement in guard cell regulation was the fact that release of caged $InsP_3$ was found to increase cytosolic Ca^{2+} and promote stomatal closure (Gilroy et al. 1990). Subsequently, it was shown that PLC is important for this process, since inhibition of PLC activity reduced ABA-mediated stomatal closure (Staxen et al. 1999) and similarly plants with reduced PLC did not fully respond to ABA (Hunt et al. 2003; Mills et al. 2004). It should be noted that changes in PLC will have broader consequences on PI turnover, in addition to reducing $InsP_3$. Inhibition of PLC could lead to a build up of $PtdInsP_2$, and also have an impact on PA either via DAG or by activating specific phospholipase D isoforms (Qin and Wang 2002). $PtdInsP_2$ was recently implicated in regulating stomatal opening (Lee et al. 2007) and PA is involved in ABA-mediated guard cell closure by the interaction with both ABI1, a negative regulator of ABA, (Zhang et al. 2004, 2005) and the G protein α subunit Gpa1 (Mishra et al. 2006). Lemtiri-Chlieh have proposed that $InsP_6$, rather than $InsP_3$, is the effector in ABA-mediated stomatal closure because $InsP_6$ was shown to increase in response to ABA in guard cells and be more effective than $InsP_3$ in inhibiting the inward-rectifying K^+ channel (Lemtiri-Chlieh et al. 2000, 2003). Since $InsP_6$ may be derived from $InsP_3$, these results suggest that the role of $InsP_3$ in guard cells may be either direct or indirect as an intermediate in $InsP_6$ synthesis. Clearly, more work is needed to resolve the contribution of the different pathway intermediates, but the evidence is strong for their involvement in guard cell signaling.

4.3 Altered PI Metabolism in Plants

More clues on the possible roles of $InsP_3$ in plants have come from mutants with altered levels of $InsP_3$. Many of these mutants belong to the inositol 5-phosphatase family (see also chapter "Signaling and the Polyphosphoinositide Phosphatases from Plants"). In *Arabidopsis*, these enzymes appear to play a role in diverse aspects of plant growth and development (Xue et al. 2007) including vascular patterning in leaves and cotyledons (Carland and Nelson 2004; Lin et al. 2005), secondary cell wall, and fiber cell development (Zhong et al. 2004), root hair initiation (Jones et al. 2006), ABA-regulated processes (Sanchez and Chua 2001;

Burnette et al. 2003), and hypocotyl elongation in the dark (Gunesekera et al. 2007) and in blue light (Chen et al. 2008).

The *fry1* mutant (which has increased $InsP_3$) is hypersensitive to many different abiotic stresses, including cold, salt, and drought stresses (Xiong et al. 2001). However, the *fry1* mutant phenotype may not be solely due to loss of 1-phosphatase activity since FRY1 is a multifunctional enzyme. The lipid phosphatase mutant *sac9* has elevated levels of both $PtdInsP_2$ and $InsP_3$ and exhibits a systemic stressed phenotype (Williams et al. 2005). In the antisense mutant of the Ca^{2+} receptor (CAS), basal $InsP_3$ levels were slightly reduced and increases in $InsP_3$ in response to external Ca^{2+} were attenuated. CAS was originally proposed to be an extracellular Ca^{2+} receptor localized in the plasma membrane (Han et al. 2003; Tang et al. 2007) but more recent work has shown that it is chloroplast-localized and is involved in regulating stomatal closure in response to high external Ca^{2+} (Weinl et al. 2008).

In our work, we have taken a molecular approach to perturb the PI pathway by the constitutive expression of heterologous enzymes in plant cells. Expression of the mammalian type 1 inositol polyphosphate 5-phosphatase in tobacco cells resulted in a significant lowering of basal $InsP_3$ and $PtdInsP_2$ levels (Perera et al. 2002). Conversely, the expression of the mammalian type 1 PIP 5-kinase α resulted in cells with increased basal $InsP_3$ and $PtdInP_2$ levels (Im et al. 2007). In vivo labeling of the transgenic phosphatase (InsP 5-ptase) and kinase (PIP5K Iα) tobacco lines has revealed that inositol phosphate metabolism is altered. Notably, $InsP_6$ levels were increased in the kinase lines and decreased in the phosphatase lines (Table 1), indicating that altering the $InsP_3$ levels affects $InsP_6$ and supporting a route for $InsP_6$ synthesis from $InsP_3$.

Stable transgenic *Arabidopsis* lines expressing the mammalian type 1 InsP 5-ptase have also been generated and characterized. The plants showed normal growth and morphology although basal $InsP_3$ levels were drastically reduced. Interestingly, the response to gravistimulation was delayed and reduced ~30% in roots, hypocotyls, and inflorescence stems compared with wild-type plants (Perera et al. 2006). Transcript profiling of dark-grown wild-type *Arabidopsis* roots during a time course of gravistimulation has revealed that several transcripts increase within 5 min of gravistimulation (Kimbrough et al. 2004). Curiously, in the InsP 5-ptase roots, the expression of five of the fastest responding genes does not change with gravistimulation (Salinas-Mondragon et al. 2005). Furthermore, the expression of these genes is also rapidly and transiently upregulated by light in the wild type but not in the InsP 5-ptase plants (Salinas-Mondragon et al. 2005). These results suggest a hitherto undiscovered role for $InsP_3$ and/or $InsP_3$-derived downstream components in the negative regulation of gene expression in plants. Transcript profiling of different mutants with altered $InsP_3$ metabolism should help identify clusters of genes that are coordinately regulated and provide valuable clues on putative downstream targets.

In yeast, $InsP_3$ is considered to be the backbone for the formation of higher phosphorylated inositol phosphates and many of these are implicated in nuclear physiology and regulation of gene expression (reviewed in Alcázar-Román and

Wente 2008). It will be of great interest to see if this type of regulation is functional in plants.

5 Conclusions and Future Prospects

As outlined in this chapter, there is substantial data from plants indicating that $InsP_3$ plays a role in propagating a signal either directly as a second messenger or as a key intermediate in the signal transduction cascade. These signals may result in rapid responses or more long-term responses, such as changes in growth. Identification of the plant-specific $InsP_3$ receptor and the development of more sensitive methods for in vivo monitoring of $InsP_3$ at the cellular level would greatly advance the field of plant PI-mediated signaling. We anticipate that new and exciting insights will emerge in the future with the advances in technology and with the characterization of novel mutant and transgenic plant lines.

Acknowledgments Funding from the National Science Foundation and the US Department of Agriculture is gratefully acknowledged.

References

Aducci P, Marra M (1990) IP_3 levels and their modulation FY fusicoccin measured by a novel [^{3}H] IP_3 binding assay. Biochem Biophys Res Commun 168:1041–1046

Alcázar-Román A, Wente S (2008) Inositol polyphosphates: a new frontier for regulating gene expression. Chromosoma 117:1–13

Alexandre J, Lassalles JP, Kado RT (1990) Opening of Ca^{2+} channels in isolated red beet root vacuole membrane by inositol 1,4,5-trisphosphate. Nature 343:567–570

Barker CJ, Wright J, Kirk CJ, Michell RH (1995) Inositol 1,2,3-trisphosphate is a product of InsP6 dephosphorylation in WRK-1 rat mammary epithelial cells and exhibits transient concentration changes during the cell cycle. Biochem Soc Trans 23:169S

Barker CJ, Wright J, Hughes PJ, Kirk CJ, Michell RH (2004) Complex changes in cellular inositol phosphate complement accompany transit through the cell cycle. Biochem J 380:465–473

Berdy SE, Kudla J, Gruissem W, Gillaspy GE (2001) Molecular characterization of At5PTase1, an inositol phosphatase capable of terminating inositol trisphosphate signaling. Plant Physiol 126:801–810

Berridge MJ (1993) Inositol trisphosphate and calcium signalling. Nature 361:315–325

Berrie CP, Iurisci C, Piccolo E, Bagnati R, Corda D (2007) Analysis of phosphoinositides and their aqueous metabolites. Methods Enzymol 434:187–232

Biswas S, Dalal B, Sen M, Biswas BB (1995) Receptor for myo-inositol trisphosphate from the microsomal fraction of *Vigna radiata*. Biochem J 306(Pt 3):631–636

Brearley CA, Hanke DE (2000) Metabolic relations of inositol 3,4,5,6-tetrakisphosphate revealed by cell permeabilization. Identification of inositol 3,4,5,6-tetrakisphosphate 1-kinase and inositol 3,4,5,6-tetrakisphosphate phosphatase activities in mesophyll cells. Plant Physiol 122:1209–1216

Brearley CA, Parmar PN, Hanke DE (1997) Metabolic evidence for PtdIns(4,5)P_2-directed phospholipase C in permeabilized plant protoplasts. Biochem J 324:123–131
Burnette RN, Gunesekera BM, Gillaspy GE (2003) An Arabidopsis inositol 5-phosphatase gain-of-function alters abscisic acid signaling. Plant Physiol 132:1011–1019
Canut H, Carrasco A, Rossignol M, Ranjeva R (1993) Is vacuole the richest store of IP_3-mobilizable calcium in plant-cells. Plant Sci 90:135–143
Carland FM, Nelson T (2004) COTYLEDON VASCULAR PATTERN2-mediated inositol (1,4,5) triphosphate signal transduction is essential for closed venation patterns of Arabidopsis foliar organs. Plant Cell 16:1263–1275
Carlsson NG, Bergman EL, Skoglund E, Hasselblad K, Sandberg AS (2001) Rapid analysis of inositol phosphates. J Agric Food Chem 49:1695–1701
Challiss RA, Batty IH, Nahorski SR (1988) Mass measurements of inositol(1,4,5)trisphosphate in rat cerebral cortex slices using a radioreceptor assay: effects of neurotransmitters and depolarization. Biochem Biophys Res Commun 157:684–691
Chattaway JA, Drobak BK, Watkins PAC, Dawson AP, Letcher AJ, Stephens LR, Irvine RF (1992) An inositol 1,4,5-trisphosphate-6-kinase activity in Pea roots. Planta 187:542–545
Chen QC, Li BW (2003) Separation of phytic acid and other related inositol phosphates by high-performance ion chromatography and its applications. J Chromatogr A 1018:41–52
Chen X, Lin W-H, Wang Y, Luan S, Xue H-W (2008) An inositol polyphosphate 5-phosphatase functions in PHOTOTROPIN1 signaling in Arabidopis by altering cytosolic Ca^{2+}. Plant Cell 20:353–366
Cho MH, Tan Z, Erneux C, Shears SB, Boss WF (1995) The effects of mastoparan on the carrot cell plasma membrane polyphosphoinositide phospholipase C. Plant Physiol 107:845–856
Cockcroft S (2006) The latest phospholipase C, PLCeta, is implicated in neuronal function. Trends Biochem Sci 31:4–7
DeWald DB, Torabinejad J, Jones CA, Shope JC, Cangelosi AR, Thompson JE, Prestwich GD, Hama H (2001) Rapid accumulation of phosphatidylinositol 4,5-bisphosphate and inositol 1,4,5-trisphosphate correlates with calcium mobilization in salt-stressed arabidopsis. Plant Physiol 126:759–769
Dijken P, Bergsma JCT, Haastert PJM (1997) Phospholipase-C-independent inositol 1,4,5-trisphosphate formation in dictyostelium cells – activation of a plasma-membrane-bound phosphatase by receptor-stimulated Ca2+ influx. Eur J Biochem 244:113–119
Dowd PE, Coursol S, Skirpan AL, Kao T-h, Gilroy S (2006) Petunia phospholipase C1 is involved in pollen tube growth. Plant Cell 18:1438–1453
Drobak BK, Ferguson IB (1985) Release of Ca2+ from plant hypocotyl microsomes by inositol-1,4,5-trisphosphate. Biochem Biophys Res Commun 130:1241–1246
Drobak BK, Watkins PA, Chattaway JA, Roberts K, Dawson AP (1991) Metabolism of inositol (1,4,5)trisphosphate by a soluble enzyme fraction from Pea (Pisum sativum) roots. Plant Physiol 95:412–419
Drøbak BK, Dewey RE, Boss WF (1998) Phosphoinositide kinases and the synthesis of polyphosphoinositides in higher plant cells. In: Jeon KW (ed) International review of cytology. Academic, New York, pp 95–130
Engstrom EM, Ehrhardt DW, Mitra RM, Long SR (2002) Pharmacological analysis of nod factor-induced calcium spiking in Medicago truncatula. Evidence for the requirement of type IIA calcium pumps and phosphoinositide signaling. Plant Physiol 128:1390–1401
Field J, Wilson MP, Mai ZM, Majerus PW, Samuelson J (2000) An Entamoeba histolytica inosital 1,3,4-trisphosphate 5/6-kinase has a novel 3-kinase activity. Mol Biochem Parasitol 108:119–123
Foskett JK, White C, Cheung K-H, Mak D-OD (2007) Inositol trisphosphate receptor Ca^{2+} release channels. Physiol Rev 87:593–658
Franklin-Tong VE, Drobak BK, Allan AC, Watkins P, Trewavas AJ (1996) Growth of pollen tubes of Papaver rhoeas is regulated by a slow-moving calcium wave propagated by inositol 1,4,5-trisphosphate. Plant Cell 8:1305–1321

Gil-Mascarell R, López-Coronado JM, Serrano BR, Rodríguez PL (1999) TheArabidopsis HAL2-like gene family includes a novel sodium-sensitive phosphatase. Plant J 17:373–383

Gilroy S, Read ND, Trewavas AJ (1990) Elevation of cytoplasmic calcium by caged calcium or caged inositol trisphosphate initiates stomatal closure. Nature 346:769–771

Gunesekera B, Torabinejad J, Robinson J, Gillaspy GE (2007) Inositol polyphosphate 5-phosphatases 1 and 2 are required for regulating seedling growth. Plant Physiol 143:1408–1417

Han S, Tang R, Anderson LK, Woerner TE, Pei Z-M (2003) A cell surface receptor mediates extracellular Ca^{2+} sensing in guard cells. Nature 425:196–200

Heilmann I, Perera IY, Gross W, Boss WF (2001) Plasma membrane phophotidylinositol 4,5-bisphosphate decreases with time in culture. Plant Physiol 126:1507–1518

Helling D, Possart A, Cottier S, Klahre U, Kost B (2006) Pollen tube tip growth depends on plasma membrane polarization mediated by tobacco PLC3 activity and endocytic membrane recycling. Plant Cell 18:3519–3534

Hirose K, Kadowaki S, Tanabe M, Takeshima H, Iino M (1999) Spatiotemporal dynamics of inositol 1,4,5-trisphosphate that underlies complex Ca^{2+} mobilization patterns. Science 284:1527–1530

Hunt L, Mills LN, Pical C, Leckie CP, Aitken FL, Kopka J, Mueller-Roeber B, McAinsh MR, Hetherington AM, Gray JE (2003) Phospholipase C is required for the control of stomatal aperture by ABA. Plant J 34:47–55

Hunt L, Otterhag L, Lee JC, Lasheen T, Hunt J, Seki M, Shinozaki K, Sornmarin M, Gilmour DJ, Pical C, Gray JE (2004) Gene-specific expression and calcium activation of Arabidopsis thaliana phospholipase C isoforms. New Phytol 162:643–654

Im YJ, Perera IY, Brglez I, Davis AJ, Stevenson-Paulik J, Phillippy BQ, Johannes E, Allen NS, Boss WF (2007) Increasing plasma membrane phosphatidylinositol(4,5)bisphosphate biosynthesis increases phosphoinositide metabolism in Nicotiana tabacum. Plant Cell 19:1603–1616

Jones MA, Raymond MJ, Smirnoff N (2006) Analysis of the root-hair morphogenesis transcriptome reveals the molecular identity of six genes with roles in root-hair development in Arabidopsis. Plant J 45:83–100

Josefsen L, Bohn L, Sorensen MB, Rasmussen SK (2007) Characterization of a multifunctional inositol phosphate kinase from rice and barley belonging to the ATP-grasp superfamily. Gene 397:114–125

Joseph SK, Esch T, Bonner WD Jr (1989) Hydrolysis of inositol phosphates by plant cell extracts. Biochem J 264:851–856

Kimbrough JM, Salinas-Mondragon R, Boss WF, Brown CS, Sederoff HW (2004) The fast and transient transcriptional network of gravity and mechanical stimulation in the arabidopsis root apex. Plant Physiol 136:2790–2805

Kost B (2008) Spatial control of Rho (Rac-Rop) signaling in tip-growing plant cells. Trends Cell Biol 18:119–127

Krinke O, Novotna Z, Valentova O, Martinec J (2007) Inositol trisphosphate receptor in higher plants: is it real? J Exp Bot 58:361–376

Kukis A (2003) Inositol phospholipid metabolism and phosphatidyl inositol kinases. Elsevier, Amsterdam

Kusano H, Testerink C, Vermeer JEM, Tsuge T, Shimada H, Oka A, Munnik T, Aoyama T (2008) The arabidopsis phosphatidylinositol phosphate 5-Kinase PIP5K3 is a key regulator of root hair tip growth. Plant Cell 20:367–380

Laxminarayan KM, Matzaris M, Speed CJ, Mitchell CA (1993) Purification and characterization of a 43-kDa membrane-associated inositol polyphosphate 5-phosphatase from human placenta. J Biol Chem 268:4968–4974

Laxminarayan KM, Chan BK, Tetaz T, Bird PI, Mitchell CA (1994) Characterization of a cDNA encoding the 43-kDa membrane-associated inositol-polyphosphate 5-phosphatase. J Biol Chem 269:17305–17310

Lee Y, Choi YB, Suh S, Lee J, Assmann SM, Joe CO, Kelleher JF, Crain RC (1996) Abscisic acid-induced phosphoinositide turnover in guard cell protoplasts of *Vicia faba*. Plant Physiol 110:987–996

Lee Y, Kim YW, Jeon BW, Park KY, Suh SJ, Seo J, Kwak JM, Martinoia E, Hwang I, Lee Y (2007) Phosphatidylinositol 4,5-bisphosphate is important for stomatal opening. Plant J 52:803–816

Legendre L, Yueh YG, Crain R, Haddock N, Heinstein PF, Low PS (1993) Phospholipase C activation during elicitation of the oxidative burst in cultured plant cells. J Biol Chem 268:24559–24563

Lemtiri-Chlieh F, MacRobbie EAC, Brearley CA (2000) Inositol hexakisphosphate is a physiological signal regulating the K^+-inward rectifying conductance in guard cells. Proc Natl Acad Sci USA 97:8687–8692

Lemtiri-Chlieh F, MacRobbie EAC, Webb AAR, Manison NF, Brownlee C, Skepper JN, Chen J, Prestwich GD, Brearley CA (2003) Inositol hexakisphosphate mobilizes an endomembrane store of calcium in guard cells. Proc Natl Acad Sci USA 100:10091–10095

Li S, Assmann SM, Albert R (2006) Predicting essential components of signal transduction networks: a dynamic model of guard cell abscisic acid signaling. PLoS Biol 4:e312

Lin W-H, Wang Y, Mueller-Roeber B, Brearley CA, Xu Z-H, Xue H-W (2005) At5PTase13 modulates cotyledon vein development through regulating auxin homeostasis. Plant Physiol 139:1677–1691

Liu HT, Gao F, Cui SJ, Han JL, Sun DY, Zhou RG (2006) Primary evidence for involvement of IP_3 in heat-shock signal transduction in Arabidopsis. Cell Res 16:394–400

Loewus FA, Murthy PPN (2000) *myo*-Inositol metabolism in plants. Plant Sci 150:1–19

Majerus PW, Kisseleva MV, Norris FA (1999) The role of phosphatases in inositol signaling reactions. J Biol Chem 274:10669–10672

Martinoia E, Locher R, Vogt E (1993) Inositol trisphosphate metabolism in subcellular fractions of barley (*Hordeum vulgare L.*) mesophyll cells. Plant Physiol 102:101–105

Meijer HJ, Munnik T (2003) Phospholipid-based signaling in plants. Annu Rev Plant Biol 54:265–306

Mikoshiba K (2007) IP_3 receptor/Ca^{2+} channel: from discovery to new signaling concepts. J Neurochem 102:1426–1446

Mills LN, Hunt L, Leckie CP, Aitken FL, Wentworth M, McAinsh MR, Gray JE, Hetherington AM (2004) The effects of manipulating phospholipase C on guard cell ABA-signalling. J Exp Bot 55:199–204

Mishra G, Zhang W, Deng F, Zhao J, Wang X (2006) A bifurcating pathway directs abscisic acid effects on stomatal closure and opening in Arabidopsis. Science 312:264–266

Morse MJ, Crain RC, Satter RL (1987a) Phosphatidylinositol Cycle Metabolites in *Samanea saman* Pulvini. Plant Physiol 83:640–644

Morse MJ, Crain RC, Satter RL (1987b) Light-stimulated inositolphospholipid turnover in *Samanea saman* leaf pulvini. Proc Natl Acad Sci USA 84:7075–7078

Mueller-Roeber B, Pical C (2002) Inositol phospholipid metabolism in Arabidopsis. Characterized and putative isoforms of inositol phospholipid kinase and phosphoinositide-specific phospholipase C. Plant Physiol 130:22–46

Muir SR, Sanders D (1997) Inositol 1,4,5-trisphosphate-sensitive Ca^{2+} release across nonvacuolar membranes in cauliflower. Plant Physiol 114:1511–1521

Perera IY, Heilmann I, Boss WF (1999) Transient and sustained increases in inositol 1,4,5-trisphosphate precede the differential growth response in gravistimulated maize pulvini. Proc Natl Acad Sci USA 96:5838–5843

Perera IY, Heilmann I, Chang SC, Boss WF, Kaufman PB (2001) A role for inositol 1,4,5-trisphosphate in gravitropic signaling and the retention of cold-perceived gravistimulation of oat shoot pulvini. Plant Physiol 125:1499–1507

Perera IY, Love J, Heilmann I, Thompson WF, Boss WF (2002) Up-regulation of phosphoinositide metabolism in tobacco cells constitutively expressing the human type I inositol polyphosphate 5-phosphatase. Plant Physiol 129:1795–1806

Perera IY, Davis AJ, Galanopoulou D, Im YJ, Boss WF (2005) Characterization and comparative analysis of Arabidopsis phosphatidylinositol phosphate 5-kinase 10 reveals differences in Arabidopsis and human phosphatidylinositol phosphate kinases. FEBS Lett 579:3427–3432

Perera IY, Hung CY, Brady S, Muday GK, Boss WF (2006) A universal role for inositol 1,4,5-trisphosphate-mediated signaling in plant gravitropism. Plant Physiol 140:746–760

Pottosin II, Schonknecht G (2007) Vacuolar calcium channels. J Exp Bot 58:1559–1569

Qin C, Wang X (2002) The Arabidopsis phospholipase D family. Characterization of a calcium-independent and phosphatidylcholine-selective PLD zeta 1 with distinct regulatory domains. Plant Physiol 128:1057–1068

Quintero FJ, Garciadeblas B, Rodriguez-Navarro A (1996) The SAL1 gene of arabidopsis, encoding an enzyme with 3'(2'),5'-bisphosphate nucleotidase and inositol polyphosphate 1-phosphatase activities, increases salt tolerance in yeast. Plant Cell 8:529–537

Raboy V (2003) myo-Inositol-1,2,3,4,5,6-hexakisphosphate. Phytochemistry 64:1033–1043

Rebecchi MJ, Pentyala SN (2000) Structure, function, and control of phosphoinositide-specific phospholipase C. Physiol Rev 80:1291–1335

Remus TP, Zima AV, Bossuyt J, Bare DJ, Martin JL, Blatter LA, Bers DM, Mignery GA (2006) Biosensors to measure inositol 1,4,5-trisphosphate concentration in living cells with spatio-temporal resolution. J Biol Chem 281:608–616

Saiardi A, Nagata E, Luo HR, Sawa A, Luo X, Snowman AM, Snyder SH (2001) Mammalian inositol polyphosphate multikinase synthesizes inositol 1,4,5-trisphosphate and an inositol pyrophosphate. Proc Natl Acad Sci USA 98:2306–2311

Salinas-Mondragon R, Brogan A, Ward N, Perera I, Boss W, Brown CS, Sederoff HW (2005) Gravity and light: integrating transcriptional regulation in roots. Gravit Space Biol Bull 18:121–122

Sanchez J-P, Chua N-H (2001) Arabidopsis PLC1 is required for secondary responses to abscisic acid signals. Plant Cell 13:1143–1154

Sanders D, Brownlee C, Harper JF (1999) Communicating with calcium. Plant Cell 11:691–706

Sanders D, Pelloux J, Brownlee C, Harper JF (2002) Calcium at the crossroads of signaling. Plant Cell 14:S401–S417

Schumaker KS, Sze H (1987) Inositol 1,4,5-trisphosphate releases Ca^{2+} from vacuolar membrane vesicles of oat roots. J Biol Chem 262:3944–3946

Shears SB (2004) How versatile are inositol phosphate kinases? Biochem J 377:265–280

Shigaki T, Bhattacharyya MK (2000) Decreased inositol 1,4,5-trisphosphate content in pathogen-challenged soybean cells. Mol Plant Microbe Interact 13:563–567

SmolenskaSym G, Kacperska A (1996) Inositol 1,4,5-trisphosphate formation in leaves of winter oilseed rape plants in response to freezing, tissue water potential and abscisic acid. Physiol Plant 96:692–698

Staxen I, Pical C, Montgomery LT, Gray JE, Hetherington AM, McAinsh MR (1999) Abscisic acid induces oscillations in guard-cell cytosolic free calcium that involve phosphoinositide-specific phospholipase C. Proc Natl Acad Sci USA 96:1779–1784

Stenzel I, Ischebeck T, Konig S, Holubowska A, Sporysz M, Hause B, Heilmann I (2008) The type B phosphatidylinositol-4-phosphate 5-kinase 3 is essential for root hair formation in Arabidopsis thaliana. Plant Cell 20:124–141

Stevenson JM, Perera IY, Heilmann I, Persson S, Boss WF (2000) Inositol signaling and plant growth. Trends Plant Sci 5:252–258

Stevenson-Paulik J, Odom AR, York JD (2002) Molecular and biochemical characterization of two plant inositol polyphosphate 6-/3-/5-kinases. J Biol Chem 277:42711–42718

Stevenson-Paulik J, Bastidas RJ, Chiou ST, Frye RA, York JD (2005) Generation of phytate-free seeds in Arabidopsis through disruption of inositol polyphosphate kinases. Proc Natl Acad Sci USA 102:12612–12617

Takahashi S, Katagiri T, Hirayama T, Yamaguchi-Shinozaki K, Shinozaki K (2001) Hyperosmotic stress induces a rapid and transient increase in inositol 1,4,5-trisphosphate independent of abscisic acid in Arabidopsis cell culture. Plant Cell Physiol 42:214–222

Takazawa K, Perret J, Dumont JE, Erneux C (1991) Molecular cloning and expression of a human brain inositol 1,4,5-trisphosphate 3-kinase. Biochem Biophys Res Commun 174:529–535

Tang RH, Han S, Zheng H, Cook CW, Choi CS, Woerner TE, Jackson RB, Pei ZM (2007) Coupling diurnal cytosolic Ca^{2+} oscillations to the CAS-IP_3 pathway in Arabidopsis. Science 315:1423–1426

Tanimura A, Nezu A, Morita T, Turner RJ, Tojyo Y (2004) Fluorescent biosensor for quantitative real-time measurements of inositol 1,4,5-trisphosphate in single living cells. J Biol Chem 279:38095–38098

Tasma IM, Brendel V, Whitham SA, Bhattacharyya MK (2008) Expression and evolution of the phosphoinositide-specific phospholipase C gene family in Arabidopsis thaliana. Plant Physiol Biochem 46(7):627–637

Testerink C, Munnik T (2005) Phosphatidic acid: a multifunctional stress signaling lipid in plants. Trends Plant Sci 10:368–375

Tucker EB, Boss WF (1996) Mastoparan-induced intracellular Ca^{2+} fluxes may regulate cell-to-cell communication in plants. Plant Physiol 111:459–467

van der Wal J, Habets R, Varnai P, Balla T, Jalink K (2001) Monitoring agonist-induced phospholipase C activation in live cells by fluorescence resonance energy transfer. J Biol Chem 276:15337–15344

van Leeuwen W, Vermeer JEM, Gadella TWJ, Munnik T (2007) Visualization of phosphatidylinositol 4,5-bisphosphate in the plasma membrane of suspension-cultured tobacco BY-2 cells and whole Arabidopsis seedlings. Plant J 52:1014–1026

Varnai P, Balla T (1998) Visualization of phosphoinositides that bind pleckstrin homology domains: calcium- and agonist-induced dynamic changes and relationship to *Myo*-[^{3}H]inositol-labeled phosphoinositide pools. J Cell Biol 143:501–510

Wang X (2004) Lipid signaling. Curr Opin Plant Biol 7:329–336

Weinl S, Held K, Schlucking K, Steinhorst L, Kuhlgert S, Hippler M, Kudla J (2008) A plastid protein crucial for Ca^{2+}-regulated stomatal responses. New Phytol 179(3):675–686

Williams RS, Eames M, Ryves WJ, Viggars J, Harwood AJ (1999) Loss of a prolyl oligopeptidase confers resistance to lithium by elevation of inositol (1,4,5) trisphosphate. EMBO J 18:2734–2745

Williams ME, Torabinejad J, Cohick E, Parker K, Drake EJ, Thompson JE, Hortter M, Dewald DB (2005) Mutations in the Arabidopsis phosphoinositide phosphatase gene SAC9 lead to over-accumulation of PtdIns(4,5)P2 and constitutive expression of the stress-response pathway. Plant Physiol 138:686–700

Xiong L, Lee B, Ishitani M, Lee H, Zhang C, Zhu JK (2001) FIERY1 encoding an inositol polyphosphate 1-phosphatase is a negative regulator of abscisic acid and stress signaling in Arabidopsis. Genes Dev 15:1971–1984

Xue H, Chen X, Li G (2007) Involvement of phospholipid signaling in plant growth and hormone effects. Curr Opin Plant Biol 10:483–489

Yu J, Leibiger B, Yang SN, Caffery JJ, Shears SB, Leibiger IB, Barker CJ, Berggren PO (2003) Cytosolic multiple inositol polyphosphate phosphatase in the regulation of cytoplasmic free Ca^{2+} concentration. J Biol Chem 278:46210–46218

Zhang X, Coté G, Crain R (2002) Involvement of phosphoinositide turnover in tracheary element differentiation in *Zinnia elegans* L. cells. Planta 215:312–318

Zhang W, Qin C, Zhao J, Wang X (2004) Phospholipase D alpha 1-derived phosphatidic acid interacts with ABI1 phosphatase 2C and regulates abscisic acid signaling. Proc Natl Acad Sci USA 101:9508–9513

Zhang W, Yu L, Zhang Y, Wang X (2005) Phospholipase D in the signaling networks of plant response to abscisic acid and reactive oxygen species. Biochim Biophys Acta 1736:1–9

Zhong R, Ye Z-H (2004) Molecular and biochemical characterization of three WD-repeat-domain-containing inositol polyphosphate 5-phosphatases in Arabidopsis thaliana. Plant Cell Physiol 45:1720–1728

Zhong R, Burk DH, Morrison WH III, Ye Z-H (2004) FRAGILE FIBER3, an Arabidopsis gene encoding a type ii inositol polyphosphate 5-phosphatase, is required for secondary wall synthesis and actin organization in fiber cells. Plant Cell 16:3242–3259

Inositol Polyphosphates and Kinases

Jill Stevenson-Paulik and Brian Q. Phillippy

Abstract Since the turn of the century, an advent of discoveries has revealed novel functions of specific inositol phosphates in eukaryotic cells. Forward- and reverse-genetic approaches in plant systems have allowed for the identification of many of the genes involved in plant inositol phosphate metabolism. There appears to be a relatively small group of promiscuous kinases responsible for a plethora of reactions that potentially generate multiple and branched inositol phosphate pathways leading to phytate synthesis. This chapter describes the known plant inositol polyphosphate kinases, the inositol phosphates that are modified by them, and the possibility for emerging roles of these small molecules in plant biology.

1 Introduction

The drive to understand the biology of inositol polyphosphate (IP) metabolism in plants has been fueled not only by the practical desire to manipulate phytate production in seeds for environmental and nutritional purposes, but also by intriguing discoveries in nonplant eukaryotes regarding the essential roles of IPs in the regulation of basic cellular functions. Because *myo*-inositol hexakisphosphate ($InsP_6$), or phytate, is indigestible to monogastrics, the majority (~80%) of seed phosphorus is not available to the animal (Raboy 2003). Phosphate is among the most expensive ingredients in feed supplementation. In 2008, the cost of mono- and dicalcium phosphate has surged, leading to intensified interest in developing crops with low phytate and highly bioavailable forms of phosphate. Forward-genetic approaches in maize and reverse-genetic approaches in *Arabidopsis* have been fruitful in identifying major genes involved in phytate synthesis, but a deeper

J. Stevenson-Paulik (✉)
BASF Plant Science, Research Triangle Park, L.P. 26 Davis Drive, NC 27709, USA
e-mail: jill.stevenson@basf.com

T. Munnik (ed.), *Lipid Signaling in Plants*, Plant Cell Monographs 16,
DOI 10.1007/978-3-642-03873-0_11, © Springer-Verlag Berlin Heidelberg 2010

understanding of the function and regulation of these genes and other members of the pathways is needed to rationally manipulate phytate metabolism in crops used as feed sources without compromising the yield and agronomic performance.

Perhaps more fascinating, from a basic cell biology point-of-view, is that there have been many surprising and provocative discoveries within the last 10 years uncovering the various roles of phytate and higher IP species in regulating many fundamental cellular processes that extend beyond the classical paradigm of $InsP_3$-mediated calcium signaling. These discoveries have revealed an enormous toolbox of small molecules available to the cell that are derived from one molecule, *myo*-inositol. As an asymmetric cyclitol harboring six hydroxyls available for phosphorylation, *myo*-inositol (Fig. 1a) provides a chemical prop from which an array of at least 64 different inositol phosphates may be generated and used as specific signals (York 2006), and the numbers increase as these IPs are conjugated to lipids or are pyrophosphorylated. Many of these functions (reviewed in York 2006), which include transcriptional regulation, mRNA export, DNA repair, chromatin remodeling, and telomere maintenance, have not been explored yet in plants, but the conservation of the inositol polyphosphate kinome from yeast to animals to plants suggests primordial roles in cell function that have been evolutionarily conserved. This chapter describes the state of current understanding of plant inositol polyphosphates, the kinases that control them, and their role in plant biology.

2 Inositol Polyphosphates and Phytate Synthesis

Though not definitively identified, plant inositol polyphosphate synthesis has been postulated to occur through two general routes, as described in Fig. 1b. Both pathways derive from de novo inositol synthesis initiated by the conversion of glucose 6-phosphate to inositol-3-phosphate [Ins(3)P] via the enzyme *myo*-inositol phosphate synthase (MIPS or INO1). The inositol lipid-independent pathway then extends through the sequential phosphorylation of Ins(3)P to eventually generate $InsP_6$. This pathway was described biochemically through short-term nonequilibrium $^{32}P–P_i$ labeling studies on developing storage tissues of the duckweed, *Spirodela polyrhiza* (Brearley and Hanke 1996). A similar pathway was also described earlier in slime mold by Stephens and Irvine (1990), thus suggesting a conservation across species. Since then, inositol polyphosphate kinases (IPKs) have been discovered, but the characterization of their substrate selectivity has often been limited to include only those IPs that are commercially available, which is a small number compared with the complete theoretical cellular IP portfolio, and unfortunately excludes many of the intermediates identified in this pathway. For this reason, neither the recapitulation of this pathway in vitro nor the mapping of the recently identified kinases to some of the intermediate steps (from Ins(3)P through Ins(3,4,5,6)P_4) has been completely accomplished. Nonetheless, the last two steps of this pathway can be explained based on in vitro assays of known plant IP kinases and are discussed in later sections of this chapter.

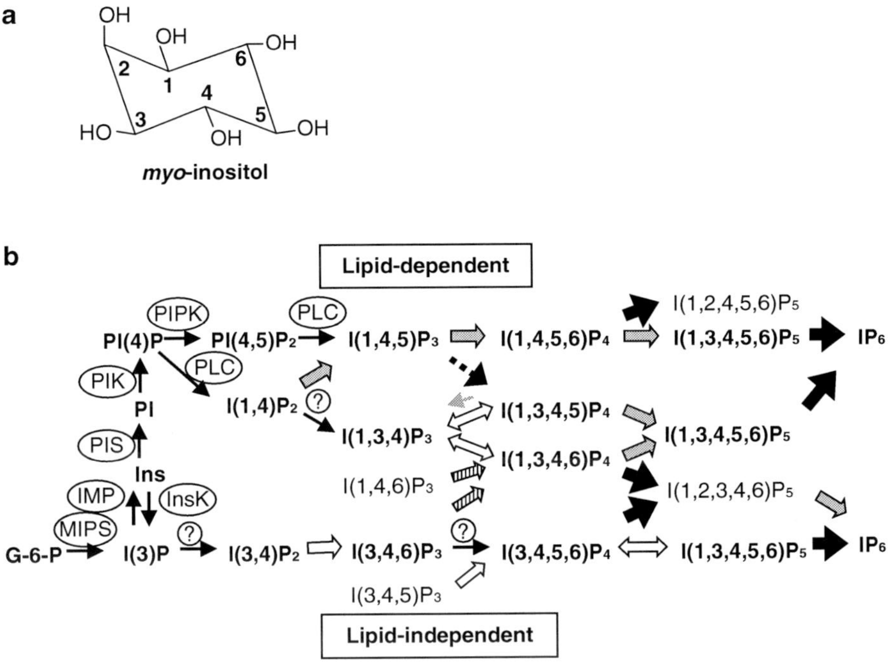

Fig. 1 Inositol structure and potential plant inositol polyphosphate pathways. (**a**) Inositol is an asymmetric 6-carbon cyclitol harboring six hydroxyl groups available for phosphorylation. Five of the six hydroxyls fall within the equatorial plane of the structure, while the single 2-OH is axial, thus conferring asymmetry to the molecule. Inositol polyphosphates are named and numbered based on the corresponding hydroxyl groups that are phosphorylated. (**b**) Proposed lipid-dependent and -independent phytate synthesis pathways in plants, based on biochemical and molecular evidence as described in this chapter. *Thin arrows* indicate reactions catalyzed by the designated circled enzyme: *MIPS*, *myo*-inositol phosphate synthase; *IMP*, inositol monophosphatase; *InsK*, inositol kinase; *PIS*, phosphatidylinositol synthase; *PIK*, phosphatidylinositol 4-kinase; *PIPK*, phosphatidylinositol 4-phosphate kinase. *Thick black arrows* indicate reactions catalyzed by Ipk1, inositol polyphosphate kinase 1 *Thick gray arrows* are reactions by Ipk2, inositol polyphosphate kinase 2 *Thick white arrows* are reactions catalyzed by ITPKs, inositol(1,3,4)P_3 5/6-kinase. *Thick striped arrows* indicate reactions catalyzed by either Ipk2 or ITPK. *Dotted arrow* indicates IP_3 3-kinase activity which has not been described in plants. *Thin gray arrow* indicates inositol polyphosphate 5-phosphatase activity. *Question marks* indicate unknown kinases

The second major pathway of plant phytate synthesis occurs through the inositol phospholipids whereby Ins(3)P is dephosphorylated by inositol monophosphatase to generate free *myo*-inositol, which is then transferred to CDP-diacylglycerol to make the phospholipid, phosphatidylinositol (PtdIns), via PtdIns synthase. PtdIns is sequentially phosphorylated to PtdIns(4,5)P_2 by the action of PI 4- and PIP 5-kinases and then hydrolyzed by phospholipase C (PLC) to generate Ins(1,4,5)P_3. Ins(1,4,5)P_3 is then sequentially phosphorylated to $InsP_6$ through the activity of two or three different IP kinases. This pathway and its variants are the major routes of phytate synthesis in yeast and animals. One major difference in plants is the higher concentration of PtdIns4P than PtdIns(4,5)P_2. Because some plant PLCs prefer PtdIns4P over PtdIns(4,5)P_2 as a substrate (Park et al. 2002), Ins(1,4)P_2 could also have a significant role. In yeast, Ins(1,4,5)P_3 is phosphorylated twice by a dual-specificity $InsP_3$/$InsP_4$ 6-/3-kinase, Ipk2, to produce Ins(1,3,4,5,6)P_5 via an Ins (1,4,5,6)P_4 intermediate (Odom et al. 2000). The $InsP_5$ is then phosphorylated by Ipk1, which specifically recognizes the single axial hydroxyl group at the 2nd position of the *myo*-inositol ring (York et al. 1999). In mammalian cells, this pathway diverges from the straightforward yeast route at the step of Ins(1,4,5)P_3 phosphorylation. In these systems, Ins(1,4,5,)P_3 is phosphorylated by an Ins(1,4,5) P_3 3-kinase, which is structurally similar yet distinct from Ipk2. This reaction generates Ins(1,3,4,5)P_4, which is then dephosphorylated by an IP 5-phosphatase to produce Ins(1,3,4)P_3. Ins(1,3,4)P_3 then becomes the substrate of a multidimensional Ins(1,3,4)P_3 5/6-kinase (IP56K or ITPK). Formation of Ins(1,3,4,6)P_4 by the ITPK is described as the rate-limiting step to higher phosphoinositols in humans (Verbsky et al. 2005). Interestingly, plant genomes do not appear to contain functional homologs of the $InsP_3$ 3-kinase (Stevenson-Paulik et al. 2002), yet they do have IP 5-phosphatase homologs that can use Ins(1,3,4,5)P_4 as a substrate (Berdy et al. 2001), and they possess a gene family of ITPK homologs. Proof that the lipid-dependent pathway(s) exists in plants is indirect and based on enzymatic properties of the plant IP kinases and their similarities to the yeast system (Stevenson-Paulik et al. 2002; Xia et al. 2003). Indeed, *Arabidopsis* Ipk2 and Ipk1 can recapitulate $InsP_6$ synthesis in vitro from a pure Ins(1,4,5)P_3 substrate and restore $InsP_6$ synthesis in yeast cells lacking endogenous Ipk1 and Ipk2 (Stevenson-Paulik et al. 2002, 2005). However, due to the probability that other inositol phosphate isomers, in addition to those already identified, may participate in the lipid independent pathway, it cannot be ruled out that Ipk2 also participates in that network. In support of this, Caddick et al. (2007) describe a potato Ipk2 that not only phosphorylates the same IPs as the *Arabidopsis* orthologs, but also acts upon Ins(1,4)P_2, a racemic mixture of Ins(1/3,4,6)P_3, and Ins(3,4,5,6)P_4.

It is quite possible that both pathways exist in plants and it is tempting to speculate that the lipid-dependent pathway(s) that is conserved from yeast to animals, which has roles in regulating basic nuclear and cellular processes that are common to all eukaryotes, occurs in plant vegetative tissues as well. Indeed, phytate regulation appears to be similar in vegetative tissues as the concentration of phytate is akin to that in yeast and animals ($<100\ \mu M$) (Irvine and Schell 2001; Bentsink et al. 2003). Conversely, the lipid-independent pathway may occur in seed (Suzuki et al. 2007) and storage tissues such as in the Duckweed turion, where massive amounts of phytate accumulate for the future use of subsequently developing tissues. In this scenario, the kinases may be the same as those involved in the lipid-dependent pathway because their activities have been shown to be promiscuous (Stevenson-Paulik et al. 2002; Josefsen et al. 2007). If this is the case, then the regulation of the pathways would have to be quite dissimilar to account for the major differences in phytate levels found in vegetative versus storage tissues.

One step where the regulation of phytate synthesis may be dramatically different in vegetative versus seed tissues most likely occurs early in the pathway and has been recently suggested by the molecular identification of the maize *lpa1* gene (Shi et al. 2007). Shi et al. have identified the lesion of *lpa1*, originally described by Raboy et al. (2000), as an ABC transporter, and have shown that the seed-specific downregulation of this gene results in approximately a 90% reduction of phytate and corresponding gains in free phosphate. Clearly, this gene plays a major role in phytate synthesis. Lpa1 is an MRP-type ABC transporter, with unknown substrates and function. Although the majority of *lpa1*, also called *ZmMRP4*, is expressed in the seed, it is also expressed at much lower levels in vegetative tissues (Shi et al. 2007), where it likely plays a role in maintaining phytate anabolism as well. The expression differential of this one gene alone might account for the dramatic tissue variation observed for phytate levels.

Interestingly, the *Arabidopsis* ortholog of *ZmMRP4* is *AtMRP5* (At1g04120) and the expression of this gene is also coincident with seed development and phytate accumulation (for visual, go to http://bbc.botany.utoronto.ca/efp/cgi-bin/efpWeb.cgi and enter At1g04120). Despite the striking expression of *AtMRP5* in developing seed, the role of this gene in seed physiology has been unexplored thus far. Instead, AtMRP5 has been the focus of several studies that describe it as a plasma membrane-resident protein involved in the ABA- and calcium-mediated regulation of stomatal movements (Gaedeke et al. 2001; Klein et al. 2003; Suh et al. 2007). Given its likely role in phytate metabolism, and its localization to vacuolar membranes (Dunkley et al. 2006) where phytate accumulation occurs, it may function in regulating stomatal control by providing $InsP_6$, which has been described to regulate guard cell K_{in} channels in a calcium-dependent manner (Lemtiri-Chlieh et al. 2000).

Not indicated in the pathway sehematic diagram is the metabolism of inositol pyrophosphates. Although emerging data from yeast systems identify these hyper-phosphorylated small molecules as important regulators of cell growth, morphology, and the key players in phosphate signaling (Mulugu et al. 2007; Lee et al. 2007), there is little evidence yet of their existence in plants. That said, it is likely that these IP species will be found to conduct communication in plant cells. Flores and Smart (2000) reported the identification of $InsP_7$ and $InsP_8$ peaks in IP profiles of labeled Duckweed turions. The recently described Vip1 and Asp1 $InsP_6$- and PP-$InsP_5$ ($InsP_7$)-kinases belong to a family of kinases that are conserved across species (Mulugu et al. 2007) and are represented in the *Arabidopsis* genome (At5g15070, At3g01310).

3 The Plant Inositol Polyphosphate Kinases

3.1 Inositol Polyphosphate Kinase 2 (Ipk2)/Inositol Phosphate Multikinase

The identification of yeast *IPK2* (Odom et al. 2000) provided the basis for the molecular discovery of its orthologs in plants. Two *Arabidopsis* genes were recognized as having deduced amino acid sequence similar to yeast Ipk2, and were subsequently cloned, expressed as fusion proteins, and characterized for enzymatic activity. The gene products of *AtIPK2α* and *AtIPK2β* are 70% identical to each other and display indistinguishable enzymatic properties (Stevenson-Paulik et al. 2002; Xia et al. 2003). Each protein phosphorylates Ins(1,4,5)P_3 to Ins(1,3,4,5,6)P_5 via an Ins(1,4,5,6)P_4 intermediate, thus they are described as having InsP_3/InsP_4 6-/3-kinase activities. It is important to label these kinases by their order of phosphorylation so as not to confuse them with the InsP_3 3-kinase family, which have some sequence similarity to the Ipk2s, but are not dual kinases and specifically phosphorylate Ins(1,4,5)P_3 on the D-3 position. The InsP_3 3-kinase family does not appear to exist in nonmetazoan species. An important discovery was that the Ipk2 proteins could not only phosphorylate the inositol ring on the 6th and 3rd position, but could also phosphorylate the 5th position (Stevenson-Paulik et al. 2002; Chang et al. 2002). This 5-kinase activity was found to be conserved in all Ipk2 orthologs and allowed for the mapping of this kinase onto the mammalian IP pathway where, until then, it was not known what gene-product was responsible for catalyzing the conversion of Ins(1,3,4,6)P_4 to Ins(1,3,4,5,6)P_5 (Chang et al. 2002).

Recent attempts to understand the function of the Ipk2 proteins and their IP products in plant biology have been made by altering the expression of the genes. A T-DNA insertion that disrupts *AtIPK2β* expression was characterized and found to have a role in phytate synthesis in *Arabidopsis* seeds. Interestingly, although the *AtIPK2α* gene was still functional and was expressed during seed development of these mutants, there was an inositol phosphate phenotype in the seeds but not in vegetative tissues. This result suggested that there are distinct roles for the two homologs in seed IP synthesis but redundant activities in other tissues (Stevenson-Paulik et al. 2005). Although the seeds of the *AtIPK2β* knockout plants had a 30% reduction in phytate and an accumulation of InsP_4 and InsP_5 species, there was no other seed or vegetative growth phenotype. Xu et al. (2005) downregulated *AtIPK2α* using antisense technology and characterized several plant growth phenotypes. They described transformants with enhanced pollen germination and pollen tube growth in the presence of low calcium, increased root growth and root hair development in the presence of high calcium, and accelerated seed germination and early seedling growth. The authors suggested that the downregulation of the gene caused a disruption of inositol polyphosphate production that led to the growth phenotypes and that the plants had an altered InsP_3-mediated calcium signaling. Unfortunately, there were no measurements of inositol phosphates done in the transgenic material, and thus there were no data to support this hypothesis. Given the presence of the other ubiquitously expressed Ipk2 in *Arabidopsis*, *AtIPK2β*, which could very well have redundant activities (indeed the biochemistry of each is the same), it is quite possible that the observed phenotypes may have had little to do with inositol phosphate metabolism.

Yang et al. (2008) overexpressed *AtIPK2β* in tobacco plants and found that this conferred a slight advantage over wild type to tolerating abiotic stresses such as salinity, drought, and oxidation, but no inositol phosphate measurements were performed to correlate the growth and stress phenomenology to inositol phosphate metabolism. Zhang et al. (2007) overexpressed *AtIPK2β* in *Arabidopsis* and reported decreased apical dominance and the altered regulation of some auxin-responsive genes, thus concluding that this gene product is involved in mediating auxin-responses. Since the high-level constitutive ectopic expression driven by the 35S promoter confers such an artificial state, it is difficult to conclude that this gene is involved in any of these processes. Again, no inositol phosphates were measured in these studies, thus it is not clear if the inositol phosphate pathway is altered in the mutants. Given that these ubiquitously expressed kinases do not appear to be rate-limiting in the pathway, it seems unlikely that their overexpression would have a profound effect on the inositol phosphate profile.

3.2 Inositol (1,3,4)P_3 5/6-Kinases/Inositol Trisphosphate Kinases (IP56K; ITPK)

Another family of multifunctional inositol kinases are a class of enzymes that were originally isolated from mammalian sources and named inositol (1,3,4)P_3 5/6-kinases, because they were first determined to phosphorylate the Ins(1,3,4)P_3 isomer on either the 5th or 6th position of the inositol ring, with a greater preference for the latter reaction (Shears 1989). The conversion of Ins(1,3,4)P_3 to Ins(1,3,4,6)P_4 is the pathway branching step that leads to the formation of higher phosphorylated IPs in animals (Verbsky et al. 2005). Subsequent to its initial kinase activity identification, this class of IP kinases has been shown to also harbor Ins(3,4,5,6)P_4 1-kinase activity (Tan et al. 2007; Yang et al. 1999), as well as phosphatase activity towards Ins(1,3,4,5)P_4 and Ins(1,3,4,6)P_4 to produce Ins(1,3,4)P_3, and towards Ins(1,3,4,5,6)P_5 to produce Ins(3,4,5,6)P_4 (Ho et al. 2002). To add to its portfolio of multidimensional activities, the mammalian IP56K also has been described as having protein kinase activity (Wilson et al. 2001) and association with the COP9 signalosome (CSN) (Sun et al. 2002), which is a protein complex that was first identified in plants as a repressor of photomorphogenesis. These additional roles are not without controversy as the protein kinase activity has been refuted by Qian et al. (2005). The *Arabidopsis* At156K-1 protein has also been reported to harbor protein kinase activity and association with the COP9 signalosome, as well as its mutant mildly phenocopying CSN mutants in red light-induced photomorphogenesis (Qin et al. 2005).

ITPKs were first described in plants by Wilson and Majerus (1997) who isolated a partial cDNA from *Arabidopsis* and showed that the recombinant protein had Ins(1,3,4)P_3 5/6-kinase activity, and by Phillippy (1998) who identified the activity in immature soybean seeds. Since then multiple reports have accumulated that identify ITPKs in various plant species. ITPKs exist as multi-gene families in plants, as *Arabidopsis*, soybean, corn, and rice have at least 4–6 members in their genomes (Sweetman et al. 2007; Stiles et al. 2008; Shi et al. 2003; Josefsen et al. 2007). Recent cloning and recombinant protein characterization of the plant ITPKs has revealed that, like their mammalian counterparts, the plant kinases also possess multiple substrate selectivities. Although the majority of these kinases from rice (OsITPK), barley (HvITPK), *Arabidopsis* (AtITPK), soybean (GmITPK), maize (ZmITPK), and potato (StITPK) phosphorylate Ins (1,3,4)P_3 to produce either Ins(1,3,4,5)P_4 or Ins(1,3,4,6)P_4, they appear to prefer Ins(3,4,5,6)P_4 as a substrate to produce Ins(1,3,4,5,6)P_5 (Josefsen et al. 2007; Caddick et al. 2008; Stiles et al. 2008; Shi et al. 2003; Sweetman et al. 2007). The fact that these enzymes prefer this $InsP_4$ substrate and participate in the reverse reaction to cycle between Ins(3,4,5,6)P_4 and Ins(1,3,4,5,6)P_5, suggest that they may play an integral role in pollen tube growth by regulating Cl^- channel conductance via Ins(3,4,5,6)P_4 (Zonia et al. 2002). Of interest, these enzymes also catalyze a range of reactions including the following: Ins(3,4)P_2 $\rightarrow$ Ins(3,4,6) P_3 (StITPK1); Ins(3,4,6)P_3 $\rightarrow$ Ins(1,3,4,6)P_4 (StITPK1, GmITPKs, AtITPKs); Ins (3,4,5)P_3 $\rightarrow$ Ins(3,4,5,6)P_4 (StITPK1, OsITPK, HvITPK); and Ins(1,4,6)P_3 $\rightarrow$ Ins (1,3,4,6)P_4 (StITPK1, AtITPKs). Additionally, the OsITPK and HvITPK enzymes were found to phosphorylate a multitude of IP isomers ranging from Ins(1)P to $InsP_6$. In several cases, the kinases were described as harboring isomerase and phosphatase activities, and thus resulted in cycling between substrate and product formation. The interpretation that an IP kinase also has phosphatase activity is sometimes based on a result whereby the ^{32}P-labeled product of the reaction is the same IP species as the unlabeled substrate, e.g., unlabeled $InsP_4$ + [^{32}P]ATP $\rightarrow$ [^{32}P]$InsP_4$ + ADP. Although this may be the result of initial phosphatase action followed by rephosphorylation, another more trivial explanation is that this is the result of an impure substrate, e.g., $InsP_3$ contamination that is phosphorylated to make $InsP_4$. Thus, care should be taken in interpreting substrate selectivity and phosphatase/kinase cycling behaviors when using [^{32}P]ATP and unlabeled IP substrates for in vitro assays.

AtITPK4, which is more divergent in primary amino acid sequence from other family members, had the unique inability to act as an Ins(3,4,5,6)P_4 1-kinase, and appeared to favor dephosphorylating and rephosphorylating Ins(1,3,4,6)P_4 (Sweetman et al. 2007). The potato ITPK also displayed phosphotransferase activity as indicated by its ability to transfer a phosphate from Ins(3,4,5,6)P_4 to Ins(1,3,4)P_3 only in the presence of ADP. Together, these data indicate that this class of IP kinases likely participates in multiple steps and potentially divergent IP pathways in plants. In particular, these data suggest that the ITPKs would most likely participate in the lipid-independent pathways and that a subset of the family members may phosphorylate earlier steps while others may participate in later steps. Of course, the transferability of the behavior of the enzymes in an isolated in vitro environment to that of the complex cellular environment has to be considered and it is possible that these enzymes have a prevailing activity within the cell, depending on location and substrate presentation. It will be enlightening to see the effect of altered expression of these genes on the cellular IP profile in different plant tissues and species. Clearly, at least some of the members of this IP kinase family are necessary for phytate synthesis pathways, as indicated by the null mutation of maize ITPK, originally identified as *lpa2*, which results in a 30% loss of seed phytate in corn kernels (Shi et al. 2003).

3.3 Inositol Polyphosphate Kinase 1

The terminal reaction of phytate synthesis is catalyzed by Ipk1, which specifically phosphorylates the 2-OH position of the inositol ring. IP 2-kinase activity was first described in mung bean (Biswas et al. 1978) and purified from immature soybean seeds (Phillippy et al. 1994). The molecular identity of the $InsP_5$ 2-kinase remained unknown until York et al. (1999) performed a screen that revealed the PLC → $InsP_6$ pathway requirement for efficient mRNA export in yeast. Through in silico approaches to identify homologous sequences to the fungal genes, four genes have been identified in *Arabidopsis* that appear to encode $InsP_5$ 2-kinases, and two of those genes are expressed (Stevenson-Paulik et al. 2005; Sweetman et al. 2006). Comparison of the expression of each of these genes indicates that both are ubiquitously expressed, but At5g42810 is much more predominant and increases in strength during seed development, while At5g59900 decreases during seed development (http://bbc.botany.utoronto.ca/efp/cgi-bin/efpWeb.cgi), thus suggesting that At5g42810 plays the major role in seed phytate synthesis. This conclusion is supported by the knockdown of At5g42810 by >70% which resulted in >80% reduction in seed phytate and >90% reduction in vegetative tissues (Stevenson-Paulik et al. 2005). Without the characterization of a complete knockout of At5g42810 and/or a knockout of At5g59900, it is not possible to know the relative contribution of At5g59900 to overall phytate synthesis. The enzymatic properties of At5g59900 remain to be characterized and it is possible that this gene-product has different substrate selectivities than the predominant form, or that it contributes to phytate synthesis in very specialized cells or responses.

Enzyme assays using recombinant protein generated from the At5g42810 gene confirmed that it is, indeed, an IP 2-kinase with a *K*m for $InsP_5$ of 4.3 μM (Stevenson-Paulik et al. 2005, supplemental data), which is similar to that of the soybean $InsP_5$ 2-kinase at 2.3 μM (Phillippy et al. 1994). These studies also reveal that this kinase, unlike Ipk2 or ITPKs, specifically phosphorylates only the D-2 position of the inositol ring, yet it does maintain some level of promiscuity in that it can use various $InsP_3$, $InsP_4$, and $InsP_5$ species, providing that the 2-OH position of inositol can accept a phosphate. Maize ZmIpk1 phosphorylates $Ins(1,3,4,5,6)P_5$ with a K_m of 119 μM, and also may use $Ins(1,4,6)P_3$, $Ins(1,4,5,6)P_4$, and $Ins(3,4,5,6)P_4$ as substrates (Sun et al. 2007). It was shown that AtIpk1, along with AtIpk2 could recapitulate $InsP_6$ synthesis in vitro from a single $Ins(1,4,5)P_3$ precursor and that the combination of AtIpk1, AtIpk2, and a mammalian ITPK (with similar activities to the plant ITPKs) can produce $InsP_6$ from an $Ins(1,3,4)P_3$ precursor (Stevenson-Paulik et al. 2005). Additionally, the end-products can be altered based on the stoichiometry of the enzymes in the tube (Stevenson-Paulik et al. 2005, supplemental data), thus highlighting the potentially complex set of higher IPs that may exist in the cell based on substrate availability and enzyme dosage.

Recently emerging roles of inositol phosphates have been to regulate the aspects of phosphate responses in yeast (Steger et al. 2003) and the knockdown of *AtIPK1* also suggests a role for the higher IPs in plant phosphate signaling. A knockdown of

AtIPK1 (At5g42810) resulted in *Arabidopsis* plants that were insensitive to external phosphate as indicated by increased root hair length in the presence of adequate media phosphate concentrations. Additionally, the *atipk1-1* plants were unable to control intracellular phosphate concentrations such that they reached toxic levels when exposed to phosphate concentrations that are typical of standard fertilizers (Stevenson-Paulik et al. 2005). These phenotypes indicate an inability of the plants to sense and respond to phosphate in order to maintain intracellular homeostasis, thus suggesting that the inositol phosphates that are effected by the *AtIPK1* gene product (Ins(1,4,5,6)P_4 and Ins(1,3,4,5,6)P_5, InsP_6, and potentially InsP_7 and InsP_8), or the protein itself, are important for regulating these processes. There was no seed defect in these plants, despite a >80% phytate reduction. This supports the hypothesis that the phytate pathway serves specialized, tissue-specific roles, e. g., that the purpose for seed phytate is for sequestering phosphate, and the purpose of the vegetative pathway is like that in other higher eukaryotes, where it serves a signaling role. It will be interesting to assess the impact of the *ipk1-1* mutation on phosphate-responsive genes and their respective gene-product activities and to dissect the specific IP effectors by studying mutants with lesions in different steps of the phytate pathway.

3.4 Other Kinases

Forward-genetic approaches have been invaluable for identifying genes involved in plant phytate synthesis. Already mentioned are the maize *lpa1* and *lpa2* genes which were first generated by producing mutagenized populations that were screened for high kernel phosphate (Raboy et al. 2000) followed by mapping of the lesions. *lpa3* was also recently identified through a high seed phosphate phenotype screen of a large collection of Mu-insertion lines. *lpa3* was found to be an inositol kinase, with homology to the pfkB carbohydrate kinase family, and phosphorylates *myo*-inositol to produce inositol monophosphates (Shi et al. 2005). Interestingly, the *lpa3* null results in a 45% reduction of seed phytate, and the *lpa2* null results in a 31% reduction, while the *lpa2lpa3* double mutant results in a 66% reduction. Since the double mutant has a greater phytate reduction than the singles, this indicates that they participate, at least partially, in separate pathways that independently contribute to phytate synthesis.

Very recently, the rice *lpa1* mutant, which confers a 45% loss of rice seed phytate, was mapped and named *OsLpa1*. The gene product has homology to 2-phosphoglycerate kinases, which have been described in hyperthermophilic methanogens and produce 2,3-bisphosphoglycerate (Kim et al. 2008). Explanations for the role of this gene in phytate synthesis are twofold. First, previous reports indicate that 2,3-bisphosphoglycerate is an inhibitor of inositol polyphosphate 5-phosphatases (Downes et al. 1982; Fischback et al. 2006), thus by reducing this inhibitor in the mutant, inositol phosphatase activity may be enhanced and lead to decreased phytate. The second possibility is that the putative 2-phosphoglycerate kinase uses inositol phosphates as substrates and participates in early steps of the lipid-independent pathway. It will certainly be interesting to learn what the exact role is for this new gene in phytate metabolism and whether it is conserved in other plant species and nonplant eukaryotes as well.

4 Closing Remarks

We are really at the beginning of understanding the basic components of the inositol polyphosphate pathways and the important gene-products involved in plant biology. This is an area rich in opportunity for revealing new mechanisms of regulation and likely will lead to surprising discoveries for roles of orphan IPs. In immediate terms, there is a need to bridge the gap between the characterization of the knockout and overexpression of the kinases *in planta* and the in vitro enzyme characterization so that conclusions can be drawn about the major pathways in different tissues and environmental conditions. Further questions remain: is the pathway in the cytosol, nucleus, or vacuole? and does this vary depending on tissue and cell type? How do globoids, which store crystalline phytate, develop from protein storage vacuoles and how are phytate and/or its precursors transported into these organelles? Perhaps, these processes are orchestrated by the recently identified MRP-type ABC transporter (Shi et al. 2007) and facilitated possibly by a vacuolar ATPase (Bentsink et al. 2003). What are the functions of orphan IPs, for instance the Ins(1,2,4,5,6)P_5 that is consistently identified in *Arabidopsis* seedlings (Stevenson-Paulik et al. 2005)? Are specific IPs important regulators of fundamental cellular processes, as have been described in fungi and metazoans, or are the detected IPs simply just intermediates on the way to phytate synthesis or degradation? Do the inositol kinases and inositol phosphates impinge on other metabolic and signaling networks? The answer to these questions is likely yes, especially as $InsP_6$ emerges as a surprising cofactor of signaling proteins, as exemplified by the recent discoveries that the *Arabidopsis* auxin receptor, TIR1 (Tan et al. 2007), and human ADAR2 involved in RNA editing (MacBeth et al. 2005) both crystallize with phytate at their core. Systems biology approaches that combine transcriptional, proteomic, and metabolomic data, as well as flux analyses on a complete set of IP kinase mutants is an attractive approach to answering these and other burning questions. Indeed, it is an exciting time for plant inositol phosphate research and discovery as new tools become available to enhance these efforts.

References

Bentsink L, Yuan K, Koornneef M, Vreugdenhil D (2003) The genetics of phytate and phosphate accumulation in seeds and leaves of *Arabidopsis thaliana*, using natural variation. Theor Appl Genet 106:1234–1243

Berdy SE, Kudla J, Gruissem W, Gillaspy GE (2001) Molecular characterization of At5PTase1, an inositol phosphatase capable of terminating inositol trisphosphate signaling. Plant Physiol 126:801–810

Biswas S, Maity IB, Chakrabarti S, Biswas BB (1978) Purification and characterization of *myo*-inositol hexaphosphate-adenosine diphosphate phosphotransferase from Phaseolus aureus. Arch Biochem Biophys185:557–566
Brearley CA, Hanke DE (1996) Metabolic evidence for the order of addition of individual phosphate esters to the *myo*-inositol moiety of inositol hexakisphosphate in the duckweed *Spirodela polyrhiza* L. Biochem J 314:227–233
Caddick SE, Harrison CJ, Stavridou I, Johnson S, Brearley CA (2007) A lysine accumulation phenotype of ScIpk2Delta mutant yeast is rescued by Solanum tuberosum inositol phosphate multikinase. Biochem J 403:381–389
Caddick SEK, Harrison CJ, Stavridou I, Mitchell JL, Hemmings AM, Brearley CA (2008) A Solanum tuberosum inositol phosphate kinase (StITPK1) displaying inositol phosphate-inositol phosphate and inositol phosphate-ADP phosphotransferase activities. FEBS Lett 582:1731–1737
Chang SC, Miller AL, Feng Y, Wente SR, Majerus PW (2002) The human homolog of the rat inositol phosphate multikinase is an inositol 1,3,4,6-tetrakisphosphate 5-kinase. J Biol Chem 277:43836–43843
Downes CP, Mussat MC, Michell RH (1982) The inositol trisphosphate phosphomonoesterase of the human erythrocyte membrane. Biochem J 203:169–177
Dunkley TP, Hester S, Shadforth IP, Runions J, Weimar T, Hanton SL, Griffin JL, Bessant C, Brandizzi F, Hawes C, Watson RB, Dupree P, Lilley KS (2006) Mapping the *Arabidopsis* organelle proteome. Proc Natl Acad Sci USA 103:6518–6523
Fischback A, Adelt S, Muller A, Vogel G (2006) Disruption of inositol biosynthesis through targeted mutagenesis in *Dictyostelium discoideum*: generation and characterization of inositol-auxotrophic mutants. Biochem J 397:509–518
Flores S, Smart CC (2000) Abscisic acid-induced changes in inositol metabolism in Spirodela polyrrhiza. Planta 211:823–832
Gaedeke N, Klein M, Kolukisaoglu HU, Forestier C, Muller A, Ansorge M, Becker D, Mamnun Y, Kuchler K, Schulz B, Mueller-Roeber B, Martinoia E (2001) The *Arabidopsis thaliana* ABC transporter AtMRP5 controls root development and stomata movement. EMBO J 20:1875–1887
Ho MW, Yang X, Carew MA, Zhang T, Hua L, Kwon YU, Chung SK, Adelt S, Vogel G, Riley AM, Potter BV, Shears SB (2002) Regulation of Ins(3,4,5,6)P(4) signaling by a reversible kinase/phosphatase. Curr Biol 12:477–482
Irvine RF, Schell MJ (2001) Back in the water: the return of the inositol phosphates. Nat Rev 2:327–338
Josefsen L, Bohn L, Sorensen MB, Rasmussen SK (2007) Characterization of a multifunctional inositol phosphate kinase from rice and barley belonging to the ATP-grasp superfamily. Gene 397:114–125
Kim SI, Andaya CB, Goyal SS, Tai TH (2008) The rice OsLpa1 gene encodes a novel protein involved in phytic acid metabolism. Theor Appl Genet 117:769–779
Klein M, Perfus-Barbeoch L, Frelet A, Gaedeke N, Reinhardt D, Mueller-Roeber B, Martinoia E, Forestier C (2003) The plant multidrug resistance ABC transporter AtMRP5 is involved in guard cell hormonal signaling and water use. Plant J 33:119–129
Lee Y-S, Mulugu S, York JD, O'Shea EK (2007) Regulation of a cyclin-CDK-CDK inhibitor complex by inositol pyrophosphates. Science 316:109–112
Lemtiri-Chlieh F, MacRobbie EA, Brearley CA (2000) Inositol hexakisphosphate is a physiological signal regulating the K + -inward rectifying conductance in guard cells. Proc Natl Acad Sci USA 97:8687–8692
Macbeth MR, Schubert HL, Vandemark AP, Lingam AT, Hill CP, Bass BL (2005) Inositol hexakisphosphate is bound in the ADAR2 core and required for RNA editing. Science 309:1534–1539
Mulugu S, Bai W, Fridy PC, Bastidas RJ, Otto JC, Dollins DE, Haystead TA, Ribeiro AA, York JD (2007) A conserved family of enzymes that phosphorylate inositol hexakisphosphate. Science 316:106–109

Odom AR, Stahlberg A, Wente SR, York JD (2000) A role for nuclear inositol 1,4,5-trisphosphate kinase in transcriptional control. Science 287:2026–2029

Park SK, Lee JR, Lee SS, Son HJ, Yoo JY, Moon JC, Kwon HY, Lim CO, Bahk JD, Cho MJ, Lee SY (2002) Partial purification and properties of a phosphatidylinositol 4,5-bisphosphate hydrolyzing phospholipase C from the soluble fraction of soybean sprouts. Mol Cells 13:377–384

Phillippy BQ (1998) Identification of inositol 1,3,4-trisphosphate 5-kinase and inositol 1,3,4,5-tetrakisphosphate 6-kinase in immature soybean seeds. Plant Physiol 116:291–297

Phillippy BQ, Ullah AH, Ehrlich KC (1994) Purification and some properties of inositol 1,3,4,5,6-Pentakisphosphate 2-kinase from immature soybean seeds. J Biol Chem 269:28393–28399

Qian X, Mitchell J, Wei SJ, Williams J, Petrovich RM, Shears SB (2005) The Ins(1,3,4)P3 5/6-kinase/Ins(3,4,5,6)P4 1-kinase is not a protein kinase. Biochem J 389:389–395

Qin Z-X, Chen Q-J, Tong Z, Wang X-C (2005) The Arabidopsis inositol 1,3,4-trisphosphate 5/6 kinase, AtItpk-1, is involved in plant photomorphogenesis under red light conditions, possibly via interaction with COP9 signalosome. Plant Physiol Biochem 43:947–954

Raboy V (2003) myo-Inositol-1,2,3,4,5,6-hexakisphosphate. Phytochemistry 64:1033–1043

Raboy V, Gerbasi PF, Young KA, Stoneberg SD, Pickett SG, Bauman AT, Murthy PP, Sheridan WF, Ertl DS (2000) Origin and seed phenotype of maize low phytic acid 1–1 and low phytic acid 2–1. Plant Physiol 124:355–368

Shears SB (1989) Metabolism of the inositol phosphates produced upon receptor activation. Biochem J 260:313–324

Shi J, Wang H, Wu Y, Hazebroek J, Meeley RB, Ertl DS (2003) The maize low-phytic acid mutant *lpa2* is caused by mutation in an inositol phosphate kinase gene. Plant Physiol 131:507–515

Shi J, Wang H, Hazebroek J, Ertl DS, Harp T (2005) The maize low-phytic acid 3 encodes a *myo*-inositol kinase that plays a role in phytic acid biosynthesis in developing seeds. Plant J 42:708–719

Shi J, Wang H, Schellin K, Li B, Faller M, Stoop JM, Meeley RB, Ertl DS, Ranch JP, Glassman K (2007) Embryo-specific silencing of a transporter reduces phytic acid content of maize and soybean seeds. Nat Biotechnol 25:930–937

Steger DJ, Haswell ES, Miller AL, Wente SR, O'Shea EK (2003) Regulation of chromatin remodeling by inositol polyphosphates. Science 299:114–116

Stephens LR, Irvine RF (1990) Stepwise phosphorylation of *myo*-inositol leading to *myo*-inositol hexakisphosphate in *Dictyostelium*. Nature 346:580–583

Stevenson-Paulik J, Odom AR, York JD (2002) Molecular and biochemical characterization of two plant inositol polyphosphate 6-/3-/5-kinases. J Biol Chem 277:42711–42718

Stevenson-Paulik J, Bastidas RJ, Chiou S-T, Frye RA, York JD (2005) Generation of phytate-free seeds in *Arabidopsis* through disruption of inositol polyphosphate kinases. Proc Natl Acad Sci USA 102:12612–12617

Stiles AR, Qian X, Shears SB, Grabau EA (2008) Metabolic and signaling properties of an Itpk gene family in Glycine max. FEBS Lett 582:1853–1858

Suh SJ, Wang YF, Frelet A, Leonhardt N, Klein M, Forestier C, Mueller-Roeber B, Cho MH, Martinoia E, Schroeder JI (2007) The ATP binding cassette transporter AtMRP5 modulates anion and calcium channel activities in *Arabidopsis* guard cells. J Biol Chem 282:1916–1924

Sun Y, Wilson MP, Majerus PW (2002) Inositol 1,3,4-trisphosphate 5/6-kinase associates with the COP9 signalosome by binding to CSN1. J Biol Chem 277:45759–45764

Sun Y, Thompson M, Lin G, Butler H, Gao Z, Thornburgh S, Yau K, Smith DA, Shukla VK (2007) Inositol 1,3,4,5,6-pentakisphosphate 2-kinase from maize: molecular and biochemical characterization. Plant Physiol 144:1278–1291

Suzuki M, Tanaka K, Kuwano M, Yoshida KT (2007) Expression pattern of inositol phosphate-related enzymes in rice (Oryza sativa L.): implications for the phytic acid biosynthetic pathway. Gene 405:55–64

Sweetman D, Johnson S, Caddick SEK, Hanke DE, Brearley CA (2006) Characterization of an Arabidopsis inositol 1,3,4,5,6-pentakisphosphate 2-kinase (AtIPK1). Biochem J 394:95–103

Sweetman D, Stavridou I, Johnson S, Green P, Caddick SEK, Brearley CA (2007) Arabidopsis thaliana inositol 1,3,4-trisphosphate 5/6-kinase 4 (AtITPK4) is an outlier to a family of ATP-grasp fold proteins from Arabidopsis. FEBS Lett 581:4165–4171

Tan X, Calderon-Villalobos IA, Sharon M, Zheng C, Robinson CV, Estelle M, Zheng N (2007) Mechanism of auxin perception by the TIR1 ubiquitin ligase. Nature 446:640–645

Verbsky JW, Chang SC, Wilson MP, Mochizuki Y, Majerus PW (2005) The pathway for the production of inositol hexakisphosphate in human cells. J Biol Chem 280:1911–1920

Wilson MP, Majerus PW (1997) Characterization of a cDNA encoding Arabidopsis thaliana inositol 1,3,4-trisphosphate 5/6-kinase. Biochem Biophys Res Commun 232:678–681

Wilson MP, Sun Y, Cao L, Majerus PW (2001) Inositol 1,3,4-trisphosphate 5/6-kinase is a protein kinase that phosphorylates the transcription factors c-Jun and ATF-2. J Biol Chem 276:40998–41004

Xia H-J, Brearley C, Elge S, Kaplan B, Fromm H, Mueller-Roeber B (2003) *Arabidopsis* inositol polyphosphate 6-/3-kinase is a nuclear protein that complements a yeast mutant lacking a functional ArgR-Mcm1 transcription complex. Plant Cell 15:449–463

Xu J, Brearley CA, Lin W-H, Wang Y, Ye R, Mueller-Roeber B, Xu Z-H, Xue H-W (2005) A role of *Arabidopsis* inositol polyphosphate kinase, AtIPK2α, in pollen germination and root growth. Plant Physiol 137:94–103

Yang X, Rudolf M, Carew MA, Yoshida M, Nerreter V, Riley AM, Chung SK, Bruzik KS, Potter BV, Schultz C, Shears SB (1999) Inositol 1,3,4-trisphosphate acts in vivo as a specific regulator of cellular signaling by inositol 3,4,5,6-tetrakisphosphate. J Biol Chem 274: 18973–18980

Yang L, Tang R, Zhu J, Liu H, Mueller-Roeber XH, Zhang H (2008) Enhancement of stress tolerance in transgenic tobacco plants constitutively expressing *AtIpk2β*, an inositol polyphosphate 6-/3-kinase from *Arabidopsis* thaliana. Plant Mol Biol 66:329–343

York JD (2006) Regulation of nuclear processes by inositol polyphosphates. Biochim Biophys Acta 1761:552–559

York JD, Odom AR, Murphy R, Ives EB, Wente SR (1999) A phospholipase C-dependent inositol polyphosphate kinase pathway required for efficient messenger RNA export. Science 285:96–100

Zhang Z-B, Yang G, Arana F, Chen Z, Li Y, Xia H-J (2007) *Arabidopsis* inositol polyphosphate 6-/3-kinase (AtIpk2β) is involved in axillary shoot branching via auxin signaling. Plant Physiol 144:942–951

Zonia L, Cordeiro S, Tupý J, Feijó JA (2002) Oscillatory chloride efflux at the pollen tube apex has a role in growth and cell volume regulation and is targeted by inositol 3,4,5,6-tetrakisphosphate. Plant Cell 14:2233–2249

Phosphoinositides and Plant Cell Wall Synthesis

Ruiqin Zhong, Ryan L. McCarthy, and Zheng-Hua Ye

Abstract Phosphoinositides are lipid second messengers known to be important for many cellular processes in yeast, including actin cytoskeletal organization, vesicle transport, and cell wall assembly. In plant cells, studies on phosphoinositide phosphatases and kinases suggest that phosphoinositides are involved in the regulation of actin cytoskeletal organization, cell wall synthesis, and cell morphogenesis. It is hypothesized that phosphoinositides may regulate the transport of vesicles carrying cell wall biosynthetic enzymes and wall components, thereby influencing cell wall synthesis and cell morphogenesis.

1 Introduction

Plant cell walls are extracellular matrices that define cell shapes and are important for cell growth and development. They provide mechanical strength to the plant body as well as physical barriers for protection from biotic and abiotic stresses. Based on the timing of their synthesis during cell growth, cell walls are generally classified into two types, primary cell walls and secondary cell walls. Primary walls are formed at the cell plates in newly divided cells and continuously remodeled during cell elongation. They typically consist of cellulose, hemicellulose, pectin, and structural proteins. Secondary walls are deposited after cessation of cell elongation and constitute the bulk part of biomass produced by plants. They are generally composed of cellulose, xylan, and lignin. The making of cell walls involves the coordinated processes of the biosynthesis, secretion, and assembly of wall components. It is conceivable that various signals and signal transduction pathways are involved in the regulation of cell wall synthesis. Recent studies have

Z.-H. Ye (✉)
Department of Plant Biology, University of Georgia, 2502 Plant Sciences, Athens, GA, 30602, USA
e-mail: zhye@plantbio.uga.edu

T. Munnik (ed.), *Lipid Signaling in Plants*, Plant Cell Monographs 16,
DOI 10.1007/978-3-642-03873-0_12, © Springer-Verlag Berlin Heidelberg 2010

revealed an essential role of phosphoinositides in secondary cell wall synthesis (Zhong et al. 2004, 2005) and root hair tip growth (Preuss et al. 2006; Kusano et al. 2008; Thole et al. 2008), probably through regulation of the secretion of vesicles containing wall components, which is the focus of this review.

2 Phosphoinositide Metabolism in Plants

Phosphoinositides are lipid second messengers present ubiquitously in all eukaryotic cell membranes. There are seven forms of phosphoinositides that differ from each other by the presence of a phosphate group on the 3-, 4-, or 5-hydroxyl position of the inositol head of phosphatidylinositol (PtdIns). Among them, PtdIns 4,5-bisphosphate [PtdIns(4,5)P_2] is the best studied phosphoinositide and is the precursor of the second messengers, inositol 1,4,5-triphopsphate and diacylglycerol. In the last decade, phosphoinositides have been demonstrated in yeast and animals to be the key regulators in a number of cellular processes, including cell signaling and proliferation, vesicle trafficking, actin cytoskeletal organization, Golgi and vacuole membrane dynamics, and activation of proteins such as phosphoinositide-dependent kinase 1 and phospholipase D (PLD; Takenawa and Itoh 2001). In plants, all forms of phosphoinositides, except PtdIns(3,4,5)P_3, have been detected and they have been shown to be important for vesicle trafficking, actin cytoskeletal organization, the tip growth of pollen tubes and root hairs, secondary wall synthesis, cell morphogenesis, and stress and hormonal responses (Mueller-Roeber and Pical 2002; Meijer and Munnik 2003).

Phosphoinositides undergo dynamic metabolism to maintain specific levels at different membrane compartments, which is crucial for regulation of specific cellular functions. This dynamic change of phosphoinositide pools is mediated by kinases and phosphatases that phosphorylate and dephosphorylate them, respectively, and by phospholipases (Fig. 1). In plants, a number of genes encoding phosphoinositide kinases, including PI 3-kinases (see chapter, "Plant Phosphatidylinositol 3-Kinase"), PI 4-kinases (see chapter, "Phosphatidylinositol 4-Phosphate is Required for Tip Growth in Arabidopsis thaliana"), and PIP 5-kinases (see chapter, "PIP-Kinases as Key Regulators of Plant Function"), have been identified (Mueller-Roeber and Pical 2002; Meijer and Munnik 2003), and analyses of their corresponding mutants have revealed their crucial roles in the tip growth of root hairs (Preuss et al. 2006; Kusano et al. 2008; Stenzel et al. 2008; see chapters, "Phosphatidylinositol 4-Phosphate is Required for Tip Growth in Arabidopsis thaliana, PIP-Kinases as Key Regulators of Plant Function, Plant Phosphatidylinositol 3-Kinase").

Dephosphorylation of phosphoinositides is carried out by two groups of phosphatases, SAC domain phosphoinositide phosphatases and inositol polyphosphate phosphatases. SAC domain phosphatases from yeast and animals have been shown to exhibit differential activities toward different forms of phosphoinositides.

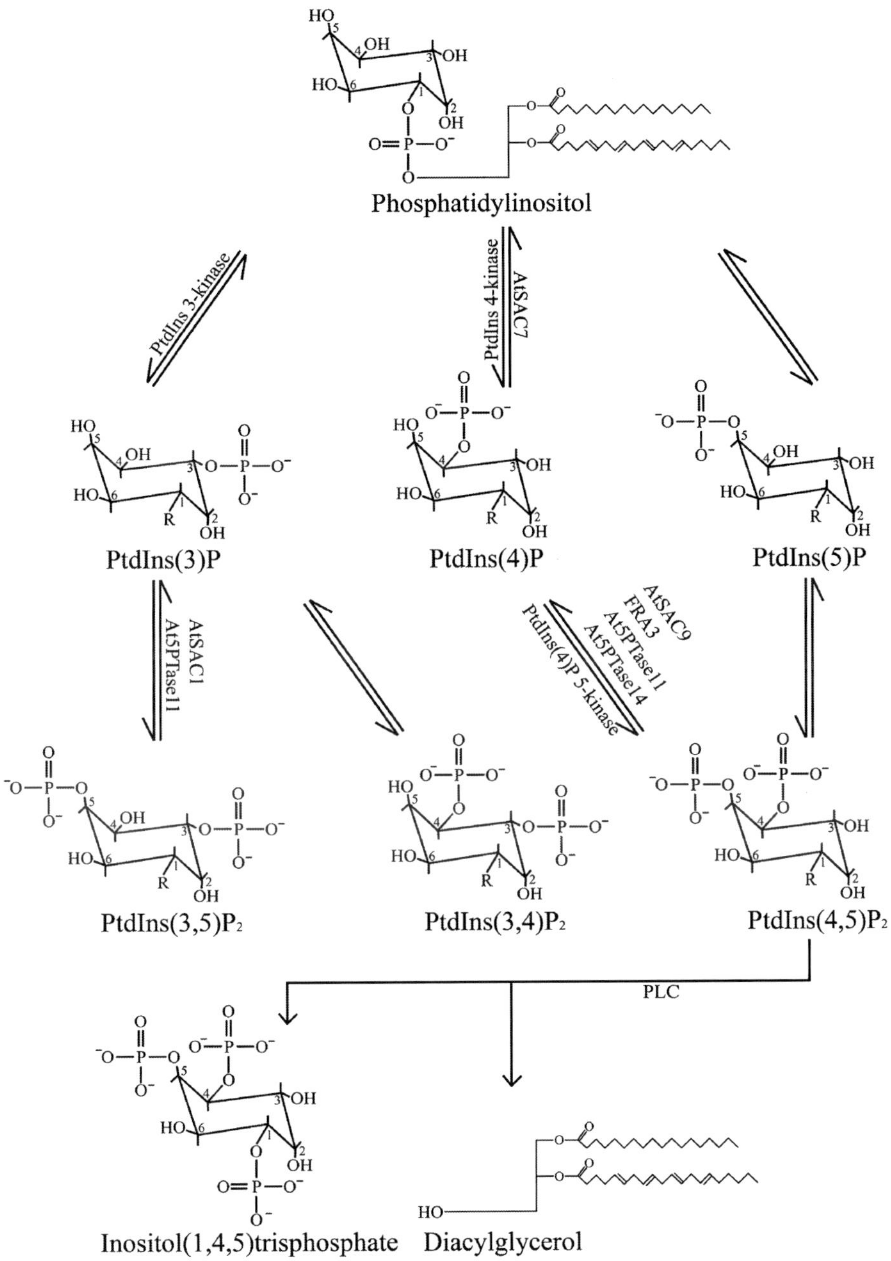

Fig. 1 Phosphoinositide metabolism in plant cells. PtdIns(3,4,5)P_3 has not been detected in plant cells. Genes encoding phosphoinositide kinases and phosphatases that have been functionally characterized in plants are shown in the diagram

For example, the SAC domains of yeast Sac1p, yeast synaptojanin homologs Inp52p and Inp53p, human synaptojanins, and rat rSAC1 are able to hydrolyze phosphate from multiple positions of the inositol head of phosphoinositides, including PtdIns3P, PtdIns4P, and PtdIns(3,5)P_2 (Guo et al. 1999; Hughes et al. 2000; Nemoto et al. 2000). By contrast, another yeast SAC domain protein, *pheromone-induced gene4* (Fig4), hydrolyzes phosphate only from PtdIns(3,5)P_2 (Rudge et al. 2004), and the human hSac2 exhibits 5-phosphatase activity toward PtdIns(4,5)P_2 and PtdIns(3,4,5)P_3 (Minagawa et al. 2001). In plants, nine *Arabidopsis* genes encoding SAC domain proteins, *AtSAC1–AtSAC9*, were identified (Despres et al. 2003; Zhong and Ye 2003) and three of them have been functionally characterized. AtSAC1 exhibits phosphatase activity toward PtdIns(3,5)P_2 (Zhong et al. 2005), whereas AtSAC7/RDH4 predominantly hydrolyzes phosphate from PtdIns4P (Thole et al. 2008). Another *Arabidopsis* SAC domain protein, AtSAC9, whose mutation causes an accumulation of PtdIns(4,5)P_2 likely possesses a phosphatase activity toward PtdIns(4,5)P_2 (Williams et al. 2005). In addition to SAC domain phosphatases, it has been demonstrated that two *Arabidopsis* type II inositol polyphosphate 5-phosphatases hydrolyze phosphate from PtdIns(4,5)P_2 (Zhong et al. 2004; Zhong and Ye 2004), and another 5-phosphatase hydrolyzes phosphate from both PtdIns(3,5)P_2 and PtdIns(4,5)P_2 (Ercetin and Gillaspy 2004).

3 Phosphoinositides Regulate Golgi-to-Plasma Membrane Transport and Cell Wall Assembly in Yeast

PtdIns4P has been shown to be essential for normal actin cytoskeletal organization, vesicle trafficking, and cell wall integrity in yeast (Schorr et al. 2001; Simonsen et al. 2001). The level of PtdIns4P in yeast is regulated by Sac1p and the PtdIns 4-kinase PIK1, which terminates and generates the PtdIns4P signal, respectively. Sac1p is an integral membrane phosphoinositide phosphatase localized at the endoplasmic reticulum (ER) and the Golgi. Mutation of Sac1p causes an accumulation of PtdIns4P, indicating its essential role in the regulation of the PtdIns4P pool (Foti et al. 2001; Schorr et al. 2001). While the Sac1p located at ER regulates the ATP uptake into the ER lumen (Kochendorfer et al. 1999), the Sac1p at Golgi is required for proper cell wall assembly (Schorr et al. 2001). The latter was demonstrated by the inability of the sac1p mutant to grow on plates containing chitin-binding dye Calcofluor White, a widely used indicator for detecting cell wall defects in yeast mutants. Detailed analysis revealed that the sac1p mutation causes aberrant deposition of chitin, which is accompanied with an alteration in the trafficking of chitin synthase (Schorr et al. 2001). The role of PtdIns4P in regulating the trafficking of chitin synthase was further confirmed by overexpression of PIK1, which also caused an increased level of PtdIns4P (Schorr et al. 2001). Furthermore, studies of a kinase-deficient form of PIK1 showed an impaired secretion of cargos from the Golgi to the plasma membrane (Hama et al. 1999; Walch-Solimena and Novick 1999; Audhya et al. 2000; Simonsen et al. 2001). These findings demonstrate that PtdIns4P is a crucial signal involved in the regulation of the Golgi-to-plasma membrane trafficking of vesicles carrying cell wall synthesizing enzymes and/or wall components in yeast.

4 Phosphoinositides are Involved in Regulation of Secondary Wall Synthesis in Plants

Plant cell walls are mainly composed of cellulose, hemicellulose, pectin, and lignin. While cellulose is synthesized by the cellulose synthase machinery located at the plasma membrane, hemicellulose and pectin are produced at the Golgi and subsequently secreted into the walls. Similar to yeast, cell wall biosynthesis in plant cells also requires the transport of vesicles carrying cell wall components and cellulose biosynthetic enzymes from the Golgi to the plasma membrane. Recent studies on phosphoinositide phosphatases in *Arabidopsis* have revealed an essential role of phosphoinositide signals in plant cell wall synthesis, probably through regulation of actin cytoskeletal organization and vesicle trafficking.

The first link of phosphoinositide metabolism and plant cell wall synthesis came from the study of the *FRAGILE FIBER3* (*FRA3*) gene in *Arabidopsis* (Zhong et al. 2004). *FRA3* encodes a type II inositol polyphosphate 5-phosphatase that hydrolyzes phosphate from PtdIns(4,5)P_2, PtdIns(3,4,5)P_3, and Ins(1,4,5)P_3. Mutation of the *FRA3* gene causes an increased accumulation of PtdIns(4,5)P_2 and Ins(1,4,5)P_3, aberrant organization of actin cytoskeleton, and severely reduced thickening of secondary walls in fibers and vessels (Fig. 2). Alterations in actin cytoskeletal organization and secondary wall thickening were also observed in the *fra7* mutant with a mutation in the *AtSAC1* gene that encodes a PtdIns(3,5)P_2-specific SAC domain phosphatase (Zhong et al. 2005; Fig. 2). The *fra7* mutation also led to a severe change in cell morphogenesis. It was found that the AtSAC1 protein is associated with the Golgi (Fig. 2) and truncation of its C-terminus by the *fra7* mutation resulted in its mislocation in the cytoplasm. The fact that the *fra7* mutation did not affect the phosphatase activity of AtSAC1 toward PtdIns(3,5)P_2 indicates that the correct localization of AtSAC1 is required for its normal cellular functions.

Although it is not known how the FRA3- and AtSAC1-regulated phosphoinositide metabolism affects secondary wall synthesis, several possible explanations could be envisioned. First, the actin cytoskeleton in plants is believed to be important in the transport of vesicles carrying cell wall polysaccharides and cellulose synthase machinery from the Golgi to the plasma membrane (Bannigan and Baskin 2005). An alteration in the actin cytoskeletal organization caused by treatment with the actin-disrupting drug cytochalasin D has been shown to affect secondary wall synthesis in fiber cells (Hu et al. 2003). Therefore, it is likely that the reduced cell wall synthesis in the *fra3* and *fra7* mutants is at least in part due to

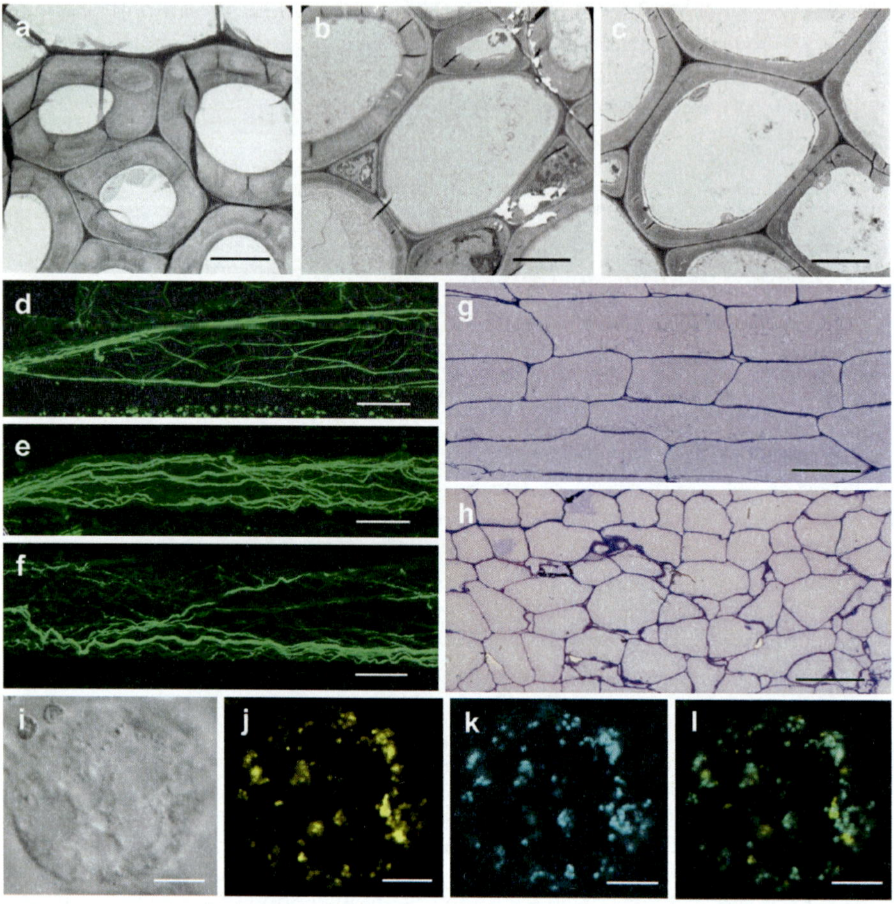

Fig. 2 Phosphoinositide phosphatases are required for normal secondary wall synthesis, actin cytoskeletal organization and cell morphogenesis. (**a**)–(**c**) The *fra3* (**b**) and *fra7* (**c**) mutations cause a severe reduction in secondary wall thickening in fiber cells compared with the wild type (**a**). Bars = 5.3 μm. (**d**)–(**f**) The *fra3* (**e**) and *fra7* (**f**) mutations result in an alteration in actin cytoskeletal organization in fiber cells compared with the wild type (**d**). *Bars* = 18 μm. (**g**) and (**h**) The *fra7* mutation (**h**) leads to a drastic change in cell morphology compared with the wild type (**g**). *Bars* = 85 μm. (**i**)–(**l**) A carrot protoplast (**i**) expressing yellow fluorescent protein tagged-AtSAC1 (**j**) and a cyan fluorescent protein-tagged Golgi marker (**k**). The signals of AtSAC1 and the Golgi marker are overlapped, indicating that AtSAC1 is associated with the Golgi. *Bars* = 22 μm. Adapted from Zhong et al. (2004, 2005)

the altered actin cytoskeletal organization, which may impede the process of vesicle trafficking. Second, the alterations in the phosphoinositide metabolism in the *fra3* and *fra7* mutants may directly affect vesicle trafficking independent of actin cytoskeleton, thus leading to cell wall defects. It has been shown in yeast that expression of a mutant form of PIK1, a PtdIns 4-kinase, causes impaired Golgi-to-plasma membrane trafficking of vesicles independent of the PtdIns4P regulation of actin cytoskeleton, indicating a direct role of PtdIns4P in vesicle trafficking (Walch-Solimena and Novick 1999). Third, the alterations in the phosphoinositide metabolism in the *fra3* and *fra7* mutants may affect cortical microtubule dynamics that is known to be important for cellulose microfibril deposition and the overall cell wall biosynthesis (Burk and Ye 2002; Burk et al. 2006). This possibility is substantiated by the finding that a PLD, whose activity could be regulated by PtdIns(4,5)P_2 (Qin et al. 2002), is involved in the organization of the microtubular cytoskeleton (Dhonukshe et al. 2003; Gardiner et al. 2003). Although the orientation of cortical microtubules in the fiber cells is not visibly altered by the *fra3* and *fra7* mutations, these mutations may affect the phosphoinositide-dependent PLD activity, which could lead to subtle alterations in cortical microtubule reorganization and subsequent cell wall synthesis.

5 Phosphoinositides are Essential for Root Hair Tip Growth in Plants

The tip growth of root hairs in plants requires the polarized secretion of new cell wall materials at the root hair tip. Several recent reports have implicated an important role of phosphoinositides in root hair tip growth. Mutations of AtSAC7/RHD4, a PtdIns4P phosphatase, and PI4K, a PtdIns 4-kinase, have been shown to cause aberrant root hair morphology (Preuss et al. 2006; Thole et al. 2008; see chapter "Phosphatidylinositol 4-Phosphate is Required for Tip Growth in Arabidopsis thaliana"). Both AtSAC7 and PI4K are targeted to tip-localized membranes in growing root hairs, which is consistent with their possible roles in regulation of membrane trafficking. In addition, PtdIns4P 5-kinase, a key enzyme involved in converting PtdIns4P to PtdIns(4,5)P_2, is predominantly localized at the plasma membrane in the tip of root hair cells, and its mutation results in root hair growth defects (Kusano et al. 2008; Stenzel et al. 2008). Using lipid biosensors, a tip-localized gradient for both PtdIns4P and PtdIns(4,5)P_2 on the plasma membrane was found (van Leeuwen et al. 2007; Vermeer et al. 2009; see chapter, "Imaging Lipids in Living Plants"). Together, these studies demonstrate that PtdIns4P and PtdIns(4,5)P_2 are essential for the tip growth of root hairs, probably by regulating the polarized secretion of cell wall components.

6 Future Perspectives

Studies on phosphoinositide phosphatases and kinases have provided a first glimpse into the important roles of phosphoinositide signals in plant cell wall synthesis. The next challenging step is to unveil the molecular mechanisms underlying the regulation of cell wall synthesis by phosphoinositide signals. First, it will be important to investigate whether phosphoinositides affect cell wall synthesis by regulating the trafficking of vesicles carrying cell wall components from the Golgi to the plasma membrane and, if they do, whether they directly affect vesicle trafficking or through regulation of the actin cytoskeleton. Fluorescent protein-tagged markers for membrane trafficking from ER to Golgi and from Golgi to the plasma membrane (Samalova et al. 2008) and mutants affecting the phosphoinositide metabolism are valuable tools in addressing this issue. Second, phosphoinositides generally relay their signals through effector proteins that bind specific phosphoinositides at particular membrane compartments. Identification and functional characterization of the effector proteins in plants will provide a molecular link between phosphoinositide signals and the cellular processes they regulate. Third, studies in yeast and animals have indicated that highly localized changes in phosphoinositide levels are mediated through the action of phosphatases and kinases at specific regions of membranes, which provides a means for temporal and spatial regulation of specific cellular functions (Simonsen et al. 2001). With the identification of many phosphoinositide phosphatase and kinase genes in plants (Mueller-Roeber and Pical 2002; Meijer and Munnik 2003), it will be imperative to determine their subcellular localization and the locations of individual forms of phosphoinositides. Imaging the dynamic changes in the levels of phosphoinositides is technically challenging in plant cells, but recent applications of the phosphoinositide-binding domain fused with fluorescent proteins have been successful in detecting the dynamic changes of phosphoinositides at specific membrane compartments (Kim et al. 2001; Jung et al. 2002; Vermeer et al. 2006, 2009; van Leeuwen et al. 2007; Thole et al. 2008; see chapter, "Imaging Lipids in Living Plants"). It is expected that further studies on phosphoinositide signaling will undoubtedly contribute to our understanding of the molecular mechanisms regulating the cell wall synthesis.

References

Audhya A, Foti M, Emr SD (2000) Distinct roles for the yeast phosphatidylinositol 4-kinases, Stt4p and Pik1p, in secretion, cell growth, and organelle membrane dynamics. Mol Biol Cell 11:2673–2689

Bannigan A, Baskin TI (2005) Directional cell expansion – turning toward actin. Curr Opin Plant Biol 8:619–624

Burk DH, Ye Z-H (2002) Alteration of oriented deposition of cellulose microfibrils by mutation of a katanin-like microtubule severing protein. Plant Cell 14:2145–2160

Burk DH, Zhong R, Morrison WHIII, Ye Z-H (2006) Disruption of cortical microtubules by overexpression of green fluorescent protein-tagged α-tubulin 6 causes a marked reduction in cell wall synthesis. J Integr Plant Biol 48:85–98

Despres B, Bouissonnié F, Wu H-J, Gomord V, Guilleminot J, Grellet F, Berger F, Delseny M, Devic M (2003) Three *SAC1*-like genes show overlapping patterns of expression in *Arabidopsis* but are remarkably silent during embryo development. Plant J 34:293–306

Dhonukshe P, Laxalt AM, Goedhart J, Gadella TWJ, Munnik T (2003) Phospholipase D activation correlates with microtubule reorganization in living plant cells. Plant Cell 15:2666–2679

Ercetin ME, Gillaspy GE (2004) Molecular characterization of an *Arabidopsis* gene encoding a phospholipid-specific inositol polyphosphate 5-phosphatase. Plant Physiol 135:938–946

Foti M, Audhya A, Emr SD (2001) Sac1 lipid phosphatase and Stt4 phosphatidylinositol 4-kinase regulate a pool of phosphatidylinositol 4-phosphate that functions in the control of the actin cytoskeleton and vacuole morphology. Mol Biol Cell 12:2396–2411

Gardiner J, Collings DA, Harper JDI, Marc J (2003) The effects of the phospholipase D-antagonist 1-butanol on seedling development and microtubule organisation in *Arabidopsis*. Plant Cell Physiol 44:687–696

Guo S, Stolz LE, Lemrow SM, York JD (1999) SAC1-like domains of yeast SAC1, INP52, INP53 and of human synaptojanin encode polyphosphoinositide phosphatases. J Biol Chem 274:12990–12995

Hama H, Schnieders EA, Thoener J, Takemoto JY, DeWald DB (1999) Direct involvement of phosphatidylinositol 4-phosphate in secretion in the yeast *Saccharomyces cerevisiae*. J Biol Chem 274:34294–34300

Hu Y, Zhong R, Morrison WH, Ye Z-H (2003) The *Arabidopsis RHD3* gene is required for cell wall biosynthesis and actin organization. Planta 217:912–921

Hughes WE, Woscholski R, Cooke FT, Patrick RS, Dove SK, McDonald NQ, Parker PJ (2000) *SAC1* encodes a regulated lipid phosphoinositide phosphatase, defects in which can be suppressed by the homologous Inp52p and Inp53p phosphatases. J Biol Chem 275:801–808

Jung J-Y, Kim Y-W, Kwak JM, Hwanga J-U, Young J, Schroeder JI, Hwang I, Lee Y (2002) Phosphatidylinositol 3- and 4-phosphate are required for normal stomatal movements. Plant Cell 14:2399–2412

Kim DH, Eu Y-J, Yoo CM, Kim Y-W, Pih KT, Jin JB, Kim SJ, Stenmark H, Hwang I (2001) Trafficking of phosphatidylinositol 3-phosphate from the trans-Golgi network to the lumen of the central vacuole in plant cells. Plant Cell 13:287–301

Kochendorfer KU, Then AR, Kearns BG, Bankaitis VA, Mayinger P (1999) Sac1p plays a crucial role in microsomal ATP transport, which is distinct from its function in Golgi phospholipid metabolism. EMBO J 18:1506–1515

Kusano H, Testerink C, Vermeer JEM, Tsuge T, Shimada H, Oka A, Munnik T, Aoyama T (2008) The *Arabidopsis* phosphatidylinositol phosphate 5-kinase PIP5K3 is a key regulator of root hair tip growth. Plant Cell 20:367–380

Meijer HJG, Munnik T (2003) Phospholipid-based signaling in plants. Annu Rev Plant Biol 54:265–306

Minagawa T, Ijuin T, Mochizuki Y, Takenawa T (2001) Identification and characterization of a Sac domain-containing phosphoinositide 5-phosphatase. J Biol Chem 276:22011–22015

Mueller-Roeber B, Pical C (2002) Inositol phospholipid metabolism in *Arabidopsis*. Characterization and putative isoforms of inositol phospholipid kinase and phosphoinositide-specific phospholipase C. Plant Physiol 130:22–46

Nemoto Y, Kearns BG, Wenk MR, Chen H, Mori K, Alb JG, Camilli PD, Bankaitis VA (2000) Functional characterization of a mammalian Sac1 and mutants exhibiting substrate-specific defects in phosphoinositide phosphatase activity. J Biol Chem 275:34293–34305

Preuss ML, Schmitz AJ, Thole JM, Bonner HKS, Otegui MS, Nielsen E (2006) A role for the RabA4b effector protein PI-4Kbeta1 in polarized expansion of root hair cells in *Arabidopsis thaliana*. J Cell Biol 172:991–998

Qin C, Wang C, Wang X (2002) Kinetic analysis of Arabidopsis phospholipase Dδ. J Biol Chem 277:49685–49690

Rudge SA, Anderson DM, Emr SD (2004) Vacuole size control: regulation of PtdIns(3, 5)P2 levels by the vacuole-associated Vac14-Fig4 complex, a PtdIns(3, 5)P2-specific phosphatase. Mol Biol Cell 15:24–36

Samalova M, Fricker M, Moore I (2008) Quantitative and qualitative analysis of plant membrane traffic using fluorescent proteins. Methods Cell Biol 85:353–380

Schorr M, Then A, Tahirovic S, Hug N, Mayinger P (2001) The phosphoinositide phosphatase Sac1p controls trafficking of the yeast Chs3p chitin synthase. Curr Biol 11:1421–1426

Simonsen A, Wurmser AE, Emr SD, Stenmark H (2001) The role of phosphoinositides in membrane transport. Curr Opin Cell Biol 13:485–492

Stenzel I, Ischebeck T, Konig S, Holubowska A, Sporysz M, Hause B, Heilmann I (2008) The type B phosphatidylinositol 4-phosphate 5-kinase 3 is essential for root hair formation in *Arabidopsis thaliana*. Plant Cell 20:124–141

Takenawa T, Itoh T (2001) Phosphoinositides, key molecules for regulation of actin cytoskeletal organization and membrane traffic from the plasma membrane. Biochim Biophys Acta 1533:190–206

Thole JM, Vermeer JE, Zhang Y, Gadella TWJ, Nielsen E (2008) *Root hair defective4* encodes a phosphatidylinositol-4-phosphate phosphatase required for proper root hair development in *Arabidopsis thaliana*. Plant Cell 20:381–395

van Leeuwen W, Vermeer JE, Gadella TW, Munnik T (2007) Visualization of phosphatidylinositol 4, 5-bisphosphate in the plasma membrane of suspension-cultured tobacco BY-2 cells and whole Arabidopsis seedlings. Plant J 52:10141026

Vermeer JE, van Leeuwen W, Tobena-Santamaria R, Laxalt AM, Jones DR, Divecha N, Gadella TW, Munnik T (2006) Visualization of PtdIns3P dynamics in living plant cells. Plant J 47:687–700

Vermeer JEM, Thole JM, Goedhart J, Nielsen E, Munnik T, Gadella TWJ Jr (2009) Visualisation of PtdIns4*P* dynamics in living plant cells. Plant J. 57:356–372

Walch-Solimena C, Novick P (1999) The yeast phosphatidylinositol-4-OH kinase pik1 regulates secretion at the Golgi. Nat Cell Biol 1:523–525

Williams ME, Torabinejad J, Cohick E, Parker K, Drake EJ, Thompson JE, Hortter M, DeWald DB (2005) Mutations in the Arabidopsis phosphoinositide phosphatase gene SAC9 lead to overaccumulation of PtdIns(4, 5)P2 and constitutive expression of the stress-response pathway. Plant Physiol 138:686–700

Zhong R, Ye Z-H (2003) The SAC domain-containing protein gene family in *Arabidopsis*. Plant Physiol 132:544–555

Zhong R, Ye Z-H (2004) Molecular and biochemical characterization of three WD-repeat domain-containing inositol polyphosphate 5-phosphatases in *Arabidopsis thaliana*. Plant Cell Physiol 45:1720–1728

Zhong R, Burk DH, Morrison WHIII, Ye Z-H (2004) *FRAGILE FIBER3*, an *Arabidopsis* gene encoding a type II inositol polyphosphate 5-phosphatase, is required for secondary wall synthesis and action organization in fiber cells. Plant Cell 16:3242–3259

Zhong R, Burk DH, Nairn CJ, Wood-Jones A, Morrison WHIII, Ye Z-H (2005) Mutation of SAC1, an *Arabidopsis* SAC domain phosphoinositide phosphatase, causes alterations in cell morphogenesis, cell wall synthesis, and actin organization. Plant Cell 17:1449–1466

Imaging Lipids in Living Plants

Joop E.M. Vermeer and Teun Munnik

Abstract Phospholipids are important constituents of biological membranes, most of them fulfilling a structural role. However, it has become clear that in plants, just as in mammalian and yeast cells, some minor phospholipids, e.g. phosphoinositides, are important regulators of cellular function, providing docking sites for target proteins via lipid-binding domains, and/or modulating their enzymatic activity. The application of fluorescent proteins fused to lipid-binding domains to create the so-called, lipid biosensors, sheds new light on lipid molecules in living plant cells. Here, an overview is presented regarding their application.

1 Introduction

Phospholipids are basic structures crucial for the cell's integrity. More recently, they have been implicated in signaling plant stress (e.g. hyperosmotic shock and plant defence) and development (e.g. polarised growth of pollen tubes and root hairs) (for reviews see (Meijer and Munnik 2003; Munnik et al. 1998a; Wang 2004)), functioning as local effectors of protein targets or as signaling precursors. For example, phospholipase A_1 and A_2 enzymes (PLA_1 and PLA_2) hydrolyse the sn-1 and sn-2 position, respectively, of their substrate phospholipids, resulting in free fatty acids and lysophospholipids, which have profound roles in signaling (see Ryu 2004; Wang 2004; Munnik and Testerink 2009; see chapter, "Phospholipase A in Plant Signal Transduction"). Phospholipase D (PLD) enzymes hydrolyse structural lipids such as phosphatidylcholine (PtdCho) and phosphatidylethanolamine (PtdEtn) to produce the lipid-second messenger, phosphatidic acid (PA) (Munnik 2001; Testerink and Munnik 2005; Wang 2005). In contrast to mammalian cells, which

T. Munnik (✉)
Section Plant Physiology, Swammerdam Institute for Life Sciences, University of Amsterdam, Sciencepark 904, 1098 XH, Amsterdam, The Netherlands
e-mail: T.Munnik@uva.nl

T. Munnik (ed.), *Lipid Signaling in Plants*, Plant Cell Monographs 16,
DOI 10.1007/978-3-642-03873-0_13, © Springer-Verlag Berlin Heidelberg 2010

only have two PLDs, Arabidopsis contains 12, involved in many different processes (Bargmann et al. 2006; Wang 2000, 2002, 2005; see chapter, "Plant Phospholipase D"). Classically, phospholipase C (PLC) hydrolyses phosphatidylinositol 4,5-bisphosphate (PtdIns(4,5)P_2) into inositol 1,4,5-trisphosphate (InsP_3) and diacylglycerol (DAG) (see chapter, "The Emerging Roles of Phospholipase C in Plant Growth and Development"). In mammalian cells, InsP_3 diffuses into the cytosol where it promotes Ca^{2+}-release from an intracellular store, while DAG remains in the membrane to activate protein kinase C. However, plant cells seem to lack an InsP_3 receptor and instead may use its conversion to InsP_6 to release Ca^{2+} (Munnik and Testerink 2009; Zonia and Munnik 2006). Similarly, plants lack the DAG target, protein kinase C (PKC), and seem to phosphorylate the resulting DAG into PA (Munnik 2001; Testerink and Munnik 2005, see chapters, "Plant Phosphatidylinositol 3-kinase, Phosphatidic acid – an Electrostatic/Hydrogen- Bond Switch?, Nitric Oxide and Phosphatidic Acid Signaling in Plants, 3-Phosphoinositide-Dependent Protein Kinase is a Switchboard from Signaling Lipids to Protein Phosphorylation Cascades, Diacylglycerol Pyrophosphate, A Novel Plant Signaling Lipid"). PLC may also hydrolyse PtdIns4P and generate InsP_2 to form InsP_6 (Munnik and Testerink 2009; Vermeer et al. 2009).

Alternatively, evidence is accumulating that polyphosphoinositides (PPIs), like PtdIns4P and PtdIns(4,5)P_2, function as lipid-second messengers themselves, and as such, PLC would function as an attenuator of PPI signalling (Munnik and Testerink 2009). PPIs represent a dynamic group of inositol lipids that are involved in a wide range of membrane signaling and trafficking events (Fig. 1; see chapters "The Emerging Roles of Phospholipase C in Plant Growth and Development, Phosphatidylinositol 4-Phosphate is Required for Tip Growth in *Arabidopsis thaliana*, PIP-kinases as key regulators of plant function, Plant Phosphatidylinositol 3-kinase, Signaling and the Polyphosphoinositide Phosphatases from Plants, InsP3 in Plant Cells, Phosphoinositides and Plant Cell Wall Synthesis"). The basic structure is PtdIns which can be phosphorylated at three different positions, yielding different isomers (Fig. 1). These include: three mono-phosphorylated PtdIns$P$ isomers (PtdIns3P, PtdIns4P and PtdIns5P), three bis-phosphorylated isomers [PtdIns(3,4)P_2, PtdIns(3,5)P_2, PtdIns(4,5)P_2] and one tris-phosphorylated isomer, PtdIns(3,4,5)P_3 (Fig. 1). Of these seven PPIs, all except PtdIns(3,4,5)P_3 have been detected in plants (Meijer and Munnik 2003). The absence of PtdIns(3,4,5)P_3 is striking, since this is an important signaling lipid in mammalian cells (Rameh and Cantley 1999). Although PtdIns(3,4)P_2 has been identified in plants (Brearley and Hanke 1993), in animal cells, it is a typical PtdIns(3,4,5)P_3 metabolite and may in fact represent PtdIns(3,5)P_2, which was discovered only a few years later (Meijer et al. 1999; Munnik and Testerink 2009).

Studies in yeast and mammalian cells have shown that PPIs play a prominent role in membrane trafficking (Corvera et al. 1999; Di Paolo and De Camilli 2006; Krauss and Haucke 2007a, b; Simonsen et al. 2001; Lemmon 2003, 2008). Recent reports suggest that PPIs have similar functions in plants (Jaillais et al. 2008, 2006; Kim et al. 2001; Preuss et al. 2006; Thole et al. 2008; van Leeuwen et al. 2007; Vermeer et al. 2009, 2006). While different isomers have been identified and shown

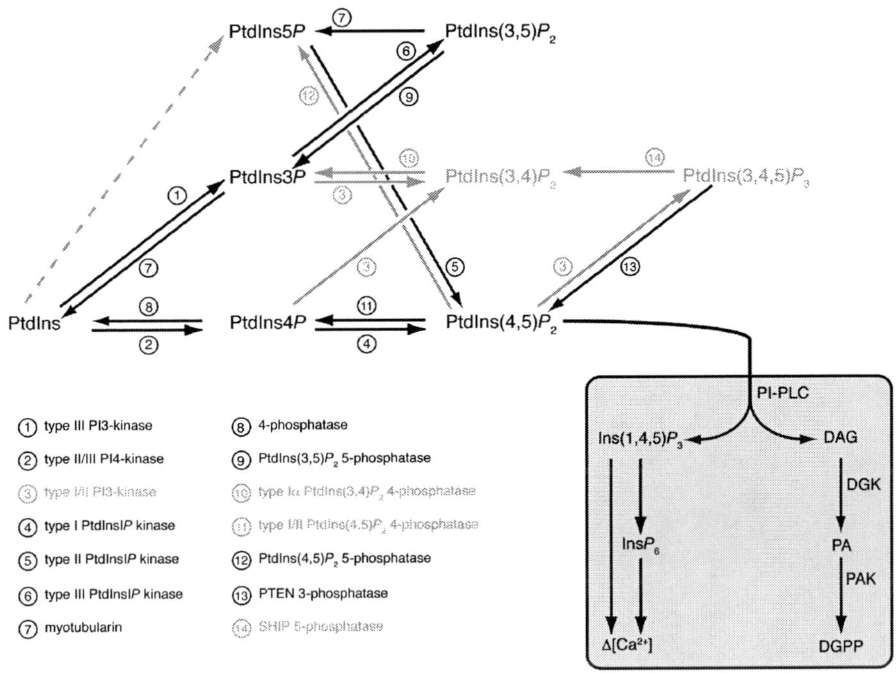

Fig. 1 Possible pathways for phosphoinositides synthesis. Pathways indicated by *solid lines* have been established in mammalian and yeast cells. *Dashed lines* indicates nonconfirmed pathways. Enzymes corresponding for the indicated activity in vitro or in vivo are indicated by *numbers* associated with the lines. Pathways and PPIs indicated by *black solid lines* and numbers are established in plants. *Boxed grey area* indicates a schematic representation of the PtdIns(4,5)P_2-PLC signaling pathway leading to Ca^{2+} release and production of DAG, PA and DGPP. *DGK* diacylglycerol kinase, *PAK* PA kinase, *DGPP* diacylglycerolpyrophosphate

to exhibit a relatively rapid turnover, still little is known about their biological role and subcellular localisation. The vast arsenal of lipid kinases, phosphatases and hydrolases, present in the Arabidopsis genome, suggests a complex network regulating the distribution and levels of these molecules.

In this chapter, we focus on the methodology to visualise lipids in living plant cells. In the end, we will also discuss the recent results in perspective and some new approaches to study the biological role of PPIs in living plant cells.

2 Visualising Lipids

2.1 $^{32}P_i$-Radiolabeling

Phospholipids are readily labelled by incubating plant material (e.g. seedlings, leaf discs, cell suspensions) with $^{32}P_i$, while signaling responses can be studied by following the quantitative changes in response to stimulation. Within seconds to minutes, dramatic changes in signal levels (2- to 50-fold) can be observed. The radiolabeling approach is very sensitive, meaning that signaling can be monitored within just a few cells or isolated material (e.g. leaf discs). The preferred systems are often cell suspensions in which most cells have direct contact to the medium, ensuring synchronous up-take of $^{32}P_i$ and exposure to the stimulus. However, it also works well for other material (leaf discs) or whole seedlings (Frank et al. 2000). Figure 2 shows a typical autoradiograph of ^{32}P-labelled phospholipids, extracted from tobacco BY-2 cells (Fig. 2a), mammalian (HeLa) cells (Fig. 2b) and *Arabidopsis* seedlings (Fig. 2c), which are separated by thin layer chromatography. What is striking is that, in the control situation, plants contain tiny amounts of PtdIns(4,5)P_2 compared to the HeLa cells, but they do synthesise it rapidly upon salt stress (Fig. 2). Most detailed studies on lipid signaling in plants have been obtained using such systems (den Hartog et al. 2001, 2003; Meijer et al. 2001, 1999; Munnik et al. 1994a, 1994b, 1998b; van der Luit et al. 2000). A complication this approach is that PPIs turn-over very quickly, meaning that radioactivity is incorporated faster and that maximal incorporation is reached much earlier than for the structural lipids. Therefore, when interpreting changes in metabolism, the length of pre-labeling is very important (see Munnik et al. 1998a; Arisz et al. 2009).

Another limitation of the approach is that the spatial information is lost. If a response takes place at a specific cell layer within the plant root, the signal will be swamped by the relatively large amount of non-responding cells. Alternatively, if not the amount but the location is changing (e.g. plasma membrane —> endosome), then this would also be missed.

2.2 Lipid Biosensors

To increase the resolution of the above-mentioned responses, an elegant solution has been developed, i.e. lipid biosensors. In general, a lipid biosensor consists of a genetically encoded lipid-binding domain (see Table 1), fused to a spectral variant of the green fluorescent protein (GFP). When expressed in cells and imaged by confocal microscopy, the fluorescent fusion protein provides both spatial and temporal information of the lipid molecules, in living cells at the subcellular resolution. This approach was first used to visualise PtdIns(4,5)P_2 in mammalian

Fig. 2 Imaging phospholipids by ^{32}P-radiolabeling. Autoradiogram showing the separation of phospholipids extracted from (**a**) tobacco BY-2 cells, (**b**) HeLa cells and (**c**) *Arabidopsis thaliana* seedlings by thin layer chromatography. Note that under normal conditions, plants have very low PtdInsP_2 levels compared to mammalian cells, but in response to salt stress, PtdInsP_2 levels dramatically increase (**b**, **c**). (**a**) Time-course experiment of pre-labelled tobacco BY-2 cells with 125 mM NaCl. (**b**) ^{32}P-labelled phospholipids profile from HeLa cells. (**c**) ^{32}P-phospholipid profile from pre-labelled Arabidopsis seedlings treated with buffer (C) or 250 mM NaCl for 15 min

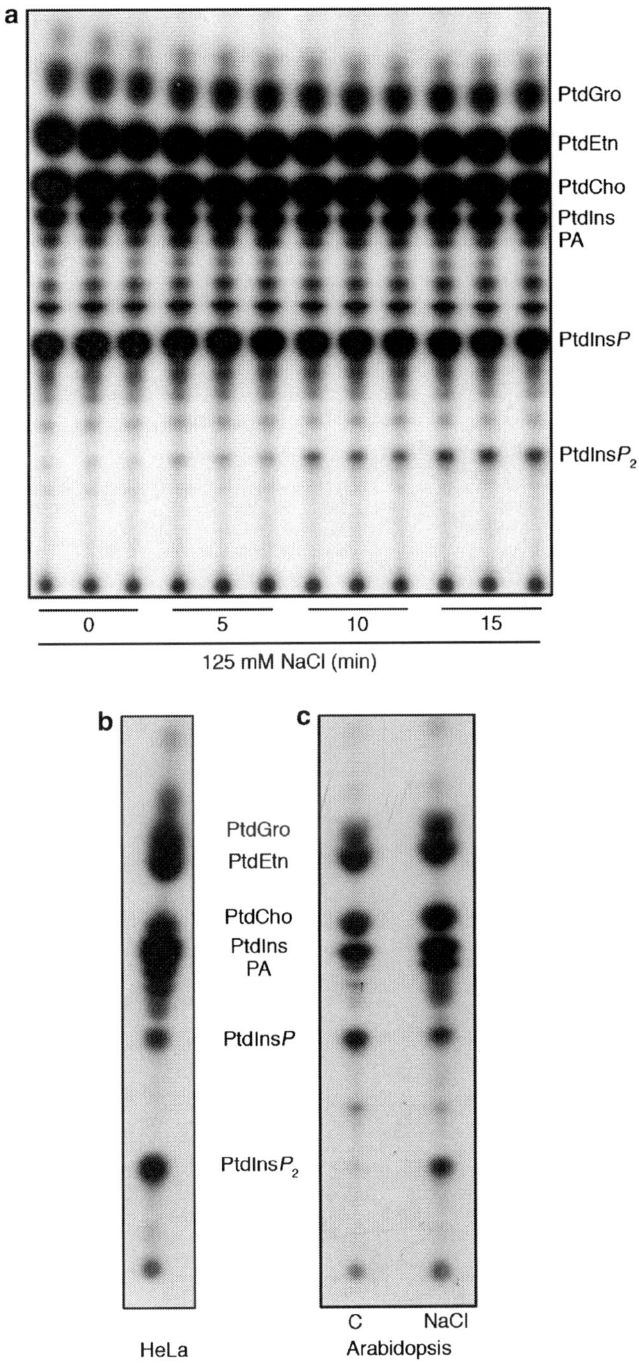
a
PtdGro
PtdEtn
PtdCho
PtdIns
PA
PtdIns*P*
PtdIns*P*2
0
5
10
15
125 mM NaCl (min)
b
c
PtdGro
PtdEtn
PtdCho
PtdIns
PA
PtdIns*P*
PtdIns*P*2
HeLa
C
NaCl
Arabidopsis

Table 1 Using lipid binding domain-GFP fusions to visualise lipids in living cells

Lipid	Protein domain	References	Localisation	Used in plants	Localisation
PtdSer	Lact-C2	Yeung et al. (2008)	PM/vesicles	–	–
DAG	C1a-PKCγ	Oancea et al. (1998)	PM	Helling et al. (2006), Vermeer (2006)	PM[b]
PtdOH[a]	NRG	Dominguez-Gonzalez et al. (2007)	Vesicles	–	–
	Raf1	Rizzo et al. (1999)	PM	–	–
	Opi1	Loewen et al. (2004)	Nucleus/ER	–	–
	PP1γ	Jones et al. (2005)	–	–	–
	PH-SOS	Zhao et al. (2007)	PM	–	–
	Spo20	Zeniou-Meyer et al. (2007)	PM	–	–
	TAPAS	Baillie et al. (2002)	Golgi	–	–
PtdIns3P	FYVE (Hrs, EEA1)	Burd and Emr (1998), Gillooly et al. (2000), Simonsen et al. (1998)	Early endosomes	Kim et al. (2001), Vermeer et al. (2006), Voigt et al. (2005)	PVC
	P40phox-PX	Ellson et al. (2001), Kanai et al. (2001)			–
PtdIns4P	PH-FAPP1	Dowler et al. (2000),	Golgi/PM	Jung et al. (2002), Thole et al. (2008), Vermeer (2006, 2008)	PM/Golgi
	PH-OSBP1	Levine and Munro (1998)	Golgi/PM		PM
	PH-OSH2	Levine and Munro (2001)	PM		–
PtdIns(4,5)P_2	PH-PLCδ1	Lemmon et al. (1995), Santagata et al. (2001), Sun et al. (2007)	PM	Dowd et al. (2006), Helling et al. (2006), Ischebeck et al. (2008) Jung et al. (2002), Kost et al. (1999), Lee et al. (2007), van Leeuwen et al. (2007), Vincent et al. (2005)	PM[c]
	Tubby-domain		PM		–
	ANTH-domain		PM		–
PtdIns(3,5)P_2	Svp1p	Dove et al. (2004)	Vacuole	–	–
PtdIns(3,4)P_2	PH-TAPP1	Dowler et al. (2000)	PM	–[d]	–
PtdIns(3,4,5)P_3	GRP1-PH	Klarlund et al. (1997)	PM	–[d]	–
	ARNH	Klarlund et al. (2000)	PM	–[d]	–
	Cytohesin-1-PH	Rameh et al. (1997)	PM	–[d]	–
	Btk-PH	Salim et al. (1996)	PM	–[d]	–

[a]No true PA-binding consensus has been identified yet. However, some PA-binding proteins have been described and used in cell biology and have therefore been included in the table

[b]YFP-C1a$_{PKC\gamma}$ is mostly localised to the cytosol of plants due to low levels of available DAG. Only in specific cells. e.g. pollen tubes, a plasma membrane localisation has been described (Helling et al. 2006)

[c]Plants have very low PtdIns(4,5)P_2 levels, hence most of the YFP-PH$_{PLC\delta1}$ fusion remains unbound in the cytosol. However, in specialised cells undergoing tip growth, such as pollen tubes and root hairs, or in cells exposed to salt stress, YFP-PH$_{PLC\delta1}$ localises to the plasma membrane (Dowd et al. 2006; Helling et al. 2006; Ischebeck et al. 2009; Kost et al. 1999; van Leeuwen et al. 2007; Vincent et al. 2005)

[d]Plants lack PtdIns(3,4,5)P_3 and the type of PI3-kinase that makes it (Munnik and Testerink 2009; see chapter, "Plant Phosphatidylinositol 3-kinase")

cells (Stauffer et al. 1998; Varnai and Balla 1998). Over the years, various lipid-binding domains with strong specificity towards distinct lipids have been identified and characterised, including Pleckstrin Homology (PH) and FYVE domains (see Lemmon 2008 for recent review), and when equipped with spectrally different fluorescent proteins, multi-parameter imaging of various PPI dynamics in living cells is possible (Schultz et al. 2005; Vermeer et al. 2009).

Table I gives an overview of the available lipid biosensors, including details on the lipid they bind to, where they localise within the cell, and whether they have been used in plants.

3 Application of Lipid Biosensors in Plants

Although the use of lipid biosensors in plants is still limited, our work as well as that from others have shown that they are excellent tools to study lipids in living cells (Dowd et al. 2006; Helling et al. 2006; Jung et al. 2002; Kim et al. 2001; Kost et al. 1999; Thole et al. 2008; van Leeuwen et al. 2007; Vermeer et al. 2009, 2006; Vincent et al. 2005). When expressed under the control of an appropriate promoter, they will allow us to follow the lipid of interest in real-time at the subcellular level. Figure 3 shows the typical fluorescence pattern of four different lipid biosensors in tobacco BY-2 cells, including YFP-2xFYVE to label PtdIns3*P* (Fig. 3a, b; Vermeer et al. 2006), YFP-PH_{FAPP1} to label PtdIns4*P* (Fig. 3c, d; Vermeer et al. 2009), YFP-$PH_{PLC\delta 1}$ to label PtdIns(4,5)P_2 (Fig. 3e, f; Van Leeuwen et al. 2007) and YFP-$C1a_{PKC\gamma}$ to label DAG (Fig. 3g, h). It is important to note that all cells grow and appear like wild-type BY-2 cells. This is an important prerequisite if conclusions about the observed localisation and dynamics are to be made. In addition, one has to be sure that the observed localisation of the biosensor reflects the localisation of the lipid. For example, most of the PtdIns(4,5)P_2 and DAG biosensor is localised in the cytosol because there is not enough of these lipids facing the cytosolic leaflet of membranes to keep it there (see Van Leeuwen et al. 2007). It is always good to determine the specificity of the biosensor, either via an in vitro lipid-binding assay (preferably as GFP-fusion), or by expressing a mutated biosensor, where amino acids critical for lipid binding are mutated and the localisation is compared with the non-mutated biosensor. It is not unlikely that the cell will compensate the lipid levels, such that there will be sufficient unbound lipid available for endogenous protein targets, as was found for PtdIns3*P* in BY2-cells overexpressing YFP-2xFYVE (Vermeer et al. 2006). Apparently, the rapid turnover of lipids allow overexpression of such domains without consequences, which is usually different when enzymes are overexpressed. Nonetheless, cell-specific overexpression of lipid binding domains has been found to interfere in stomatal movements and root hair growth, which may point to the functional roles of these PPIs (Jung et al. 2002; Lee et al. 2008, 2007).

The use of these fluorescent lipid sensors has given new insight into the dynamics and subcellular localisation of PPIs in plants. In addition, they have

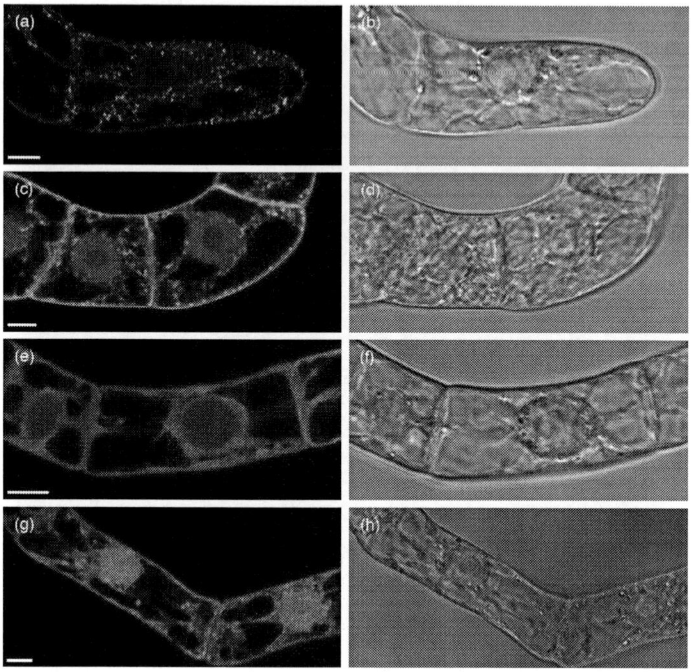

Fig. 3 Imaging PtdIns3*P*, PtdIns4*P*, PtdIns(4,5)P_2 and DAG in living tobacco BY-2 cells. (**a**, **b**) The PtdIns3*P* biosensor YFP-2xFYVE, which labels punctate structures, identified as late endosomal compartments; (**c**, **d**) the PtdIns4*P* biosensor YFP-PH_{FAPP1}, labeling the plasma membrane and Golgi stacks; (**e**, **f**) the PtdIns(4,5)P_2 biosensor YFP-$PH_{PLC\delta1}$, diffusing freely through the cytosol in an unbound state due to a very low PtdIns(4,5)P_2 concentration; (**g**, **h**) DAG biosensor YFP-$C1a_{PKC\gamma}$, also diffusing freely through the cytosol due to low concentration of DAG in the cytosolic leaflet of the membranes. (**a**, **c**, **e**, **g**) show YFP fluorescence while (**b**, **d**, **f**, **h**) show DIC. *Bars* = 10 μm

been instrumental to highlight the tightly-regulated interplay between different isomers in cells undergoing polarised growth (e.g. pollen tubes and root hairs) (Fig. 4; Dowd et al. 2006; Thole et al. 2008; van Leeuwen et al. 2007; Vermeer et al. 2006, 2009; Vincent et al. 2005). Recently, the YFP-PH_{FAPP1} fusion has been successfully used to monitor the subcellular distribution of PtdIns4*P* in root hairs of a PtdIns4*P*-specific phosphatase mutant (*rhd4-1*) (Fig. 5; Thole et al. 2008), making them attractive tools to use in other phospholipid hydrolyses, kinases and phosphatases mutants.

Knowing that the Arabidopsis genome contains many proteins with a PH-, FYVE- or PX domain (van Leeuwen et al. 2004), makes it likely that also in plants these domains could exert similar important functions as has been shown for their counterparts in other organisms. Isolating these proteins and characterise their lipid-binding properties will surely give more insights into the physiological roles of PPIs in plants.

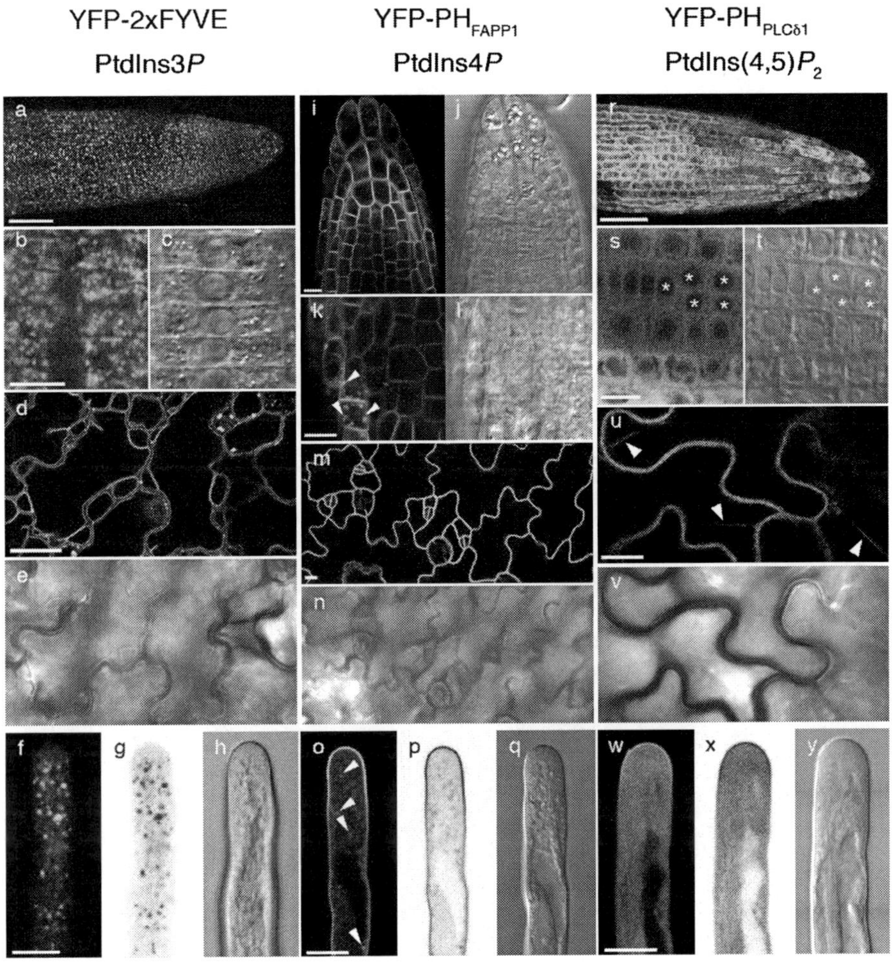

Fig. 4 Imaging PtdIns3*P*, PtdIns4*P* and PtdIns(4,5)P_2 in *Arabidopsis thaliana*. Confocal images of Arabidopsis seedlings expressing the PtdIns3*P* biosensor, YFP-2xFYVE (**a–h**), the PtdIns4*P* biosensor, YFP-PH_{FAPP1} (**i–q**) and the PtdIns(4,5)P_2 biosensor, YFP-$PH_{PLC\delta1}$ (**r–y**). (**a**) Maximal image projection of a root tip, (**b**, **c**) root cortex cells showing the vesicular localisation of YFP-2xFYVE, (**d**, **e**) leaf epidermal cells showing strong labeling of the tonoplast membrane, (**f**, **g**) growing root hair showing non-polarised distribution of YFP-2xFYVE-labelled compartments; (g) and (**f**) are identical but the image is inverted to highlight vesicular structures. Bars = 50 μm (**a**) and 10 μm (**b**, **d**, **f**). (**i**, **j**) Confocal image of a root tip, (**k**, **l**) root cortex cells and (**m**, **n**) leaf epidermal cells. *Arrowheads* in (**k**) indicate YFP-PH_{FAPP1}-labelled Golgi compartments. (**o**) Growing root hair showing a tip-focussed plasma membrane gradient of YFP-PH_{FAPP1}, which is only present in growing root hairs. (**p**) Inverted image of (**o**) to highlight the gradient. *Bars* = 50 μm (**i**), 10 μm (**k**, **o**) and 20 μm (**m**). (**r**) Maximal image projection of a root tip expressing the PtdIns(4,5)P_2 biosensor, YFP-$PH_{PLC\delta1}$. (**s**, **t**) Root cortex cells, (**u**, **v**) leaf epidermal cells. *Asterisks* in (**m**, **n**) indicate absence of YFP-$PH_{PLC\delta1}$ fluorescence in the nucleus due to a nuclear export signal present in $PH_{PLC\delta1}$. *Arrowheads* in (**u**) highlight transvacuolar strands, indicating a cytosolic localisation of the unbound probe. Most of the PtdIns(4,5)P_2 biosensor is in the cytosol as plants contain very low amounts of this lipid. (**w–y**) Growing root hair revealing a small gradient of YFP-$PH_{PLC\delta1}$ in the plasma membrane at the extreme tip of the growing root hair. (**x**) Inverted image of (**w**) to highlight the gradient. *Bars* = 50 μm (**r**), 10 μm (**s**, **u**, **w**)

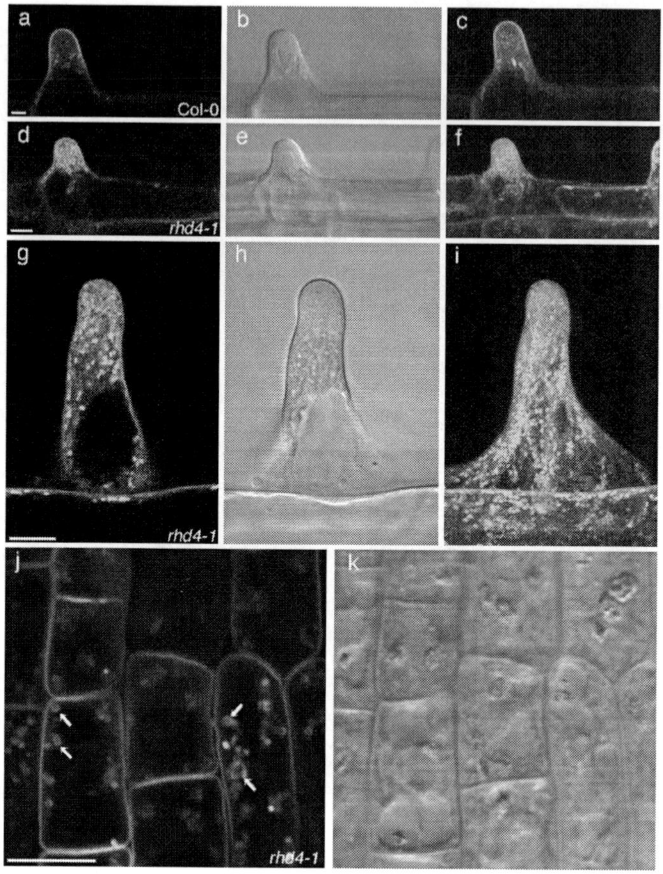

Fig. 5 Differential localisation of YFP-PH$_{FAPP1}$ in the PtdIns4*P*-phosphatase mutant *rhd4-1* (*sac7-1*). Confocal images of Arabidopsis wild-type accession, Col-0 (**a–c**) and *rhd4-1* seedlings (**d–l**) expressing the PtdIns4*P* biosensor, YFP-PH$_{FAPP1}$. (**a**, **b**) Emerging root hair, (**c**) maximal projection of an image stack taken from root hair shown in (**a**) showing strong labeling of the plasma membrane. (**d**, **e**) Emerging root hair of the *rhd4* mutant, (**f**) maximal projection of an image stack taken from root hair shown in (**d**), (**g**, **h**) growing root hair, (**i**) maximal image projection of same root hair as shown in (**g**). As clear from (**d–i**), YFP-PH$_{FAPP1}$ accumulates on internal membranes and hardly labels the plasma membrane of mutant root hairs. (**j**, **k**) Cells close to the root tip showing YFP-PH$_{FAPP1}$ labeling of the plasma membrane and "donut-like structures". (**a**, **c**, **d**, **f**, **g**, **i**, and **j**) show YFP fluorescence, (**b**, **e**, **h** and **k**) show DIC. *Bar* = 10 μm. From Vermeer et al (2009), with permission from Wiley-Blackwell publishing

4 New Tools to Study Lipid Signaling

Next to the arsenal of existing biosensors, researchers are also actively developing new tools to study lipid functions. One recent approach, developed for mammalian cell research, looks very promising. It is a two-component system, consisting of a membrane-targeted GFP, fused to the FRB domain (fragment of mammalian target of rapamycin [mTOR] that binds FKBP12), plus a customised lipid-modifying enzyme (i.e. a lipid kinase or phosphatase) fused to the FKBP12 (FK506 binding protein 12) domain. In the presence of rapamycin, both domains will heterodimerise, leading to the synthetic activation of the lipid-modifying enzyme. This approach has been very successful in elucidating the roles of PtdIns(4,5)P_2 in endocytosis and ion-channel regulation in mammalian cells (Suh et al. 2006; Varnai et al. 2006). It will be intriguing to test whether this system can be adopted to study the physiological role of lipids in living plant cells.

References

Arisz SA, Testerink C, Munnik T (2009) Plant PA signaling via diacylglycerol kinase. Biochim Biophys Acta 1791:869–875

Baillie GS, Huston E, Scotland G, Hodgkin M, Gall I, Peden AH, MacKenzie C, Houslay ES, Currie R, Pettitt TR, Walmsley AR, Wakelam MJ, Warwicker J, Houslay MD (2002) TAPAS-1, a novel microdomain within the unique N-terminal region of the PDE4A1 cAMP-specific phosphodiesterase that allows rapid, Ca^{2+}-triggered membrane association with selectivity for interaction with phosphatidic acid. J Biol Chem 277:28298–28309

Bargmann BO, Laxalt AM, Riet BT, Schouten E, van Leeuwen W, Dekker HL, de Koster CG, Haring MA, Munnik T (2006) LePLDbeta1 activation and relocalization in suspension-cultured tomato cells treated with xylanase. Plant J 45:358–368

Brearley CA, Hanke DE (1993) Pathway of synthesis of 3, 4- and 4, 5-phosphorylated phosphatidylinositols in the duckweed Spirodela polyrhiza L. Biochem J 290:145–150

Burd CG, Emr SD (1998) Phosphatidylinositol(3)-phosphate signaling mediated by specific binding to RING FYVE domains. Mol Cell 2:157–162

Corvera S, D'Arrigo A, Stenmark H (1999) Phosphoinositides in membrane traffic. Curr Opin Cell Biol 11:460–465

den Hartog M, Musgrave A, Munnik T (2001) Nod factor-induced phosphatidic acid and diacylglycerol pyrophosphate formation: a role for phospholipase C and D in root hair deformation. Plant J 25:55–65

den Hartog M, Verhoef N, Munnik T (2003) Nod factor and elicitors activate different phospholipid signaling pathways in suspension-cultured alfalfa cells. Plant Physiol 132:311–317

Di Paolo G, De Camilli P (2006) Phosphoinositides in cell regulation and membrane dynamics. Nature 443:651–657

Dominguez-Gonzalez I, Vazquez-Cuesta SN, Algaba A, Diez-Guerra FJ (2007) Neurogranin binds to phosphatidic acid and associates to cellular membranes. Biochem J 404:31–43

Dove SK, Piper RC, McEwen RK, Yu JW, King MC, Hughes DC, Thuring J, Holmes AB, Cooke FT, Michell RH, Parker PJ, Lemmon MA (2004) Svp1p defines a family of phosphatidylinositol 3, 5-bisphosphate effectors. EMBO J 23:1922–1933

Dowd PE, Coursol S, Skirpan AL, Kao TH, Gilroy S (2006) Petunia phospholipase c1 is involved in pollen tube growth. Plant Cell 18:1438–1453

Dowler S, Currie RA, Campbell DG, Deak M, Kular G, Downes CP, Alessi DR (2000) Identification of pleckstrin-homology-domain-containing proteins with novel phosphoinositide-binding specificities. Biochem J 351:19–31

Ellson CD, Gobert-Gosse S, Anderson KE, Davidson K, Erdjument-Bromage H, Tempst P, Thuring JW, Cooper MA, Lim ZY, Holmes AB, Gaffney PR, Coadwell J, Chilvers ER, Hawkins PT, Stephens LR (2001) PtdIns(3)P regulates the neutrophil oxidase complex by binding to the PX domain of p40(phox). Nat Cell Biol 3:679–682

Frank W, Munnik T, Kerkmann K, Salamini F, Bartels D (2000) Water deficit triggers phospholipase D activity in the resurrection plant Craterostigma plantagineum. Plant Cell 12:111–124.
Gillooly DJ, Morrow IC, Lindsay M, Gould R, Bryant NJ, Gaullier JM, Parton RG, Stenmark H (2000) Localization of phosphatidylinositol 3-phosphate in yeast and mammalian cells. EMBO J 19:4577–4588
Helling D, Possart A, Cottier S, Klahre U, Kost B (2006) Pollen tube tip growth depends on plasma membrane polarization mediated by tobacco PLC3 activity and endocytic membrane recycling. Plant Cell 18:3519–3534
Ischebeck T, Stenzel I, Heilmann I (2008) Type B phosphatidylinositol-4-phosphate 5-kinases mediate Arabidopsis and Nicotiana tabacum pollen tube growth by regulating apical pectin secretion. Plant Cell 20:3312–3330
Jaillais Y, Fobis-Loisy I, Miege C, Rollin C, Gaude T (2006) AtSNX1 defines an endosome for auxin-carrier trafficking in Arabidopsis. Nature 443:106–109
Jaillais Y, Fobis-Loisy I, Miege C, Gaude T (2008) Evidence for a sorting endosome in Arabidopsis root cells. Plant J 53:237–247
Jones JA, Rawles R, Hannun YA (2005) Identification of a novel phosphatidic acid binding domain in protein phosphatase-1. Biochemistry 44:13235–13245
Jung JY, Kim YW, Kwak JM, Hwang JU, Young J, Schroeder JI, Hwang I, Lee Y (2002) Phosphatidylinositol 3- and 4-phosphate are required for normal stomatal movements. Plant Cell 14:2399–2412
Kanai F, Liu H, Field SJ, Akbary H, Matsuo T, Brown GE, Cantley LC, Yaffe MB (2001) The PX domains of p47phox and p40phox bind to lipid products of PI(3)K. Nat Cell Biol 3:675–678
Kim DH, Eu YJ, Yoo CM, Kim YW, Pih KT, Jin JB, Kim SJ, Stenmark H, Hwang I (2001) Trafficking of phosphatidylinositol 3-phosphate from the trans-Golgi network to the lumen of the central vacuole in plant cells. Plant Cell 13:287–301
Klarlund JK, Guilherme A, Holik JJ, Virbasius JV, Chawla A, Czech MP (1997) Signaling by phosphoinositide-3, 4, 5-trisphosphate through proteins containing pleckstrin and Sec7 homology domains. Science 275:1927–1930
Klarlund JK, Tsiaras W, Holik JJ, Chawla A, Czech MP (2000) Distinct polyphosphoinositide binding selectivities for pleckstrin homology domains of GRP1-like proteins based on diglycine versus triglycine motifs. J Biol Chem 275:32816–32821
Kost B, Lemichez E, Spielhofer P, Hong Y, Tolias K, Carpenter C, Chua NH (1999) Rac homologues and compartmentalized phosphatidylinositol 4, 5-bisphosphate act in a common pathway to regulate polar pollen tube growth. J Cell Biol 145:317–330
Krauss M, Haucke V (2007a) Phosphoinositide-metabolizing enzymes at the interface between membrane traffic and cell signalling. EMBO Rep 8:241–246
Krauss M, Haucke V (2007b) Phosphoinositides: regulators of membrane traffic and protein function. FEBS Lett 581:2105–2111
Lee Y, Kim YW, Jeon BW, Park KY, Suh SJ, Seo J, Kwak JM, Martinoia E, Hwang I (2007) Phosphatidylinositol 4, 5-bisphosphate is important for stomatal opening. Plant J 52:803–816
Lee Y, Bak G, Choi Y, Chuang WI, Cho HT, Lee Y (2008) Roles of phosphatidylinositol 3-kinase in root hair growth. Plant Physiol 147(2):624–635
Lemmon MA (2003) Phosphoinositide recognition domains. Traffic 4:201–213
Lemmon MA (2008) Membrane recognition by phospholipid-binding domains. Nat Rev Mol Cell Biol 9:99–111
Lemmon MA, Ferguson KM, O'Brien R, Sigler PB, Schlessinger J (1995) Specific and high-affinity binding of inositol phosphates to an isolated pleckstrin homology domain. Proc Natl Acad Sci USA 92:10472–10476
Levine TP, Munro S (1998) The pleckstrin homology domain of oxysterol-binding protein recognises a determinant specific to Golgi membranes. Curr Biol 8:729–739
Levine TP, Munro S (2001) Dual targeting of Osh1p, a yeast homologue of oxysterol-binding protein, to both the Golgi and the nucleus-vacuole junction. Mol Biol Cell 12:1633–1644

Loewen CJ, Gaspar ML, Jesch SA, Delon C, Ktistakis NT, Henry SA, Levine TP (2004) Phospholipid metabolism regulated by a transcription factor sensing phosphatidic acid. Science 304:1644–1647

Meijer HJ, Munnik T (2003) Phospholipid-based signaling in plants. Annu Rev Plant Biol 54:265–306

Meijer HJG, Divecha N, van den Ende H, Musgrave A, Munnik T (1999) Hyperosmotic stress induces rapid synthesis of phosphatidyl-D-inositol 3, 5 bisphosphate in plant cells. Planta 208:294–298

Meijer HJ, Berrie CP, Iurisci C, Divecha N, Musgrave A, Munnik T (2001) Identification of a new polyphosphoinositide in plants, phosphatidylinositol 5-monophosphate (PtdIns5P), and its accumulation upon osmotic stress. Biochem J 360:491–498

Munnik T (2001) Phosphatidic acid: an emerging plant lipid second messenger. Trends Plant Sci 6:227–233

Munnik T, Testerink C (2009) Plant phospholipid signaling: "in a nutshell". J Lipid Res 50(Suppl): S260–S265

Munnik T, Irvine RF, Musgrave A (1994a) Rapid turnover of phosphatidylinositol 3-phosphate in the green alga Chlamydomonas eugametos: signs of a phosphatidylinositide 3-kinase signalling pathway in lower plants? Biochem J 298(Pt 2):269–273

Munnik T, Musgrave A, de Vrije T (1994b) Rapid turnover of polyphosphoinositides in carnation flower petals. Planta 193:89–98

Munnik T, Irvine RF, Musgrave A (1998a) Phospholipid signalling in plants. Biochim Biophys Acta 1389:222–272

Munnik T, van Himbergen JAJ, Ter Riet B, Braun F, Irvine RF, van den Ende H, Musgrave A (1998b) Detailed analysis of the turnover of polyphosphoinositides ans phosphatidic acid upon activation of phospholipases C and D in *Chlamydomonas* cells treated with non-permeabilizing concentrations of mastoparan. Planta 207:133–145

Oancea E, Teruel MN, Quest AF, Meyer T (1998) Green fluorescent protein (GFP)-tagged cysteine-rich domains from protein kinase C as fluorescent indicators for diacylglycerol signaling in living cells. J Cell Biol 140:485–498

Preuss ML, Schmitz AJ, Thole JM, Bonner HKS, Otegui MS, Nielsen E (2006) A role for the RabA4b effector protein PI-4K1 in polarized expansion of root hair cells in Arabidopsis thaliana. J Cell Biol 172(7):991–998

Rameh LE, Cantley LC (1999) The role of phosphoinositide 3-kinase lipid products in cell function. J Biol Chem 274:8347–8350

Rameh LE, Arvidsson A, Carraway KL 3 rd, Couvillon AD, Rathbun G, Crompton A, VanRenterghem B, Czech MP, Ravichandran KS, Burakoff SJ, Wang DS, Chen CS, Cantley LC (1997) A comparative analysis of the phosphoinositide binding specificity of pleckstrin homology domains. J Biol Chem 272:22059–22066

Rizzo MA, Shome K, Vasudevan C, Stolz DB, Sung TC, Frohman MA, Watkins SC, Romero G (1999) Phospholipase D and its product, phosphatidic acid, mediate agonist-dependent raf-1 translocation to the plasma membrane and the activation of the mitogen-activated protein kinase pathway. J Biol Chem 274:1131–1139

Ryu SB (2004) Phospholipid-derived signaling mediated by phospholipase A in plants. Trends Plant Sci 9:229–235

Salim K, Bottomley MJ, Querfurth E, Zvelebil MJ, Gout I, Scaife R, Margolis RL, Gigg R, Smith CI, Driscoll PC, Waterfield MD, Panayotou G (1996) Distinct specificity in the recognition of phosphoinositides by the pleckstrin homology domains of dynamin and Bruton's tyrosine kinase. EMBO J 15:6241–6250

Santagata S, Boggon TJ, Baird CL, Gomez CA, Zhao J, Shan WS, Myszka DG, Shapiro L (2001) G-protein signaling through tubby proteins. Science 292:2041–2050

Schultz C, Schleifenbaum A, Goedhart J, Gadella TW Jr (2005) Multiparameter imaging for the analysis of intracellular signaling. ChemBiochem 6:1323–1330

Simonsen A, Lippe R, Christoforidis S, Gaullier JM, Brech A, Callaghan J, Toh BH, Murphy C, Zerial M, Stenmark H (1998) EEA1 links PI(3)K function to Rab5 regulation of endosome fusion. Nature 394:494–498
Simonsen A, Wurmser AE, Emr SD, Stenmark H (2001) The role of phosphoinositides in membrane transport. Curr Opin Cell Biol 13:485–492
Stauffer TP, Ahn S, Meyer T (1998) Receptor-induced transient reduction in plasma membrane PtdIns(4, 5)P2 concentration monitored in living cells. Curr Biol 8:343–346
Suh BC, Inoue T, Meyer T, Hille B (2006) Rapid chemically induced changes of PtdIns(4, 5)P2 gate KCNQ ion channels. Science 314:1454–1457
Sun Y, Carroll S, Kaksonen M, Toshima JY, Drubin DG (2007) PtdIns(4, 5)P2 turnover is required for multiple stages during clathrin- and actin-dependent endocytic internalization. J Cell Biol 177:355–367
Testerink C, Munnik T (2005) Phosphatidic acid: a multifunctional stress signaling lipid in plants. Trends Plant Sci 10:368–375
Thole JM, Vermeer JE, Zhang Y, Gadella TW Jr, Nielsen E (2008) ROOT HAIR DEFECTIVE4 encodes a phosphatidylinositol-4-phosphate phosphatase required for proper root hair development in Arabidopsis thaliana. Plant Cell 20:381–395
van der Luit AH, Piatti T, van Doorn A, Musgrave A, Felix G, Boller T, Munnik T (2000) Elicitation of suspension-cultured tomato cells triggers the formation of phosphatidic acid and diacylglycerol pyrophosphate. Plant Physiol 123:1507–1516
van Leeuwen W, Ökrész L, Bögre L, Munnik T (2004) Learning the lipid language of plant signalling. Trends Plant Sci 9:378–384
van Leeuwen W, Vermeer JE, Gadella TW Jr, Munnik T (2007) Visualization of phosphatidylinositol 4, 5-bisphosphate in the plasma membrane of suspension-cultured tobacco BY-2 cells and whole Arabidopsis seedlings. Plant J 52:1014–1026
Varnai P, Balla T (1998) Visualization of phosphoinositides that bind pleckstrin homology domains: calcium- and agonist-induced dynamic changes and relationship to myo-[3H]inositol-labeled phosphoinositide pools. J Cell Biol 143:501–510
Varnai P, Thyagarajan B, Rohacs T, Balla T (2006) Rapidly inducible changes in phosphatidylinositol 4, 5-bisphosphate levels influence multiple regulatory functions of the lipid in intact living cells. J Cell Biol 175:377–382
Vermeer JEM (2006) Visualisation of polyphosphoinositide dynamics in living plant cells. PhD thesis, University of Amsterdam, pp 177
Vermeer JE, van Leeuwen W, Tobena-Santamaria R, Laxalt AM, Jones DR, Divecha N, Gadella TW Jr, Munnik T (2006) Visualization of PtdIns3P dynamics in living plant cells. Plant J 47:687–700
Vermeer JE, Thole JM, Goedhart J, Nielsen E, Munnik T, Gadella TW Jr (2009) Imaging phosphatidylinositol 4-phosphate dynamics in living plant cells. Plant J 57:356–372
Vincent P, Chua M, Nogue F, Fairbrother A, Mekeel H, Xu Y, Allen N, Bibikova TN, Gilroy S, Bankaitis VA (2005) A Sec14p-nodulin domain phosphatidylinositol transfer protein polarizes membrane growth of Arabidopsis thaliana root hairs. J Cell Biol 168:801–812
Voigt B, Timmers AC, Samaj J, Hlavacka A, Ueda T, Preuss M, Nielsen E, Mathur J, Emans N, Stenmark H, Nakano A, Baluska F, Menzel D (2005) Actin-based motility of endosomes is linked to the polar tip growth of root hairs. Eur J Cell Biol 84:609–621
Wang X (2000) Multiple forms of phospholipase D in plants: the gene family, catalytic and regulatory properties, and cellular functions. Prog Lipid Res 39:109–149
Wang X (2002) Phospholipase D in hormonal and stress signaling. Curr Opin Plant Biol 5:408–414
Wang X (2004) Lipid signaling. Curr Opin Plant Biol 7:329–336
Wang X (2005) Regulatory functions of phospholipase D and phosphatidic acid in plant growth, development, and stress responses. Plant Physiol 139:566–573
Yeung T, Gilbert GE, Shi J, Silvius J, Kapus A, Grinstein S (2008) Membrane phosphatidylserine regulates surface charge and protein localization. Science 319:210–213

Zeniou-Meyer M, Zabari N, Ashery U, Chasserot-Golaz S, Haeberle AM, Demais V, Bailly Y, Gottfried I, Nakanishi H, Neiman AM, Du G, Frohman MA, Bader MF, Vitale N (2007) Phospholipase D1 production of phosphatidic acid at the plasma membrane promotes exocytosis of large dense-core granules at a late stage. J Biol Chem 282:21746–21757
Zhao C, Du G, Skowronek K, Frohman MA, Bar-Sagi D (2007) Phospholipase D2-generated phosphatidic acid couples EGFR stimulation to Ras activation by Sos. Nat Cell Biol 9:706–712
Zonia L, Munnik T (2006) Cracking the green paradigm: functional coding of phosphoinositide signals in plant stress responses. Subcell Biochem 39:207–237

Part V
PA Signaling

Phosphatidic Acid: An Electrostatic/ Hydrogen-Bond Switch?

Edgar Eduard Kooijman and Christa Testerink

Abstract Phosphatidic acid (PA) has been shown to be an important bioactive lipid that is specifically recognized by various proteins. As such, it plays a crucial role in cellular signaling in all eukaryotes. An important determinant for PA's role in its diverse functions is its anionic headgroup that resides very close to the hydrophobic interior of the lipid bilayer. In this chapter, we describe a new model, the electrostatic/hydrogen-bond switch that describes PA's ionization properties and its specific interaction with proteins. Furthermore, we will allude to the broader implications of the model for all phosphomonoester moieties found in biological compounds. Recent data in support of the model, as well as biological predictions arising from it, are also discussed.

1 Introduction

Of all membrane glycero-phospholipids, phosphatidic acid (PA) is the most simple. Total cellular content is often no more than 1%. Despite its low concentration and simple chemical structure, PA is absolutely critical for the life of a cell, foremost, because of its role in glycero-phospholipid synthesis and, not the least, because of its many signaling functions inside the cell. PA's chemical structure consists of the alcohol, glycerol, to which two fatty acids (also termed acyl-chains) and a phosphate are esterified to positions 1, 2, and 3, respectively (see Fig. 1). The phosphate headgroup is attached as a phosphomonoester directly to the glycerol backbone. This feature is unique to PA and is likely to play an important role in the interaction of PA with proteins.

Edgar E. Kooijman (✉)
Department of Biological Sciences, Kent State University, Kent, OH 44242, USA
e-mail: ekooijma@kent.edu

T. Munnik (ed.), *Lipid Signaling in Plants*, Plant Cell Monographs 16,
DOI 10.1007/978-3-642-03873-0_14, © Springer-Verlag Berlin Heidelberg 2010

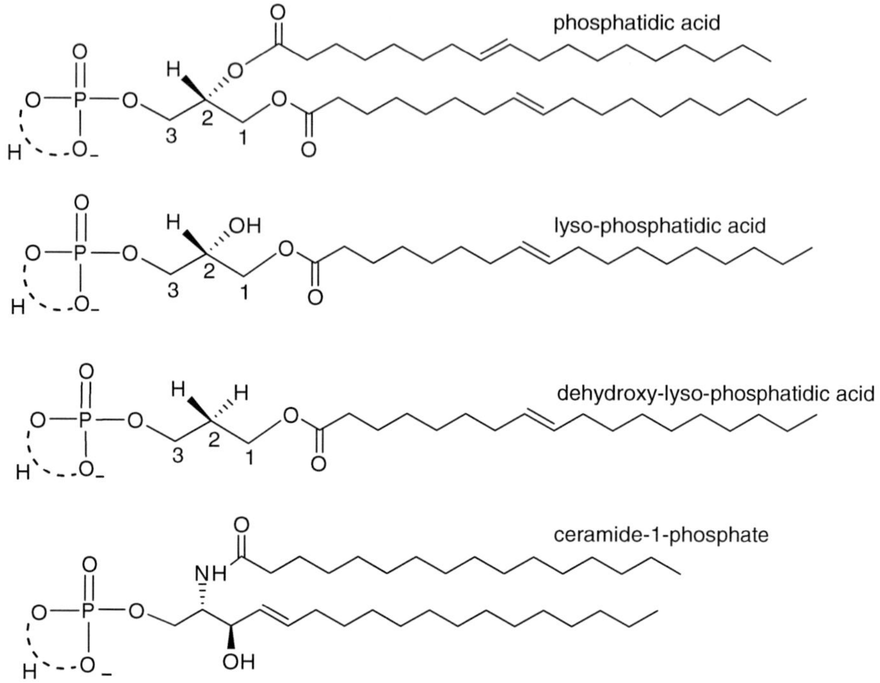

Fig. 1 Chemical structures. Chemical structure of dioleoyl (18:1c9) phosphatidic acid, lysophosphatidic acid, dehydroxy-lysophosphatydic acid, and (C16) ceramide-1-phosphate. The proton in the headgroup is indicated as being shared by the two phosphoryl oxygens, in accordance with the electrostatic/hydrogen-bond model discussed in this chapter

1.1 Biosynthesis of PA

The de novo synthesis of PA occurs via lyso PA (LPA) and can proceed via two different acylation pathways (Fig. 2). The first and main synthesis route of LPA, present in both prokaryotes and eukaryotes, is the glycerol 3-phosphate (Gro3P) pathway, in which Gro3P is acylated by a glycerol 3-phosphate acyl transferase (Coleman et al. 2000; Dircks and Sul 1997; Zheng and Zou 2001). The second pathway to LPA formation occurs only in yeast and mammals and involves the acylation of dihydroxyacetone phosphate (DHAP), via DHAP acyltransferase (Hajra 1997) and 1-acyl-DHAP reductase (Datta et al. 1990). LPA can then be acylated further by a LPA acyltransferase (LPAAT) to form PA (Nagiec et al. 1993). This PA is the major substrate for de novo synthesis of all other glycero-(phospho-)lipids and triacylglycerols (Athenstaedt and Daum 1999). The enzymes required for de novo synthesis of LPA and PA are generally localized at the endoplasmic reticulum (ER), with the notable exceptions of specialized intracellular compartments, such as mitochondria, peroxisomes in mammalian cells, lipid droplets in yeast, and chloroplasts in plants (Athenstaedt and Daum 1999).

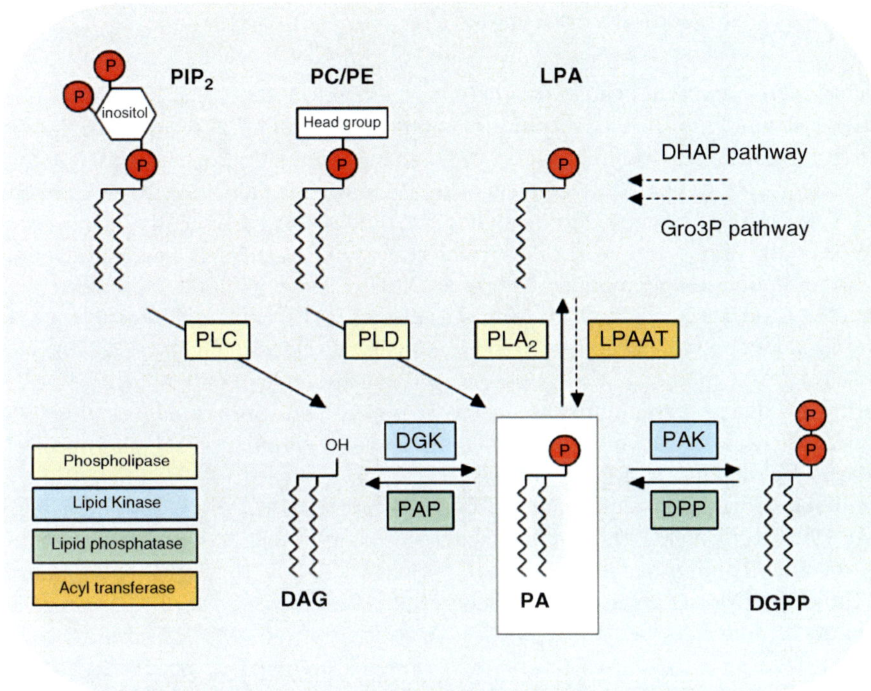

Fig. 2 Pathways to the formation and attenuation of PA. The de novo synthesis routes to the formation of PA are shown by *dashed arrows*. *Solid arrows* represent those synthesis routes most often involved in signaling processes. Note that the Gro3P pathway is localized both in the ER and in chloroplasts of plants, while the DHAP pathway is specific to yeast and mammals. PLD enzymes preferentially use PC and PE as substrates, whereas PLC hydrolyses the polyphosphoinositides PI4P and PI(4,5)P_2. Plants and yeast contain a PA kinase that converts PA into DGPP

Besides de novo biosynthesis of PA, there are several additional routes for its formation in a cell (Fig. 2). Two distinct and tightly regulated signaling pathways contribute to a separate pool of PA involved in cell signaling and vesicular trafficking. One of these is phospholipase D (PLD), which generates PA directly from structural phosholipids via the hydrolysis of the ester linkage between the primary alcohol and the phosphate (Munnik et al. 1998; see chapter, "Plant Phospholipase D"). Alternatively, PA may be formed by the sequential action of phospholipase C (PLC), which forms DAG, and diacylglycerol kinase (DGK), which phosphorylates DAG to PA (Munnik et al. 1998; see chapters "The Emerging Roles of Phospholipase C in Plant Growth and Development" and "Diacylglycerol Kinase"). Note that PA formed via either of these pathways may be rapidly phosphorylated to diacylglycerol pyrophosphate (DGPP) (van Schooten et al. 2006; see chapter, "Diacylglycerol Pyrophosphate, a Novel Plant Signaling Lipid") or dephosphorylated to DAG (Fig. 2; see chapter "Phosphatidic Acid Phosphatases in Seed Plants"), and that PA and DAG are often found in a dynamic and tightly regulated equilibrium (Nanjundan and Possmayer 2003).

1.2 Physiological Functions of PA

PA has been implicated in numerous cellular responses, as diverse as actin polymerization (Ha and Exton 1993), respiratory burst (McPhail et al. 1999), Ca^{2+} signaling (English et al. 1996; Munnik et al. 1998), and the activation of various signaling networks (Cazzolli et al. 2006; Ghosh et al. 1996; Rizzo et al. 1999, 2000; Siddiqui and Yang 1995). The important signaling function of PA has recently been underscored by the implication of PA in the recruitment of Sos and subsequent activation of Ras at the plasma membrane (Mor et al. 2007; Zhao et al. 2007) as reviewed by Hancock (2007). Exactly how PA exerts its effects is still unclear but local intracellular production via PLD appears to be responsible for these cellular responses.

A recurring theme in PA signaling is the involvement of PA in membrane trafficking events particularly along the secretory pathway (Huang et al. 2005; Roth 2008; Zeniou-Meyer et al. 2007) and in yeast sporulation (Nakanishi et al. 2006). PLD activity and PA synthesis have been proposed to be involved in the formation of transport carriers at the Golgi complex (Chen et al. 1997; Ktistakis et al. 1996; Roth et al. 1999; Siddhanta and Shields 1998). Recent work showed that PA specifically binds and activates PI4P 5-kinase and thus stimulates the formation of PI(4,5)P_2 (Moritz et al. 1992; Honda et al. 1999; Stace et al. 2007), which is required, among others, for exocytosis (De Camilli et al. 1996). Manifava et al. (2001) have recently shown that several proteins involved in vesicle trafficking bind PA in vitro. Furthermore, COPI vesicle fusion requires the BARS protein, which was recently shown to induce membrane curvature and tubulation in cooperation with PA (Yang et al. 2008).

In plants, PA is involved in development as well as in various biotic- and abiotic stress responses, including salt, drought, cold, elicitors produced by pathogens and the plant hormone abscisic acid (ABA). A few PA target proteins have been identified from plants so far. The best characterized include the protein phosphatase ABI1 and the protein kinases PDK1 and CTR1 (Anthony et al. 2004, 2006; Testerink et al. 2007; Zhang et al. 2004). For CTR1 and ABI1, which are involved in ethylene and ABA signaling, respectively, PA binding results in inhibition of their activity in vitro. In contrast, PDK1 activates the protein kinase, OXI1, in a PA-dependent manner. PDK1 contains the so-called pleckstrin homology domain that is responsible for lipid-binding (Deak et al. 1999).

Unlike other bioactive lipids, such as the phosphoinositides, no common PA-binding domain has been recognized (Stace and Ktistakis 2006; Testerink and Munnik 2005). In most cases, a number of positively charged residues are required for binding, but these can be separated in the primary structure, and therefore are quite difficult to recognize as a PA-binding site. Recently, the solution structure of the FKBP12-rapamycin-binding (FRB) domain of mTOR was determined in the presence of PA (Veverka et al. 2008). It was shown that PA's acyl chains have hydrophobic interactions with a number of residues, while the phosphate group indeed interacts with the side group of an arginine residue that had previously been implicated in PA-binding (Fang et al. 2001).

Several excellent reviews have appeared in recent years detailing the signaling properties of PA (Andresen et al. 2002; Ktistakis et al. 2003; Stace and Ktistakis 2006; Testerink and Munnik 2005; Wang et al. 2006), PLD (Cockcroft 2001; Freyberg et al. 2003; Jenkins and Frohman 2005; Roth 2008; Zambonelli and Roberts 2005; see chapters "Plant Phospholipase D" and "Nitric Oxide and Phosphatidic Acid Signaling in Plants"), PLC, and DGK (see chapters "The Emerging Roles of Phospholipase C in Plant Growth and Development", "Diacylglycerol Kinase", "Nitric Oxide and Phosphatidic Acid Signaling in Plants"). The reader is referred to these reviews for a more in-depth discussion of PA formation, function, and cell-biology.

In this chapter, we focus on the intriguing biophysics of this lipid with a particular focus on the ionization properties of its phosphomonoester headgroup. The main hypothesis will be that PA's electrostatic interaction with its molecular environment, i.e., lipids as well as proteins, forms the basis of the many roles that PA plays in living cells. We begin with a brief introduction into lipid and membrane electrostatics, followed by a detailed description of the ionization properties of PA. These results are summarized in the electrostatic/hydrogen bond-switch model. Furthermore, we make several predictions and review the recent literature that indeed appears to support our new model.

2 The Negative Charge of Biological Membranes Affects Protein Function and Organization

Biomembranes are generally negatively charged, as illustrated by the inner leaflet of the PM, which contains well over 30% of anionic phospholipids (Op den Kamp 1979; van Meer 1998), mainly PS and PI in mammalian cells (Vance and Steenbergen 2005; Yeung et al. 2008). Plant cells contain predominantly PI and PG and much less PS (Devaiah et al. 2006).

Negative charge carried by membrane lipids is an important determinant of biomembrane structure and function. It is, for example, well established that negatively charged membranes act as a site of attraction for positively charged (basic) protein domains (Kim et al. 1991; Montich et al. 1993; Roy et al. 2000). For example, the cytosolic protein myristoylated alanine-rich C-kinase substrate (MARCKS) is recruited to anionic lipid membranes by an unstructured domain, containing 13 basic amino acid residues (Arbuzova et al. 1998). Interestingly, using a biosensor that specifically recognizes PS, Yeung et al. (2008) recently showed that the PS concentration in cellular membranes also affects peripheral membrane protein localization. Positively charged residues in transmembrane proteins can similarly interact with the negatively charged lipids in the membrane (Killian and von Heijne 2000), and may guide the membrane insertion, and orientation of these proteins (van Klompenburg et al. 1997). Membrane insertion, orientation and other biophysical aspects of transmembrane protein–lipid bilayer interaction have been probed by the use of synthetic transmembrane α-helical peptides (Killian and Nyholm 2006; Nyholm et al. 2007; Ozdirekcan et al. 2007).

PA is an anionic lipid with two ionization constants; one falling in the physiological pH range 5 $<$pH $<$8. PA's negative charge is likely to play an important role in the organization of the membrane and its interaction with proteins. In general, the following features of anionic biological membranes affect the negative charge of an ionizable group in the headgroup–water interface. First, for an anionic membrane, the surface pH is lower than that of the bulk solution, an effect which is related to the electrostatic surface potential of that membrane. Increasing the overall negative charge decreases the interfacial pH and therefore decreases the degree of ionization of an ionizable group. Second, increases in ionic conditions of the medium lower the proton concentration at the membrane–medium interface and subsequently result in an increase in the degree of ionization of an ionizable group. Lastly, ionizable protons that are involved in a hydrogen bond are stabilized against dissociation.

3 PA's Phosphomonoester Headgroup Has Unique Ionization Properties

The anionic PA phosphomonoester headgroup suggests important electrostatic interactions with positively charged protein domains. Indeed, for those protein domains known to selectively bind PA, positively charged amino acid residues appear to be essential, coupled with the presence of hydrophobic residues (Stace and Ktistakis 2006; Testerink and Munnik 2005). Based on this observation, however, it is not at all clear what sets PA apart from, for example, PS and PI. In other words, how does a PA-binding domain recognize the minor lipid PA in a sea of PI, PS, and PG? Recent work has highlighted the tremendous importance of the phosphomonoester headgroup of PA in this process (Kooijman et al. 2005a, 2007), and these results are reviewed below.

3.1 Factors that Influence the Negative Charge of Ionizable Lipids

In order to understand PA–protein interactions, we need to consider the ionization properties of PA. Measurement of the ionic equilibria that describe the concentrations of $PH_2O_4 \Leftrightarrow PHO_4^{-1} \Leftrightarrow PO_4^{-2}$ (pKa_1 and pKa_2, respectively) is most conveniently achieved by using ^{31}P-NMR. The isotropic chemical shift, which is a measure of the magnetic properties of the phosphor nucleus, is very sensitive to the charge carried by the phosphomonoester headgroup. Earlier, using static ^{31}P-NMR it was demonstrated that the chemical shift of PA is sensitive to pH and that dissociation constants could be determined (Hauser 1989; Koter et al. 1978; Swairjo et al. 1994; Traikia et al. 2002). These measurements were limited to small unilamellar vesicles and/or micelles, because static ^{31}P-NMR of most lamellar lipid dispersions gives rise to a large chemical shift anisotropy that will mask any pH-dependent changes in chemical shift. In order to circumvent the intrinsic curvature associated with the unilamellar and micellar systems, Kooijman et al. (2005a) made use of a NMR technique called magic angle spinning (MAS). MAS ^{31}P NMR can obtain the isotropic chemical shift from systems with a large chemical shift anisotropy such as multi lamellar vesicle (MLV) dispersions (Watts 1998) and allows a more comprehensive understanding of the ionization behavior of PA in essentially flat bilayers.

3.2 Hydrogen Bonds Influence the Ionization Properties of PA

Using MAS ^{31}P NMR, it was established that, despite the fact that LPA and PA carry an identical phosphomonoester headgroup (Fig. 1), the ionization behavior is clearly different as was shown by their nonidentical titration curves (Kooijman et al. 2005a). The major difference between LPA, dehydroxylated LPA (dehydroxy-LPA; Fig. 1), and PA is that the former contains a hydroxyl-group at the *sn*-2 position of the glycerol backbone, while the latter two do not. The near complete overlap in titration curves for PA and dehydroxy-LPA, thus, indicated that the hydroxyl group in LPA is somehow able to lower the pKa_2 of LPA with respect to PA. A lower pKa_2 means that at an identical pH, LPA will carry more negative charge than PA. In other words, there will be more LPA lipids carrying two negative charges than PA at the same pH. These results were obtained for low concentrations of PA in a phosphatidylcholine (PC) bilayer. Additional results on the ionization properties of PA in a phosphatidylethanolamine (PE) bilayer indicated that PE decreases the pKa_2 of PA (Kooijman et al. 2005a). These seemingly unrelated results, i.e., the pKa_2 of PA is higher than that of LPA and PE decreases the pKa_2 of both PA and LPA, resulted in the following hypothesis for the ionization properties of the phosphomonoester headgroup of PA:

Upon an initial deprotonation of the first hydroxyl group, the second proton in the phosphomonoester becomes more tightly bound. In fact, it is now shared between the two hydroxyl oxygens of the phosphomonoester headgroup (Fig. 1). Hydrogen bonds formed with this phosphate headgroup will destabilize this second proton by competing for oxygen electrons, thereby facilitating its dissociation.

This is indeed what is observed for PA, LPA, and dehydroxy-LPA (Kooijman et al. 2005a). In a PC bilayer, the pKa_2 of LPA is lower than that of PA, indicating that at constant pH, there are more LPA lipids that carry two negative charges than in the case of PA. The authors proposed that the hydroxyl group on LPA forms an intramolecular hydrogen bond with the phosphomonoester headgroup, facilitating its dissociation. Such a hydrogen bond is indeed observed in the LPA crystal structure, and is apparently preserved in the fully hydrated lipid membrane (Pascher and Sundell 1985).

This model was further supported by the results of the PE bilayer. PE differs from PC in that PE carries a primary amine while PC has a quaternary amine in its headgroup. The primary amine can form a hydrogen bond with PA and thereby facilitate its dissociation, exactly as was observed. Interestingly, this hydrogen bond model, which can be thought of as an electrostatic/hydrogen-bond switch (see below), also explained the observation that LPA and PA have essentially the same pKa_2 in the bilayer rich in PE. The large amount of hydrogen bond donors provided by the amine group of PE in this bilayer will clearly overrule the hydrogen bonding ability of the hydroxyl at the *sn*-2 position of LPA.

4 PA, an Electrostatic/Hydrogen-Bond Switch?

These intriguing results raised the following question: "Do hydrogen bonds indeed facilitate the dissociation of (a proton from the headgroup of) PA?" Recently, this question was indirectly but definitively answered (Kooijman et al. 2007). Two different amphiphiles were used to distinguish between positive charge and hydrogen bond effects. Both amphiphiles carry the same positive charge but differ in their ability to form a hydrogen bond with the headgroup of PA. Only dodecylamine is capable of forming hydrogen bonds and indeed its effect on the negative charge of PA is considerably larger than that of dodecyltrimethylammonium (over 60% larger).

Moreover, these results were confirmed by pH-titration curves for pure PA, PA in the presence of DOTAP (a cationic lipid), and PA in the presence of KALP23 (a transmembrane peptide flanked by lysine residues (de Planque et al. 2001, 2002; Strandberg et al. 2002) (Kooijman et al. 2007). It is clear that the lysine residues in KALP23 that are able to form hydrogen bonds with PA, unlike DOTAP which only carries a positive charge, shift the pKa_2 of PA to lower values compared with DOTAP. Similar to the differential effects of the dodecyl amphiphiles, KALP23 causes an additional ~60% increase in negative charge of PA compared with DOTAP. These results clearly indicate that hydrogen bonds donated from positively charged chemical groups cause an increase in the negative charge of PA, i.e., shift the pKa_2 of PA to lower pH values (Kooijman et al. 2007).

As both PE and basic residues in membrane-interacting peptides can form hydrogen bonds with PA (Kooijman et al. 2005a, 2007), an interesting question that now arises is whether or not PE might hamper the binding of a PA-binding protein to PA. This does not seem to be the case, as in vitro data on the binding of PA binding domains of both Raf-1 kinase and CTR1 confirm that even in the presence of PE, binding is still specific for PA (Kooijman et al. 2007; C. Testerink, unpublished data). Electrostatic arguments also argue against this possibility. The headgroup of PE, containing a primary amine, is not a "point" charge, like the side chain of a lysine residue, but a dipole due to the presence of the negatively charged phosphate (Griffiths 1989). Although this electrostatic argument is an oversimplification, ignoring the presence of the bilayer interface, the argument above suggests that at least the initial affinity of a lysine residue for the phosphate of PA is higher than that of PE, which argues against a competition between lysine residues (or any other point charges) and PE.

4.1 The Model

Indirect but compelling evidence, thus, supports the idea that the negative charge of the phosphomonoester headgroup of PA is increased and stabilized upon the formation of hydrogen bonds. These results can be summarized in a model for the interaction of PA with proteins that at the same time captures the unique ionization properties of this unusual lipid. This model has been coined the *electrostatic/hydrogen bond-switch model* and is depicted in Fig. 3.

PA-binding proteins will initially have an electrostatic interaction with the negatively charged membrane in which PA resides in a sea of other negatively charged lipids. In Fig. 3, the headgroup of PA is shown on the left. Lit oxygens carry a negative charge. The positive charge of lysine and arginine side chains in PA-binding domains will randomly sample the membrane surroundings, i.e., interact electrostatically and form hydrogen bonds with negatively charged phosphodiesters (−1). Next, as soon as the side chain hydrogen bond donor (a primary amine group in Fig. 3, on the right) and a phosphate headgroup of PA come into close proximity (<3.5 Å), a hydrogen bond is formed, leading to a further

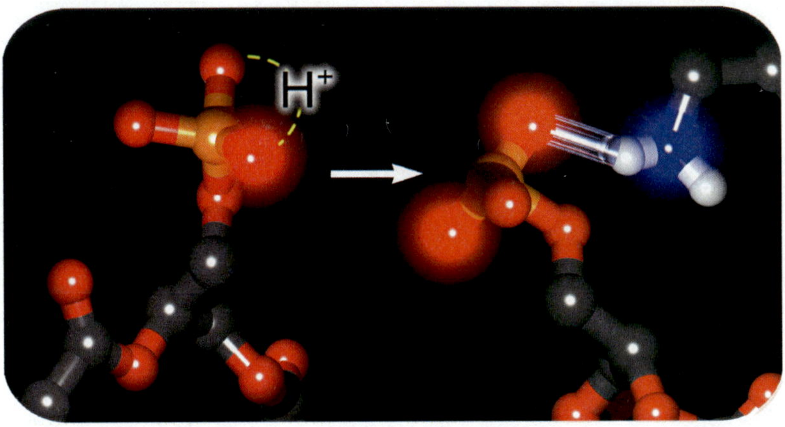

Fig. 3 The electrostatic hydrogen bond-switch model. PA carrying one negative charge is shown on the *left*, and PA interacting with a primary amine via a hydrogen bond is shown on the *right*. Carbons are *gray*, phosphate is *orange*, oxygen is *red*, nitrogen is *blue*, and explicit protons are *white*. *(High)Lighted* atoms indicate those that carry a single charge, *red* for negative, and *blue* for positive. Cartoon was provided by Christian Kandt

deprotonation (to −2) of the headgroup of PA. The increase in negative charge enhances the electrostatic attraction, as indicated by the two red lights on PA in Fig. 3. Coupled to the formation of hydrogen bonds, this now locks the positively charged lysine or arginine side chains on the headgroup of PA, and results in a docking of the membrane interacting protein on a di-anionic PA molecule.

The difference observed between lysine and arginine residues in hydrogen bond length and PA's negative charge (Kooijman et al. 2007) would suggest that lysine residues are more effective in docking on the phosphomonoester headgroup of PA in a lipid bilayer. In light of recent theoretical calculations of the hydrogen bond strength between basic amino acid residues and a phosphomonoester moiety, this is an interesting observation and one that requires further investigation (Mandell et al. 2007). The authors (Kooijman et al. 2007) proposed, and we reiterate this point here, that the electrostatic/hydrogen-bond switch is a key element of the specific recognition of PA by PA-binding proteins.

5 Experimental Support and Biological Implications of the Electrostatic/Hydrogen Bond-Switch Model

The observation that hydrogen bonds influence the negative charge of PA has led to the electrostatic/hydrogen bond switch model. Here, we review recent literature that indeed appears to validate the model and shows its broad applicability.

5.1 Computational and Experimental Evidence for the Electrostatic/Hydrogen Bond Switch Model

Recent experimental and computational work by Tigyi and Parrill further substantiate the importance of the phosphomonoester headgroup in lipid–protein interactions and subsequently illustrate the role of intramolecular hydrogen bonds in the dissociation constants of these phosphomonoesters headgroups (Naor et al. 2007). LPA and sphingosine-1-phosphate (S1P) are important signaling molecules that bind and activate G-protein-coupled receptors (GPCRs), located in the plasma-membrane and nuclear membrane of cells. The binding of LPA and S1P to their respective GPCRs critically depends on the phosphomonoester headgroup (Lynch and Macdonald 2002; Sardar et al. 2002), and the phosphate-binding region of the LPA receptors all contain two conserved basic residues (one arginine and one lysine residue, (Sardar et al. 2002)).

Interestingly, the results described above would suggest that LPA and S1P bind to their receptors in a di-anionic form. This is exactly what recent work by Tigyi and Parrill on the binding of S1P to the $S1P_1$ receptor showed (Naor et al. 2007). Importantly, this work also revealed that the intramolecular hydrogen bond between the hydroxyl group in the backbone of S1P and the phosphomonoester stabilizes the deprotonated form of S1P, exactly what was observed experimentally for the related lipid ceramide-1-phosphate (Cer-1-P; Kooijman et al. 2008; Naor et al. 2007). The ionization properties of Cer-1-P, which also contains a backbone free hydroxyl group, were similar to those of LPA than PA. The work by Tigyi and Parrill provides a strong computational basis for the validity of the electrostatic/hydrogen bond-switch model in that they show that conformations of S1P that lacked the intramolecular hydrogen bond have a higher energy. Thus, formation of the intramolecular hydrogen bond in LPA, Cer-1-P, and S1P is energetically favorable.

Additional support for the model comes from the interaction of basic residues in transmembrane proteins with anionic phospholipids. Our electrostatic/hydrogen bond-switch model predicts that regions of transmembrane proteins enriched in basic amino acids will have a special interaction with PA, and that PA plays a major part in their function. An important feature of many transmembrane proteins is indeed that they are flanked on the cytosolic side of the membrane by basic amino acids. These basic residues are thought to stabilize the transmembrane orientation of the protein and/or regulate its activity (Killian and von Heijne 2000; Lee 2004, 2006; Schmidt et al. 2006; Zimmerberg and Kozlov 2005). Clusters of these basic amino acids may specifically bind to PA. Indeed, recent evidence suggests that this is the case for several transmembrane ion channels (Deol et al. 2006; Powl et al. 2005; Raja et al. 2007). The first such special interaction between basic amino acid residues and PA was described for the mechanosensitive channel of large conductance (MscL). MscL carries a cluster of three basic amino acids (R98, K99, and K100) on its cytosolic face, and has a specific, high affinity, interaction with PA (Powl et al. 2005). Recently, computational studies of the bacterial potassium channel KcsA showed that PA appears to have a special interaction with basic residues in the interface between the monomers of the tetrameric KcsA channel (Deol et al. 2006). Indeed, a special interaction between PA and KcsA was found in TFE unfolding studies, and the authors suggested that special electrostatic and hydrogen bond interactions played a role in the stabilizing effect of PA on the KcsA tetramer (Raja et al. 2007).

5.2 *PA is the Preferred Anionic Lipid for the Interfacial Insertion of Proteins*

The docking of basic protein domains on PA may be followed by insertion of hydrophobic protein domains into the hydrophobic interior of the lipid bilayer. One example of such a favorable hydrophobic interaction has been described in vitro for the GTPase, dynamin. Dynamin binds negatively charged membranes and was found to insert considerably more in mixed-lipid monolayers containing PA rather than other negatively charged phospholipids (Burger et al. 2000).

How can we understand these hydrophobic interactions? On top of its high charge and capacity to form hydrogen bonds, (unsaturated) PA also has a special "effective molecular shape" (Kooijman et al. 2003, 2005b). The idea of effective molecular shape is nicely reviewed by Mouritsen (Mouritsen 2005). PA is the only anionic phospholipid with a pronounced cone shape under physiological conditions (Zimmerberg and Kozlov 2005), with the possible exception of Cer-1-P which occurs in even smaller (than PA) amounts in cells (Kooijman et al. 2008). Cone-shaped lipids facilitate protein penetration into the membrane by forming favorable insertion sites in the headgroup region of the lipid bilayer (van den Brink-van der Laan et al. 2004).

The effect of cone-shaped lipids has been investigated in vitro for the well-known PA-binding domain of Raf-1 kinase, RPA (Ghosh et al. 1996, 2003; Rizzo et al. 2000) and the plant protein kinase, CTR1 (Testerink et al. 2007). Indeed, the cone-shaped lipid PE was found to strongly increase PA binding of the PA-binding domain of Raf-1 (Kooijman et al. 2007). This binding depended critically on the presence of PA. Only hydrophobic insertion sites did not lead to a binding of the RPA domain and RPA bound at least 2.5-fold more to PA compared with PS, even in the presence of high concentrations of PE (Kooijman et al. 2007). Furthermore, lipid-binding experiments with the PA-binding protein CTR1 support this notion (Testerink et al. 2007). CTR1 is a key regulator of ethylene signaling (Chen et al. 2005; Guo and Ecker 2004; Kieber et al. 1993) and a close homolog of the mammalian PA-binding protein, Raf-1 kinase (Ghosh et al. 1996, 2003; Rizzo et al. 2000), and specifically binds vesicles containing PA in vitro (Testerink et al. 2007). Replacing half of the PC with PE did not prevent binding of the PA-binding domain of CTR1 but increased the binding, indicating that there is no competition between PE and this PA-binding domain (C. Testerink, unpublished data).

5.3 Interaction of the C2 Domain of PKC with PA

Recent work by Gomez-Fernandez and coworkers on the peripheral membrane protein, protein kinase C (PKC), in particular the α and ε isoforms, also provides tantalizing evidence for the electrostatic/hydrogen bond-switch model (Ochoa et al. 2002; Sanchez-Bautista et al. 2007). PKC proteins are targeted to the membrane via a C2 domain, a membrane-binding module with a large range of lipid affinities (Cho and Stahelin 2006; Corbalan-Garcia et al. 2007). In the case of the α isoform, Ochoa et al. (2002) showed that a PA molecule is bound in the crystal structure of the C2 domain of this protein. The PA molecule is located at a polybasic region of the C2 domain flanking the lipid bilayer, consistent with our switch model. In the case of PKCε, the same authors now showed that the binding of the C2 domain is specific for PA and suggested that upon binding, the charge of PA is indeed increased to -2 (Sanchez-Bautista et al. 2007), again in agreement with our model.

Finally, the observation that PA may act as a docking site for membrane-interacting peptides (Kooijman et al. 2007) very close to the hydrophobic interior of the lipid bilayer, together with the cone shape of PA, turns PA into a very effective insertion site for positively charged membrane-active proteins. A commonly overlooked possibility for negatively charged membranes may play an important role in this regard as well. The protons absorbed into the negatively charged membrane–water interface (see above) may protonate acidic residues, such as aspartic and glutamic acid, and facilitate their insertion in the hydrophobic interior of the membrane. A corollary may be that upon binding PA, a PA-binding domain liberates a proton from the headgroup of PA and this proton may act to protonate an acidic residue and facilitate its insertion. The data thus far suggest that the electrostatic/hydrogen bond-switch in the phosphate headgroup of PA, coupled to the location of the phosphate headgroup very close to the hydrophobic interior of the lipid bilayer (Kooijman et al. 2007), distinguishes PA from all other anionic membrane lipids.

6 Further Consequences of the Electrostatic/Hydrogen Bond-Switch Model of PA

Phosphomonoesters are ubiquitous in nature, occurring not only in lipids such as PA, but also in many proteins, where (de)phosphorylation often switches enzymes between active and nonactive states, as well as in many small bioactive molecules, such as glucose-6-phosphate. Recognition and binding to the phosphomonoester moiety(ies) of these compounds is likely regulated by similar principles as those presented for PA.

Indeed, a comparison of the crystal structure of 12 randomly selected proteins binding a phosphomonoester moiety(ies) shows that multiple hydrogen bonds are present between amino acid side chains and the phosphomonoester, which inevitably carries two negative charges (Aleshin et al. 1998; Baraldi et al. 1999; Bravo et al. 2001; Dumas et al. 2001; Ferguson et al. 2000; Graham Solomons et al. 2004; Kutateladze and Overduin 2001; Lee et al. 2003, 2005; Lee and Jeffery 2005; Lietzke et al. 2000; Oliva et al. 1995; Thomas et al. 2002). Lysine and arginine residues form the positively charged binding pocket and often form the hydrogen bond donors as well, but other residues donating side chain hydrogen bonds also regularly occur. The most common of these residues (after inspection of the crystal structures discussed above) are serine, threonine, and tyrosine, whereas histidine, glutamine, and asparagine are somewhat less common hydrogen bond donors, but all can donate a side chain hydrogen bond to coordinate the phosphomonoester.

One recent example of a PA molecule bound in the crystal structure of a protein is the structure of the C2 domain of PKCα (Ochoa et al. 2002). This work shows that PA is bound to a polybasic region of the protein. Interestingly, the recently identified PA-binding domain of protein phosphatase 1 contains a serine residue (in the proximity of a positively charged region) that is essential for binding to PA (Jones et al. 2005), indicating that not only lysine and arginine residues but a combination of positive charge and hydrogen bond donor are sufficient to fulfill the electrostatic requirements for the binding of PA by PA-binding proteins. Thus, it can be predicted that the electrostatic/hydrogen-bond switch is the mechanism by which phosphomonoesters are recognized and bound by proteins.

7 Perspectives

PA's role is not only that of a basic building block from which additional glycerol-phospho-lipids are synthesized but may also have formed a means for the putative cell to divide (Kooijman et al. 2003, 2005a, 2005b). Recent studies, focused on the spontaneous replication of vesicles without enzymes and on the creation of a minimal cell (Berclaz et al. 2001; Goto et al. 2000; Luisi 2006; Rasi et al. 2003), suggest that in the evolution of the first primitive cell, PA is likely to have played an important role (Kuruma 2007). Interesting work on the role of PLD in plant wounding and new biophysical studies on the effect of PA in membrane deformation also point to the fundamental role of PA in membrane deformation (Bargmann and Munnik 2006; Cambrea et al. 2007; Lamberson et al. 2007).

The electrostatic/hydrogen-bond model makes several specific predictions for the mode of action of PA. Foremost of these predictions is that the phosphomonoester headgroup acts as an effective docking site for basic amino acids in membrane proteins. These proteins can either be transmembrane- or peripheral membrane proteins. In the case of transmembrane proteins, basic amino acids that reside in the headgroup/hydrophobic interface of the lipid bilayer will act as a "hotspot" for acidic lipids. The electrostatic/hydrogen bond-switch model suggests that PA is an ideal anionic lipid for such a "hotspot," fulfilling all of the requirements of such a site, with electrostatic-, hydrogen bond-, and hydrophobic interactions acting in concert to allow for a highly specific interaction. It is likely that future studies will further highlight the importance of the often overlooked lipid, PA, in the action, stability, and function of such transmembrane proteins. The same is true for peripheral membrane proteins that use PA as a specific membrane anchor. Only a handful of PA-binding proteins are known to date, but many more are most likely to be identified in the future. Interestingly, there appears to be very little sequence homology in the binding domains of these PA binding proteins. An outstanding problem is, thus, the identification of the structure of several putative PA-binding domains to verify the key interactions in PA–protein interaction. Electrostatic-, hydrogen bond-, and hydrophobic interactions will likely be the key features of such a binding domain.

Acknowledgments The authors thank Ben de Kruijff, Koert N.J. Burger, and Teun Munnik for their interest and enthusiasm in the research that led up to this chapter. Christian Kandt is acknowledged for his superb artwork. EEK is indebted to Satyendra Kumar for creating the academic atmosphere in which the writing of this chapter was made possible.

EEK acknowledges financial support from the US National Science Foundation grants DMR-0637221 and CHE-0724082 as well as a Research Challenge award from the Ohio Board of Regents. CT acknowledges NWO for financial support through CW-Vidi grant 700.56.429.
The authors apologize to those authors whose original work is not cited here due to length considerations.

References

Aleshin AE, Zeng C, Bourenkov GP, Bartunik HD, Fromm HJ, Honzatko RB (1998) The mechanism of regulation of hexokinase: new insights from the crystal structure of recombinant human brain hexokinase complexed with glucose and glucose-6-phosphate. Structure 6:39–50

Andresen BT, Rizzo MA, Shome K, Romero G (2002) The role of phosphatidic acid in the regulation of the Ras/MEK/Erk signaling cascade. FEBS Lett 531:65–68

Anthony RG, Henriques R, Helfer A, Meszaros T, Rios G, Testerink C, Munnik T, Deak M, Koncz C, Bogre L (2004) A protein kinase target of a PDK1 signaling pathway is involved in root hair growth in Arabidopsis. EMBO J 23:572–581

Anthony RG, Khan S, Costa J, Pais MS, Bogre L (2006) The Arabidopsis protein kinase PTI1–2 is activated by convergent phosphatidic acid and oxidative stress signaling pathways downstream of PDK1 and OXI1. J Biol Chem 281:37536–37546

Arbuzova A, Murray D, McLaughlin S (1998) MARCKS, membranes, and calmodulin: kinetics of their interaction. Biochim Biophys Acta 1376:369–379

Athenstaedt K, Daum G (1999) Phosphatidic acid, a key intermediate in lipid metabolism. Eur J Biochem 266:1–16

Baraldi E, Carugo KD, Hyvonen M, Surdo PL, Riley AM, Potter BV, O'Brien R, Ladbury JE, Saraste M (1999) Structure of the PH domain from Bruton's tyrosine kinase in complex with inositol 1, 3, 4, 5-tetrakisphosphate. Structure 7:449–460

Bargmann BO, Munnik T (2006) The role of phospholipase D in plant stress responses. Curr Opin Plant Biol 9:515–522

Berclaz N, Blochliger E, Muller M, Luisi PL (2001) Matrix effect of vesicle formation as investigated by cryo transmission electron microscopy. J Phys Chem B 105:1065–1071

Bravo J, Karathanassis D, Pacold CM, Pacold ME, Ellson CD, Anderson KE, Butler PJ, Lavenir I, Perisic O, Hawkins PT, Stephens L, Williams RL (2001) The crystal structure of the PX domain from p40(phox) bound to phosphatidylinositol 3-phosphate. Mol Cell 8:829–839

Burger KNJ, Demel RA, Schmid SL, de Kruijff B (2000) Dynamin is membrane-active: lipid insertion is induced by phosphoinositides and phosphatidic acid. Biochemistry 39: 12485–12493

Cambrea LR, Haque F, Schieler JL, Rochet JC, Hovis JS (2007) Effect of ions on the organization of phosphatidylcholine/phosphatidic acid bilayers. Biophys J 93:1630–1638

Cazzolli R, Shemon AN, Fang MQ, Hughes WE (2006) Phospholipid signaling through phospholipase D and phosphatidic acid. IUBMB Life 58:457–461

Chen YG, Siddhanta A, Austin CD, Hammond SM, Sung TC, Frohman MA, Morris AJ, Shields D (1997) Phospholipase D stimulates release of nascent secretory vesicles from the trans-Golgi network. J Cell Biol 138:495–504

Chen YF, Etheridge N, Schaller GE (2005) Ethylene signal transduction. Ann Bot 95:901–915

Cho W, Stahelin RV (2006) Membrane binding and subcellular targeting of C2 domains. Biochim Biophys Acta 1761:838–849

Cockcroft S (2001) Signaling roles of mammalian phospholipase D1 and D2. Cell Mol Life Sci 58:1674–1687

Coleman RA, Lewin TM, Muoio DM (2000) Physiological and nutritional regulation of enzymes of triacylglycerol synthesis. Annu Rev Nutr 20:77–103

Corbalan-Garcia S, Guerrero-Valero M, Marin-Vicente C, Gomez-Fernandez JC (2007) The C2 domains of classical/conventional PKCs are specific PtdIns(4, 5)P(2)-sensing domains. Biochem Soc Trans 35:1046–1048

Datta SC, Ghosh MK, Hajra AK (1990) Purification and properties of acyl/alkyl dihydroxyacetone-phosphate reductase from guinea pig liver peroxisomes. J Biol Chem 265:8268–8274

De Camilli P, Emr SD, McPherson PS, Novick P (1996) Phosphoinositides as regulators in membrane traffic. Science 271:1533–1539

de Planque MR, Goormaghtigh E, Greathouse DV, Koeppe RE 2nd, Kruijtzer JA, Liskamp RM, de Kruijff B, Killian JA (2001) Sensitivity of single membrane-spanning alpha-helical peptides to hydrophobic mismatch with a lipid bilayer: effects on backbone structure, orientation, and extent of membrane incorporation. Biochemistry 40:5000–5010

de Planque MR, Boots JW, Rijkers DT, Liskamp RM, Greathouse DV, Killian JA (2002) The effects of hydrophobic mismatch between phosphatidylcholine bilayers and transmembrane alpha-helical peptides depend on the nature of interfacially exposed aromatic and charged residues. Biochemistry 41:8396–8404

Deak M, Casamayor A, Currie RA, Downes CP, Alessi DR (1999) Characterisation of a plant 3-phosphoinositide-dependent protein kinase-1 homologue which contains a pleckstrin homology domain. FEBS Lett 451:220–226

Deol SS, Domene C, Bond PJ, Sansom MS (2006) Anionic phospholipid interactions with the potassium channel KcsA: simulation studies. Biophys J 90:822–830

Devaiah SP, Roth MR, Baughman E, Li M, Tamura P, Jeannotte R, Welti R, Wang X (2006) Quantitative profiling of polar glycerolipid species from organs of wild-type Arabidopsis and a phospholipase Dalpha1 knockout mutant. Phytochemistry 67:1907–1924

Dircks LK, Sul HS (1997) Mammalian mitochondrial glycerol-3-phosphate acyltransferase. Biochim Biophys Acta 1348:17–26

Dumas JJ, Merithew E, Sudharshan E, Rajamani D, Hayes S, Lawe D, Corvera S, Lambright DG (2001) Multivalent endosome targeting by homodimeric EEA1. Mol Cell 8:947–958

English D, Cui Y, Siddiqui RA (1996) Messenger functions of phosphatidic acid. Chem Phys Lipids 80:117–132

Fang Y, Vilella-Bach M, Bachmann R, Flanigan A, Chen J (2001) Phosphatidic acid-mediated mitogenic activation of mTOR signaling. Science 294:1942–1945

Ferguson KM, Kavran JM, Sankaran VG, Fournier E, Isakoff SJ, Skolnik EY, Lemmon MA (2000) Structural basis for discrimination of 3-phosphoinositides by pleckstrin homology domains. Mol Cell 6:373–384

Freyberg Z, Siddhanta A, Shields D (2003) "Slip, sliding away": phospholipase D and the Golgi apparatus. Trends Cell Biol 13:540–546

Ghosh S, Strum JC, Sciorra VA, Daniel L, Bell RM (1996) Raf-1 kinase possesses distinct binding domains for phosphatidylserine and phosphatidic acid. Phosphatidic acid regulates the translocation of Raf-1 in 12-O-tetradecanoylphorbol-13-acetate-stimulated Madin-Darby canine kidney cells. J Biol Chem 271:8472–8480

Ghosh S, Moore S, Bell RM, Dush M (2003) Functional analysis of a phosphatidic acid binding domain in human Raf-1 kinase: mutations in the phosphatidate binding domain lead to tail and trunk abnormalities in developing zebrafish embryos. J Biol Chem 278: 45690–45696

Goto A, Suzuki A, Yoshioka H, Goto R, Imae T, Yamazaki K, Walde P (2000) Dynamic aspects of fatty acid vesicles: pH-induced vesicle-micelle transition and dilution induced formation of giant vesicles. In: Luisi PL, Walde P (eds) Giant vesicles: perspectives in supramol chemistry. Wiley, Chichester, pp 261–270

Graham Solomons JT, Zimmerly EM, Burns S, Krishnamurthy N, Swan MK, Krings S, Muirhead H, Chirgwin J, Davies C (2004) The crystal structure of mouse phosphoglucose isomerase at 1.6A resolution and its complex with glucose 6-phosphate reveals the catalytic mechanism of sugar ring opening. J Mol Biol 342:847–860

Griffiths DJ (1989) Introduction to electrodynamics. Prentice Hall, New Jersey

Guo H, Ecker JR (2004) The ethylene signaling pathway: new insights. Curr Opin Plant Biol 7:40–49

Ha KS, Exton JH (1993) Activation of actin polymerization by phosphatidic acid derived from phosphatidylcholine in IIC9 fibroblasts. J Cell Biol 123:1789–1796

Hajra AK (1997) Dihydroxyacetone phosphate acyltransferase. Biochim Biophys Acta 1348:27–34

Hancock JF (2007) PA promoted to manager. Nat Cell Biol 9:615–617

Hauser H (1989) Mechanism of spontaneous vesiculation. Proc Natl Acad Sci USA 86:5351–5355

Honda A, Nogami M, Yokozeki T, Yamazaki M, Nakamura H, Watanabe H, Kawamoto K, Nakayama K, Morris AJ, Frohman MA, Kanaho Y (1999) Phosphatidylinositol 4-phosphate 5-kinase alpha is a downstream effector of the small G protein ARF6 in membrane ruffle formation. Cell 99:521–532

Huang P, Altshuller YM, Hou JC, Pessin JE, Frohman MA (2005) Insulin-stimulated plasma membrane fusion of Glut4 glucose transporter-containing vesicles is regulated by phospholipase D1. Mol Biol Cell 16:2614–2623

Jenkins GM, Frohman MA (2005) Phospholipase D: a lipid centric review. Cell Mol Life Sci 62:2305–2316

Jones JA, Rawles R, Hannun YA (2005) Identification of a novel phosphatidic acid binding domain in protein phosphatase-1. Biochemistry 44:13235–13245

Kieber JJ, Rothenberg M, Roman G, Feldmann KA, Ecker JR (1993) CTR1, a negative regulator of the ethylene response pathway in Arabidopsis, encodes a member of the raf family of protein kinases. Cell 72:427–441

Killian JA, Nyholm TK (2006) Peptides in lipid bilayers: the power of simple models. Curr Opin Struct Biol 16:473–479

Killian JA, von Heijne G (2000) How proteins adapt to a membrane-water interface. Trends Biochem Sci 25:429–434

Kim J, Mosior M, Chung LA, Wu H, McLaughlin S (1991) Binding of peptides with basic residues to membranes containing acidic phospholipids. Biophys J 60:135–148

Kooijman EE, Chupin V, de Kruijff B, Burger KNJ (2003) Modulation of membrane curvature by phosphatidic acid and lysophosphatidic acid. Traffic 4:162–174

Kooijman EE, Carter KM, van Laar EG, Chupin V, Burger KNJ, de Kruijff B (2005a) What makes the bioactive lipids phosphatidic acid and lysophosphatidic acid so special? Biochemistry 44:17007–17015

Kooijman EE, Chupin V, Fuller NL, Kozlov MM, de Kruijff B, Burger KNJ, Rand PR (2005b) Spontaneous curvature of phosphatidic acid and lysophosphatidic acid. Biochemistry 44:2097–2102

Kooijman EE, Tieleman DP, Testerink C, Munnik T, Rijkers DT, Burger KN, de Kruijff B (2007) An electrostatic/hydrogen bond switch as the basis for the specific interaction of phosphatidic acid with proteins. J Biol Chem 282:11356–11364

Kooijman EE, Sot J, Montes LR, Alonso A, Gericke A, De Kruijff B, Kumar S, Goni FM (2008) Membrane organization and ionization properties of the minor but crucial lipid ceramide-1-phosphate. Biophys J 94:4320–4330

Koter M, de Kruijff B, van Deenen LL (1978) Calcium-induced aggregation and fusion of mixed phosphatidylcholine-phosphatidic acid vesicles as studied by 31P NMR. Biochim Biophys Acta 514:255–263

Ktistakis NT, Brown HA, Waters MG, Sternweis PC, Roth MG (1996) Evidence that phospholipase D mediates ADP ribosylation factor-dependent formation of Golgi coated vesicles. J Cell Biol 134:295–306

Ktistakis NT, Delon C, Manifava M, Wood E, Ganley I, Sugars JM (2003) Phospholipase D1 and potential targets of its hydrolysis product, phosphatidic acid. Biochem Soc Trans 31:94–97

Kuruma Y (2007) Question 7: biosynthesis of phosphatidic acid in liposome compartments – toward the self-reproduction of minimal cells. Orig Life Evol Biosph 37:409–413

Kutateladze T, Overduin M (2001) Structural mechanism of endosome docking by the FYVE domain. Science 291:1793–1796
Lamberson ER, Cambrea LR, Hovis JS (2007) Controlling the charge and organization of anionic lipid bilayers: effect of monovalent and divalent ions. J Phys Chem 111:13664–13667
Lee AG (2004) How lipids affect the activities of integral membrane proteins. Biochim Biophys Acta 1666:62–87
Lee AG (2006) Ion channels: a paddle in oil. Nature 444:697
Lee JH, Jeffery CJ (2005) The crystal structure of rabbit phosphoglucose isomerase complexed with D-sorbitol-6-phosphate, an analog of the open chain form of D-glucose-6-phosphate. Protein Sci 14:727–734
Lee DC, Cottrill MA, Forsberg CW, Jia Z (2003) Functional insights revealed by the crystal structures of Escherichia coli glucose-1-phosphatase. J Biol Chem 278:31412–31418
Lee SA, Eyeson R, Cheever ML, Geng J, Verkhusha VV, Burd C, Overduin M, Kutateladze TG (2005) Targeting of the FYVE domain to endosomal membranes is regulated by a histidine switch. Proc Natl Acad Sci USA 102:13052–13057
Lietzke SE, Bose S, Cronin T, Klarlund J, Chawla A, Czech MP, Lambright DG (2000) Structural basis of 3-phosphoinositide recognition by pleckstrin homology domains. Mol Cell 6:385–394
Luisi PL (2006) The emergence of Life: from chemical origin to synthetic biology. Cambridge University Press, Cambridge
Lynch KR, Macdonald TL (2002) Structure-activity relationships of lysophosphatidic acid analogs. Biochim Biophys Acta 1582:289–294
Mandell DJ, Chorny I, Groban ES, Wong SE, Levine E, Rapp CS, Jacobson MP (2007) Strengths of hydrogen bonds involving phosphorylated amino acid side chains. J Am Chem Soc 129:820–827
Manifava M, Thuring JW, Lim ZY, Packman L, Holmes AB, Ktistakis NT (2001) Differential binding of traffic-related proteins to phosphatidic acid- or phosphatidylinositol (4, 5)- bisphosphate-coupled affinity reagents. J Biol Chem 276:8987–8994
McPhail LC, Waite KA, Regier DS, Nixon JB, Qualliotine-Mann D, Zhang WX, Wallin R, Sergeant S (1999) A novel protein kinase target for the lipid second messenger phosphatidic acid. Biochim Biophys Acta 1439:277–290
Montich G, Scarlata S, McLaughlin S, Lehrmann R, Seelig J (1993) Thermodynamic characterization of the association of small basic peptides with membranes containing acidic lipids. Biochim Biophys Acta 1146:17–24
Mor A, Campi G, Du G, Zheng Y, Foster DA, Dustin ML, Philips MR (2007) The lymphocyte function-associated antigen-1 receptor costimulates plasma membrane Ras via phospholipase D2. Nat Cell Biol 9:713–719
Moritz A, De Graan PN, Gispen WH, Wirtz KW (1992) Phosphatidic acid is a specific activator of phosphatidylinositol-4-phosphate kinase. J Biol Chem 267:7207–7210
Mouritsen OG (2005) Life as a matter of fat. The emerging science of lipidomics. Springer, Berlin
Munnik T, Irvine RF, Musgrave A (1998) Phospholipid signaling in plants. Biochim Biophys Acta 1389:222–272
Nagiec MM, Wells GB, Lester RL, Dickson RC (1993) A suppressor gene that enables Saccharomyces cerevisiae to grow without making sphingolipids encodes a protein that resembles an Escherichia coli fatty acyltransferase. J Biol Chem 268:22156–22163
Nakanishi H, Morishita M, Schwartz CL, Coluccio A, Engebrecht J, Neiman AM (2006) Phospholipase D and the SNARE Sso1p are necessary for vesicle fusion during sporulation in yeast. J Cell Sci 119:1406–1415
Nanjundan M, Possmayer F (2003) Pulmonary phosphatidic acid phosphatase and lipid phosphate phosphohydrolase. Am J Physiol Lung Cell Mol Physiol 284:L1–L23
Naor MM, Walker MD, Van Brocklyn JR, Tigyi G, Parrill AL (2007) Sphingosine 1-phosphate pKa and binding constants: intramolecular and intermolecular influences. J Mol Graphics Model 26:519–528
Nyholm TK, Ozdirekcan S, Killian JA (2007) How protein transmembrane segments sense the lipid environment. Biochemistry 46:1457–1465

Ochoa WF, Corbalan-Garcia S, Eritja R, Rodriguez-Alfaro JA, Gomez-Fernandez JC, Fita I, Verdaguer N (2002) Additional binding sites for anionic phospholipids and calcium ions in the crystal structures of complexes of the C2 domain of protein kinase calpha. J Mol Biol 320:277–291

Oliva G, Fontes MR, Garratt RC, Altamirano MM, Calcagno ML, Horjales E (1995) Structure and catalytic mechanism of glucosamine 6-phosphate deaminase from Escherichia coli at 2.1 A resolution. Structure 3:1323–1332

Op den Kamp JA (1979) Lipid asymmetry in membranes. Annu Rev Biochem 48:47–71

Ozdirekcan S, Etchebest C, Killian JA, Fuchs PF (2007) On the orientation of a designed transmembrane peptide: toward the right tilt angle? J Am Chem Soc 129:15174–15181

Pascher I, Sundell S (1985) Interactions and space requirements of the phosphate head group in membrane lipids. The crystal structure of disodium lysophosphatidate dihydrate. Chem Phys Lipids 37:241–250

Powl AM, East JM, Lee AG (2005) Heterogeneity in the binding of lipid molecules to the surface of a membrane protein: hot spots for anionic lipids on the mechanosensitive channel of large conductance MscL and effects on conformation. Biochemistry 44:5873–5883

Raja M, Spelbrink RE, de Kruijff B, Killian JA (2007) Phosphatidic acid plays a special role in stabilizing and folding of the tetrameric potassium channel KcsA. FEBS Lett 581:5715–5722

Rasi S, Mavelli F, Luisi PL (2003) Cooperative micelle binding and matrix effect in oleate vesicle formation. J Phys Chem B 107:14068–14076

Rizzo MA, Shome K, Vasudevan C, Stolz DB, Sung TC, Frohman MA, Watkins SC, Romero G (1999) Phospholipase D and its product, phosphatidic acid, mediate agonist-dependent raf-1 translocation to the plasma membrane and the activation of the mitogen-activated protein kinase pathway. J Biol Chem 274:1131–1139

Rizzo MA, Shome K, Watkins SC, Romero G (2000) The recruitment of Raf-1 to membranes is mediated by direct interaction with phosphatidic acid and is independent of association with Ras. J Biol Chem 275:23911–23918

Roth MG (2008) Molecular mechanisms of PLD function in membrane traffic. Traffic 9:1233–1239

Roth MG, Bi K, Ktistakis NT, Yu S (1999) Phospholipase D as an effector for ADP-ribosylation factor in the regulation of vesicular traffic. Chem Phys Lipids 98:141–152

Roy MO, Leventis R, Silvius JR (2000) Mutational and biochemical analysis of plasma membrane targeting mediated by the farnesylated, polybasic carboxy terminus of K-ras4B. Biochemistry 39:8298–8307

Sanchez-Bautista S, de Godos A, Rodriguez-Alfaro JA, Torrecillas A, Corbalan-Garcia S, Gomez-Fernandez JC (2007) Interaction of the C2 domain from protein kinase C (epsilon) with model membranes. Biochemistry 46:3183–3192

Sardar VM, Bautista DL, Fischer DJ, Yokoyama K, Nusser N, Virag T, Wang DA, Baker DL, Tigyi G, Parrill AL (2002) Molecular basis for lysophosphatidic acid receptor antagonist selectivity. Biochim Biophys Acta 1582:309–317

Schmidt D, Jiang QX, MacKinnon R (2006) Phospholipids and the origin of cationic gating charges in voltage sensors. Nature 444:775–779

Siddhanta A, Shields D (1998) Secretory vesicle budding from the trans-Golgi network is mediated by phosphatidic acid levels. J Biol Chem 273:17995–17998

Siddiqui RA, Yang YC (1995) Interleukin-11 induces phosphatidic acid formation and activates MAP kinase in mouse 3T3–L1 cells. Cell Signal 7:247–259

Stace CL, Ktistakis NT (2006) Phosphatidic acid- and phosphatidylserine-binding proteins. Biochim Biophys Acta 1761:913–926

Stace CL, Manifava M, Delon C, Coadwell J, Cockroft S, Ktistakis NT (2007) PA binding of phosphatidylinositol 4-phosphate 5-kinase. Adv Enzyme Regul. doi:10.1016/j.advenzreg.2007.11.008

Strandberg E, Morein S, Rijkers DT, Liskamp RM, van der Wel PC, Killian JA (2002) Lipid dependence of membrane anchoring properties and snorkeling behavior of aromatic and charged residues in transmembrane peptides. Biochemistry 41:7190–7198

Swairjo MA, Seaton BA, Roberts MF (1994) Effect of vesicle composition and curvature on the dissociation of phosphatidic acid in small unilamellar vesicles–a 31P-NMR study. Biochim Biophys Acta 1191:354–361

Testerink C, Munnik T (2005) Phosphatidic acid: a multifunctional stress signaling lipid in plants. Trends Plant Sci 10:368–375

Testerink C, Larsen PB, van der Does D, van Himbergen JA, Munnik T (2007) Phosphatidic acid binds to and inhibits the activity of Arabidopsis CTR1. J Exp Bot 58:3905–3914

Thomas CC, Deak M, Alessi DR, van Aalten DM (2002) High-resolution structure of the pleckstrin homology domain of protein kinase b/akt bound to phosphatidylinositol (3, 4, 5)-trisphosphate. Curr Biol 12:1256–1262

Traikia M, Warschawski DE, Lambert O, Rigaud JL, Devaux PF (2002) Asymmetrical membranes and surface tension. Biophys J 83:1443–1454

van den Brink-van der Laan E, Killian JA, de Kruijff B, Killian JA, de Kruijff B (2004) Nonbilayer lipids affect peripheral and integral membrane proteins via changes in the lateral pressure profile. Biochim Biophys Acta 1666:275–288

van Klompenburg W, Nilsson I, von Heijne G, de Kruijff B (1997) Anionic phospholipids are determinants of membrane protein topology. EMBO J 16:4261–4266

van Meer G (1998) Lipids of the Golgi membrane. Trends Cell Biol 8:29–33

van Schooten B, Testerink C, Munnik T (2006) Signaling diacylglycerol pyrophosphate , a new phosphatidic acid metabolite. Biochim Biophys Acta 1761:151–159

Vance JE, Steenbergen R (2005) Metabolism and functions of phosphatidylserine. Prog Lipid Res 44:207–234

Veverka V, Crabbe T, Bird I, Lennie G, Muskett FW, Taylor RJ, Carr MD (2008) Structural characterization of the interaction of mTOR with phosphatidic acid and a novel class of inhibitor: compelling evidence for a central role of the FRB domain in small molecule-mediated regulation of mTOR. Oncogene 27:585–595

Wang X, Devaiah SP, Zhang W, Welti R (2006) Signaling functions of phosphatidic acid. Prog Lipid Res 45:250–278

Watts A (1998) Solid-state NMR approaches for studying the interaction of peptides and proteins with membranes. Biochim Biophys Acta 1376:297–318

Yang JS, Gad H, Lee SY, Mironov A, Zhang L, Beznoussenko GV, Valente C, Turacchio G, Bonsra AN, Du G, Baldanzi G, Graziani A, Bourgoin S, Frohman MA, Luini A, Hsu VW (2008) A role for phosphatidic acid in COPI vesicle fission yields insights into Golgi maintenance. Nat Cell Biol 10:1146–1153

Yeung T, Gilbert GE, Shi J, Silvius J, Kapus A, Grinstein S (2008) Membrane phosphatidylserine regulates surface charge and protein localization. Science 319:210–213

Zambonelli C, Roberts MF (2005) Non-HKD phospholipase d enzymes: new players in phosphatidic acid signaling? Prog Nucleic Acid Res Mol Biol 79:133–181

Zeniou-Meyer M, Zabari N, Ashery U, Chasserot-Golaz S, Haeberle AM, Demais V, Bailly Y, Gottfried I, Nakanishi H, Neiman AM, Du G, Frohman MA, Bader MF, Vitale N (2007) Phospholipase D1 production of phosphatidic acid at the plasma membrane promotes exocytosis of large dense-core granules at a late stage. J Biol Chem 282:21746–21757

Zhang W, Qin C, Zhao J, Wang X (2004) Phospholipase D alpha 1-derived phosphatidic acid interacts with ABI1 phosphatase 2C and regulates abscisic acid signaling. Proc Natl Acad Sci USA 101:9508–9513

Zhao C, Du G, Skowronek K, Frohman MA, Bar-Sagi D (2007) Phospholipase D2-generated phosphatidic acid couples EGFR stimulation to Ras activation by Sos. Nat Cell Biol 9:706–712

Zheng Z, Zou J (2001) The initial step of the glycerolipid pathway: identification of glycerol 3-phosphate/dihydroxyacetone phosphate dual substrate acyltransferases in Saccharomyces cerevisiae. J Biol Chem 276:41710–41716

Zimmerberg J, Kozlov MM (2005) How proteins produce cellular membrane curvature. Nat Rev Mol Cell Biol 7:9–19

Nitric Oxide and Phosphatidic Acid Signaling in Plants

Ayelen M. Distéfano, M. Luciana Lanteri, Arjen ten Have, Carlos García-Mata, Lorenzo Lamattina, and Ana M. Laxalt

Abstract Nitric oxide (NO) is an important redox-based regulator of cell physiology involved in many signaling processes in plants. The precise mechanism of how NO activates or interacts with different targets is still poorly understood. The polar lipid, phosphatidic acid (PA) is another molecule involved in plant signaling. NO and PA have been independently regarded as general, multifunctional, stress-signaling molecules in plants. Since they share common effectors, we hypothesized that NO and PA participate in the same signaling pathways. Results from our laboratory revealed that NO can induce PA formation during (1) plant-defense responses, (2) stomatal closure, and (3) adventitious root formation. Two enzymatic pathways produce PA, phospholipase D, and phospholipase C in concerted action with the diacylglycerol kinase. We discuss how NO might act on PA-generating enzymes as well as on their common downstream effectors such as Ca^{2+}, reactive oxygen species, protein kinases, and phosphatases.

1 Introduction

1.1 Chemistry of NO

Nitric oxide (NO) is a short-lived bioactive gas, able to cross biological membranes (Stamler et al. 1992). It is a well-established second messenger in animals (Davis et al. 2001) and has been shown to be involved in different physiological and developmental processes in plants (Lamattina et al. 2003). The involvement of NO, and derived reactive nitrogen species, in a wide variety of physiological processes

A.M. Laxalt (✉)
Instituto de Investigaciones Biológicas, Facultad de Ciencias Exactas y Naturales, Universidad Nacional de Mar del Plata, CC 1245, 7600, Mar del Plata, Argentina
e-mail: amlaxalt@mdp.edu.ar

T. Munnik (ed.), *Lipid Signaling in Plants*, Plant Cell Monographs 16,
DOI 10.1007/978-3-642-03873-0_15,

is achieved via a rich redox and additive chemistry toward its targets (Stamler 1994). Proteins containing transition metals, thiols, or tyrosines residues, strategically located at either allosteric or active sites, are NO targets and essential components of NO signaling.

Membrane proteins and lipoproteins are theoretically more exposed to NO because it concentrates in membranes where it more readily reacts with oxygen to generate oxidizing, nitrosating, and nitrating species (Moller et al. 2005). Membrane NO-targets have been identified in animals (Stamler 1994) but not in plants (Lindermayr et al. 2005). There are no reports of either nitrosylation or nitration of animal or plant phospholipases. Besides reacting with membrane proteins, NO can react with lipid peroxyl radicals, effectively inhibiting lipid peroxidation chain reactions, displaying an oxidant-protective and antiinflammatory role in animals (Rubbo et al. 2000). These reactions have not been studied in plants, but since they depend on the chemical characteristics of the players involved, we envisage that similar events can take place in plants.

2 NO Production and Localization in Plants

Several nonenzymatic or enzymatic sources of NO have been described in plants (Besson-Bard et al. 2008). They are either L-arginine- or nitrate/nitrite-dependent. The first involves an NO synthase (NOS)-like enzyme plus an uncharacterized process that uses polyamines as the substrate. NOS-like activities have been detected in several plant tissues and shown to be involved in NO synthesis during hormone treatments and (a)biotic stress (Neill et al. 2003). A plant NOS gene has not been identified. The nitrate/nitrite-dependent production includes a nonenzymatic NO source and the enzymes nitrate reductase (NR) and nitrite–NO reductase (Ni–NOR). *Arabidopsis* has two NR genes. NR-deficient double mutants, *nia1*/*nia2*, show low NO levels and are impaired in abscisic acid (ABA)-induced stomatal closure (Desikan et al. 2002), root hair elongation (Lombardo et al. 2006), and lateral root formation (Kolbert et al. 2007). Ni–NOR may be involved in various root processes (Besson-Bard et al. 2008). The identity of Ni–NOR is currently unknown. NO production has been localized in the cytosol, nucleus, peroxisomes, mitochondria, and chloroplasts (Neill et al. 2003). The specificity of NO signaling can be explained by this temporal and spatial NO distribution, superimposed to the spatial distribution of its effectors (Besson-Bard et al. 2008). Such spatial versatility can possibly be ascribed to the variety of NO sources.

3 Phosphatidic Acid Signaling

Phosphatidic acid (PA) has emerged as a second messenger in plants (Testerink and Munnik 2005). It is formed within minutes in response to drought stress, ABA treatments, salt stress, and during pathogenic and mutualistic interactions. Increases in PA are transient since PA can be converted to diacylglycerol pyrophosphate (DGPP) by PA kinase or to diacylglycerol (DAG) by lipid phosphate phosphatases (see chapters, "Diacylglycerol Kinase," "Phosphatidic Acid Phosphatases in Seed Plants," and "Diacylglycerol Pyrophosphate, a Novel Plant Signaling Lipid").

Different pathways generate PA. Phospholipase C (PLC) hydrolyzes phosphatidylinositol 4,5-bisphosphate (PIP_2), phosphatidylinositol 4-monophosphate (PIP), and phosphatidylinositol (PI). Hydrolysis of PIP_2 results in inositol 1,4,5-trisphosphate (IP_3), which diffuses into the cytosol and is involved in increasing cytosolic Ca^{2+} concentration, and DAG, which remains in the membrane. In plants, DAG is phosphorylated to PA by DAG kinase (DGK). This pathway is referred to as the PLC/DGK pathway (Munnik 2001; see chapter, "Diacylglycerol Kinase"). Phospholipase D (PLD) hydrolyzes structural phospholipids, such as phosphatidylcholine, generating PA and choline.

Arabidopsis has seven putative PLC genes and two pseudogenes (Mueller-Roeber and Pical 2002). PLCs contain a Ca^{2+}-dependent phospholipid binding domain (C2), known to be activated by Ca^{2+} but many aspects regarding the regulation of the various PLC isoforms remain to be elucidated.

A family of 12 PLD isoforms with specific biochemical, regulatory, and structural properties exists in *Arabidopsis* (Elias et al. 2002; Qin and Wang 2002; see chapter, "Plant Phospholipase D"). They are classified into two classes. PX/PH-PLDs class (PLDζ) has N-terminal phox homology (PX) and pleckstrin homology (PH) domains, whereas the C2-PLDs, unique to plants, have a C2 domain. The latter is further classified into PLDα, β, γ, δ, and ε classes. Genetically modified plants have been successfully used to address the role of individual members during a number of stress responses, demonstrating that some functional redundancy exists.

4 NO Signaling: Connecting to Phosphatidic Acid Signaling

NO has been shown to be involved in the IP_3- and cADPR-dependent production of Ca^{2+} (Garcia-Mata et al. 2003; Lanteri et al. 2006). Ca^{2+} is a regulator of PLD and PLC, and many other enzymes, among which one is NADPH oxidase. NADPH oxidase is responsible for reactive oxygen species (ROS) production. ROS production occurs concomitantly with NO generation in several physiological processes (Delledonne et al. 2001). PA has been shown to trigger an oxidative burst, suggesting a similar activation of NADPH oxidase (Sang et al. 2001a; de Jong et al. 2004; Park et al. 2004). This suggested a putative cross-talk among NO, Ca^{2+}, PA, and ROS. NO and PA share other downstream effectors, such as an inward rectifying-K^+ channel, the protein phosphatase ABI1, a mitogen-activated protein kinase (MAPK), and a Ca^{2+}-dependent protein kinase (CDPK). There are a number of processes in which NO- and PA-generating enzymes has been shown to act (Table 1). These instances prompted us to study whether NO- and PA-signaling are interrelated in plants.

Table 1 Involvement of PLC/PLD and NO during diverse physiological processes in plants

Signals-stimuli	Physiological process	References PLC and PLD	References NO
ABA	Stomatal movement	Lee et al. (1996); Jacob et al. (1999); Sang et al. (2001b); Hunt et al. (2003); Zhang et al. (2004)	Desikan et al. (2002); Garcia-Mata and Lamattina (2002); Neill et al. (2002)
	Germination	Katagiri et al. (2005)	Sarath et al. (2006)
Auxin	Root architecture	Li and Xue (2007)	Gouvea et al. (1997); Pagnussat et al. (2002); Correa-Aragunde et al. (2004); Huang et al. (2007); Kolbert et al. (2007); Zhao et al. (2007a)
Gibberellic acid	Germination	Villasuso et al. (2003)	Beligni and Lamattina (2000); Beligni et al. (2002)
Ethylene	Senescence	Fan et al. (1997); Mao and Huber (2004)	Leshem and Haramaty (1995); Leshem and Pinchasov (2000)
Salt		Drobak and Watkins (2000); Munnik et al. (2000); Takahashi et al. (2001); Thiery et al. (2004); Parre et al. (2007)	Gould et al. (2003); Zhao et al. (2004b)
Drought		Katagiri et al. (2001); Sang et al. (2001b); Wang et al. (2008)	Garcia-Mata and Lamattina (2001)
UV-B		Zhang et al. (2003)	Mackerness et al. (2001); Hari et al. (2003); An et al. (2005); He et al. (2005)
Wounding	JA accumulation	Lee et al. (1997); Wang et al. (2000)	Huang et al. (2004); Wang and Wu (2005)
Aluminium		Li and Fleming (1999); Pejchar et al. (2008); Ramos-Diaz et al. (2007)	Wang and Yang (2005); Illes et al. (2006); Tian et al. (2007)
Elicitation	Nodulation	den Hartog et al. (2001); Park et al. (2004)	Baudouin et al. (2006)
	Oxidative burst/HR	van der Luit et al. (2000); Sang et al. (2001a); den Hartog et al. (2003); de Jong et al. (2004); Park et al. (2004); Zhao et al. (2004a); Yamaguchi et al. (2005); Andersson et al. (2006); Bargmann et al. (2006)	Delledonne et al. (1998; 2001); Durner et al. (1998); de Pinto et al. (2002); Lamotte et al. (2004); Zhao et al. (2007b)

Since 2003, our group has studied whether a cross-talk between NO and phospholipid signaling exists in plants. We chose three processes in which NO and PA signaling were implicated (see Table 1) but for which no direct link was reported: plant defense, stomatal closure, and adventitious root formation (Fig. 1).

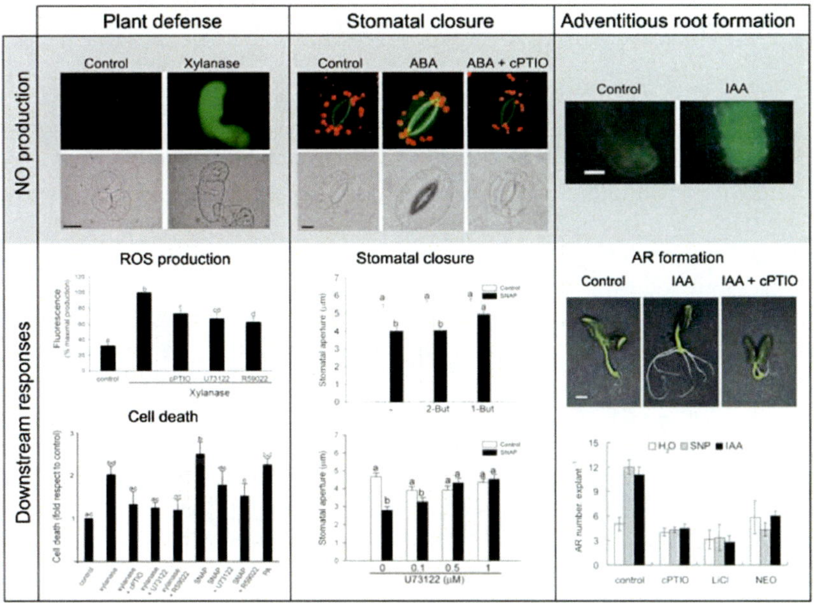

Fig. 1 Nitric oxide (NO) production and downstream physiological responses in (1) xylanase-treated tomato cells, (2) ABA-treated *Vicia faba* guard cells, and (3) auxin-treated cucumber explants. NO production was measured with the specific fluorescent probe DAF-2DA and visualized under the microscope (*green fluorescence*). *Plant defense:* NO production in tomato cells treated 15 min with xylanase (200 μg ml^{-1}) or nontreated (control). *Bar* = 5 μm. ROS production and cell death was determined in the presence or absence of either 1 mM cPTIO (NO scavenger), 5 μM U73122 (PLC inhibitor), or 10 μM R59022 (DGK inhibitor). ROS levels were determined using H2DCF-DA. Cell death was determined at 19 h after treatments with xylanase, 1 mM SNAP, or 10 μM water-soluble, synthetic PA C8:0. Adapted from Laxalt et al. (2007), with permission from the American Society for Biochemistry and Molecular Biology. *Stomatal closure:* NO production in *Vicia faba* guard cells treated 60 min with 20 μM ABA in the presence or absence of 200 μM cPTIO. Red fluorescence corresponds to chlorophyll autofluorescence. A bright field image demonstrating the difference in stomatal aperture between treatments is shown. *Bar* = 5 μm. Stomatal closure experiments were performed by treating epidermal peels for 60 min with 0.1% (v/v) of 1-Butanol (1-But), 2-Butanol (2-But) (*upper panel*), or with different concentrations of U73122 (*lower panel*) in the presence or absence of the NO donor SNAP (200 μM). Adapted from Distéfano et al. (2008) with permission from Blackwell Publishing. *Adventitious root (AR) formation:* NO production in longitudinal sections from the tip of cucumber hypocotyls treated 1 day with water (control) or 10 μM IAA. *Bar* = 0.5 mm. ARs were visualized and quantified after 5 days of treatment with 10 μM IAA or NO donor SNP in the presence or absence of cPTIO (200 μM) or of the inhibitors of IP_3 formation lithium chloride (5 mM LiCl) or neomycin sulfate (50 μM NEO). *Bar* = 5 mm. Adapted from Pagnussat et al. (2002) and Lanteri et al. (2006), with permission from the American Society of Plant Biologists and Society of Experimental Biology, respectively

5 Plant Defense

NO is involved in the plant defense response of a growing list of plant–pathogen interactions (Table 1). One downstream response to NO is the release of Ca^{2+}, following the elicitation of tobacco cells with cryptogein (Lamotte et al. 2004). Treatments with inhibitors of NO accumulation compromised the hypersensitive response (HR) (Durner and Klessig 1999). NO and ROS act together, triggering HR in plant–pathogen interactions (Delledonne et al. 2001).

PA has been shown to accumulate upon treatments with the elicitors xylanase, chitotetraose, and flagellin in tomato cells (van der Luit et al. 2000); AVR4 in Cf-4 expressing tobacco cells (de Jong et al. 2004), Nod factors in alfalfa cells (den Hartog et al. 2003), *N*-acetylchitooligosaccharide in rice cells (Yamaguchi et al. 2005), and during AvrRpm1- and AvrRpt2-induced disease resistance responses in *Arabidopsis* (Andersson et al. 2006) (Table 1).

At this moment, the only elicitor for which an interaction between PA and NO has been established is xylanase (Laxalt et al. 2007). Xylanase is a well-studied fungal elicitor. It activates a rapid PA production via both PLD and PLC/DGK pathways (van der Luit et al. 2000). Xylanase-induced PLC/DGK-derived PA is NO dependent. Pretreatment of cells with the NO scavenger cPTIO or the PLC-specific inhibitor U73122 reduced PA formation. The xylanase-induced rapid-PLD activity appears to be independent of NO. Defense-related responses downstream to PA have been studied. Scavenging of NO or inhibition of either the PLC or the DGK enzyme diminished xylanase-induced ROS production, gene expression, and cell death (Fig. 1).

In plant–pathogen interactions, the oxidative burst mediated by specific elicitors occurs in two phases. The first phase shows a rapid, transient, and low ROS response, whereas the second phase shows a prolonged and massive ROS production (Lamb and Dixon 1997). A biphasic ROS generation was observed and associated with the activation of PLC and PLD, in rice cells induced by *N*-acetylchitooligosaccharide elicitor. The activation of both enzymes was shown for the first phase of ROS generation, whereas for the second phase, only activation of PLD was evident (Yamaguchi et al. 2005). Recognition of *Pseudomonas syringae* AvrRpm1 or AvrRpt2 in *Arabidopsis* induced a biphasic accumulation of PA. The first wave was attributed to the PLC/DGK pathway and the second to PLD. Both phospholipase pathways acted upstream of ROS formation (Andersson et al. 2006). We suggest that, NO-dependent, PLC/DGK-generated PA is involved in the induction of ROS production during the first peak of the oxidative burst in xylanase-treated cells.

A model summarizing the signaling events downstream xylanase perception in tomato in relation to phospholipid signaling and NO is proposed (Fig. 2). The proposed signaling is subject to several regulatory mechanisms. Firstly, an increase in cytosolic Ca^{2+} concentration generated by PLC-released IP_3 might stimulate the NADPH oxidase. Increase in Ca^{2+} concentrations might also stimulate PLC or C2-PLDs, thereby establishing possible signal amplification via a positive feedback; however, this mechanism has not been proved so far. A second positive feedback

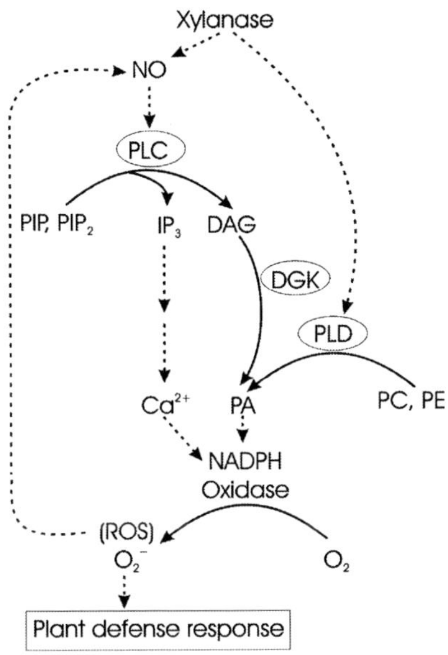

Fig. 2 Model of NO- and PA-signaling downstream of xylanase perception in suspension-cultured tomato cells. Xylanase induces the PLC/DGK pathway in an NO-dependent manner. Xylanase also induces PLD to produce PA, apparently in an NO-independent manner. PA on its turn activates NADPH oxidase, generating superoxide and related ROS. The resulting ROS form a positive feedback via a stimulation of NO production. PLC signaling also results in IP_3 formation. IP_3 induces the release of Ca^{2+} from internal stores and Ca^{2+} further stimulates NADPH oxidase leading to plant defense responses. *Solid arrows* indicate metabolic conversion, *dashed arrows* indicate activation (directly or indirectly)

can be proposed between ROS and NO, which is based on a number of observations. H_2O_2 triggers NO production in different biological systems (Lum et al. 2002; Bright et al. 2006; de Pinto et al. 2006). H_2O_2 also induces PA formation (Yamaguchi et al. 2004) and PA induces ROS production (Sang et al. 2001a; de Jong et al. 2004; Park et al. 2004) and NO (A.M. Laxalt et al. unpublished). Although all these data suggest a positive feedback between ROS and NO, mediated by PA, genetic approaches are required to unequivocally demonstrate the order of events following xylanase perception. More recent experiments have shown that yet other elicitors, chitosan and AVR4 from *Cladosporium fulvum*, induce NO-dependent PLC/DGK formation of PA (A.M. Laxalt et al. unpublished).

6 Stomatal Closure

The opening and closure of stomata is regulated by guard cell turgor and occurs rapidly in response to environmental signals such as light, temperature and humidity, or hormones, including ABA, auxin, and ethylene, which is essential to the plant's survival. The ability to integrate a wide range of stimuli makes guard cells an attractive model system to study signaling networks. During drought stress conditions, ABA regulates water loss by two different process, induction of stomatal closure and inhibition of stomatal opening (Mishra et al. 2006). Stomatal closure involves several processes, finally resulting in an inactivation of inward-rectifying K^+ (K_{in}^+) channels, activation of outward-rectifying K^+ (K_{out}^+) channels, and activation of slow and fast-anion channels (MacRobbie 2006). Here, we will focus on the processes related to NO and PA signaling.

NO is produced during the induction of stomatal closure by ABA, methyl jasmonate, UV-B light, and bicarbonate (Desikan et al. 2002; Garcia-Mata and Lamattina 2002; Neill et al. 2002; Suhita et al. 2004; He et al. 2005; Kolla et al. 2007). Exogenous NO, applied as NO donors, reduces transpirational water loss by inducing stomatal closure in both dicotyledonous and monocotyledonous plants (Garcia-Mata and Lamattina 2001). Since this first report, NO and ABA cross-talk has been thoroughly studied. In ABA-treated guard cells, endogenous NO has been shown to induce stomatal closure (Fig. 1). NO is generated enzymatically by either NR or a NOS-like activity, depending on the species studied (Neill et al. 2003). NO, in turn, participates in a subset of ABA-evoked responses by inactivating K_{in}^+ channels via a cGMP/cADPR-dependent increase of cytosolic Ca^{2+} concentration (Garcia-Mata et al. 2003). Sokolovski et al. (2005) showed that several NO-dependent signals can be modulated through protein phosphorylation upstream of intracellular Ca^{2+} release. Recently, PA was incorporated in the NO signaling cascade (Distéfano et al. 2008). We showed that NO induces PLD activation. The PLC inhibitor U73122 affects NO-induced accumulation of PA. Since DGK activity could not be measured due to methodological reasons, we do not know if this PA is generated via PLC/DGK or via a PLD activated by Ca^{2+} via PLC-released IP_3 as was proposed by Andersson et al. (2006) during plant defense responses. When either PLC or PLD activity was inhibited, NO failed to induce stomatal closure (Fig. 1). This evidence suggests that PLD and PLC are participating in the same signaling pathway and agrees with similar evidence presented for ABA-induced stomatal closure (Jacob et al. 1999). The proposed roles of NO and PA during stomatal closure are summarized in Fig. 3. After ABA recognition, NO activates PLC with the consequent IP_3 and DAG production. IP_3-induced Ca^{2+} release from intracellular stores and/or other NO-mediated pathways activates PLD resulting in PA accumulation. PA can be also produced by the activation of DGK.

Both PLC and PLD are involved in drought stress and ABA-related signaling networks. Which isoforms are activated by NO remains to be elucidated. PLDα1 (Sang et al. 2001b), PLDδ (Katagiri et al. 2001), and PLC1 (Hirayama et al. 1995) are the main candidates, since they participate in drought stress response. Preliminary data show that at least PLDα1 and PLDδ are involved in NO-induced stomatal closure (Distéfano et al, unpublished). Once the candidates have been identified, we can study how NO activates phospholipase.

Several components can be proposed as putative targets for NO-dependent PA production, among which are the protein phosphatase 2C, ABI1, and the K_{in}^+-channel activity (Fig. 3). It has been reported that PA inhibits the function of ABI1,

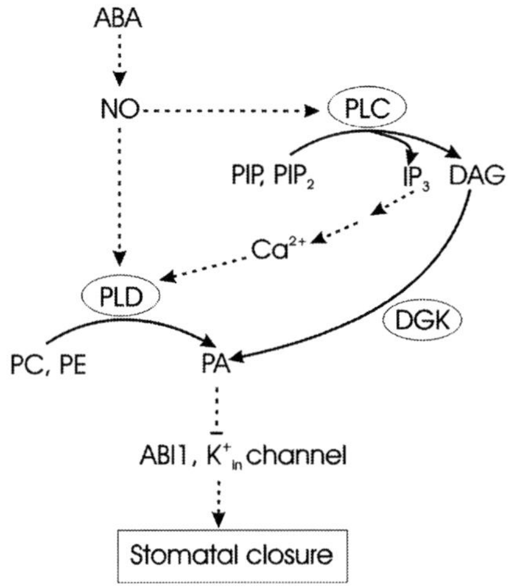

Fig. 3 Model of NO-induced PA signaling during stomatal closure in *Vicia faba* guard cells. ABA induces NO production, which in turn activates the PLC/DGK and PLD pathways generating PA. IP_3 induces the release of Ca^{2+} from internal stores and Ca^{2+} further stimulates PLD. PA inhibits K_{in}^+ channels and ABI1, thereby relieving their inhibition of stomatal closure, resulting in stomatal closure. *Solid arrows* indicate metabolic conversion, *dashed arrows* indicate activation () or inhibition (⊥) (directly or indirectly)

which is a negative regulator of ABA responses (Zhang et al. 2004). Since *abi1* mutant is able to produce NO in response to ABA, but is impaired in ABA induction of stomatal closure, NO has been proposed to act upstream of ABI1 (Desikan et al. 2002; Desikan et al. 2004). Thus, it is speculated that NO inactivates ABI1 via PA production. Guard cell K_{in}^+ channel activity is regulated by both PA and NO (Jacob et al. 1999; Garcia-Mata et al. 2003). ROS might also be pointed downstream of NO-induced PA production. PA addition to *Arabidopsis* leaves and tobacco-suspension cells induces high ROS levels (Sang et al. 2001a; de Jong et al. 2004; Park et al. 2004). NO addition also induces ROS production in guard cells (He et al. 2005). ROS induce stomatal closure and inhibit the K_{in}^+ channel (Zhang et al. 2001; Kohler et al. 2003). However, Bright et al. (2006) reported that ROS production is upstream of NO formation during ABA-induced stomatal closure. A similar unclear situation exists among NO, PA, and ROS during plant–pathogen interaction (Laxalt et al. 2007). It might be interesting to investigate NO, PA, and ROS interactions in guard cells.

It has recently been reported that both NO (Garcia-Mata and Lamattina 2007; Zhang et al. 2007) and PA (Mishra et al. 2006) are involved in the inhibition of light-induced stomatal opening. The exact roles of NO as well as PA during this process are still unknown, hence it will be interesting to see whether the link between NO and PA again exists.

7 Adventitious Root Formation

The plant hormone auxin modulates diverse aspects of root growth and development. The root system plays an essential role in the growth and survival of plants. NO is a second messenger in the auxin signal transduction leading to lateral root formation (Correa-Aragunde et al. 2004), root gravitropism (Hu et al. 2005), root hair development (Lombardo et al. 2006), and adventitious root (AR) formation (Fig. 1). In various species of dicotyledons, ARs occasionally emerge from the hypocotyls or stems, in particular upon wounding (Osmont et al. 2007). In monocotyledons, the root system is characterized by the formation of many ARs (Osmont et al. 2007).

Auxin induces a transient increase in the level of NO in the tip of cucumber (*Cucumis sativus*) hypocotyls, where ARs are formed (Fig. 1). It was proposed that auxin-induced NO triggers cGMP production by regulating the activity of the enzyme guanylate cyclase (Pagnussat et al. 2003). NO also activates a MAPK cascade in a cGMP-independent pathway (Pagnussat et al. 2004). Another pathway that is required is via the increase in cytosolic Ca^{2+} concentration and CDPK activity (Lanteri et al. 2006). It was pharmacologically demonstrated that this pathway partially depends on IP_3 formation (Fig. 1).

Recent evidence from our lab indicates that auxin-induced NO triggers a rapid and transient accumulation of PA via activation of a PLD in cucumber hypocotyls. AR formation was stimulated by treatment with PA and blocked by 1-butanol (Lanteri et al. 2008). Activation of the PLC/DGK pathway could not be demonstrated (Lanteri et al. 2008). In addition, we found that auxin induces an NO-dependent accumulation of PIP and PIP_2, although we do not know the type of isomers accumulated (Lanteri et al. 2008). The increase in PIP and PIP_2 could be the result of homeostasis restoration after PLC activation. PIP and PIP_2 not only act as substrates for PLC but also induce activities of PLDs (Chung et al. 1997; Pappan et al. 1997; Qin et al. 1997). It has been shown that both PIP and PIP_2 act as second messengers in a wide scope of processes, critical for cell survival, environmental adaptation, and growth (see chapters, "Phosphatidylinositol 4-Phosphate is Required for Tip Growth in Arabidopsis thaliana," "PIP-Kinases as Key Regulators of Plant Function," "Plant Phosphatidylinositol 3-Kinase," "Imaging Lipids in Living Plants"). Changes in IP_3, PIP, and PIP_2 levels have been shown to be regulated upon auxin treatment in a number of biological systems (Ettlinger and Lehle 1988; Zbell and Walter-Back 1989; Zbell et al. 1989; Grabowski et al. 1991).

Genetic evidence in *Arabidopsis* suggested an involvement of the PLDξ class in root architecture and auxin responses. PLDξ1 is likely to participate in both initiation and maintenance of root hair morphogenesis (Ohashi et al. 2003). PLDξ2 has a role in auxin-induced primary root elongation, hypocotyl elongation, and gravitropism (Li and Xue 2007). Inhibition of PLD with 1-butanol in wild-type *Arabidopsis* seedlings showed stronger repression of gravitropism and lateral root formation than is observed in *pldξ2* mutant, indicating that other PLD members may be involved in auxin responses (Li and Xue 2007). It will be interesting to analyze the level of PA upon auxin treatment in *pldξ2* mutant. A subsequent article by Mancuso et al. provides evidence that specifically PLDξ2, and no other PLD, is specialized in the vesicular regulation of polar auxin transport in the distal portion of the transition zone in root apices (Mancuso et al. 2007).

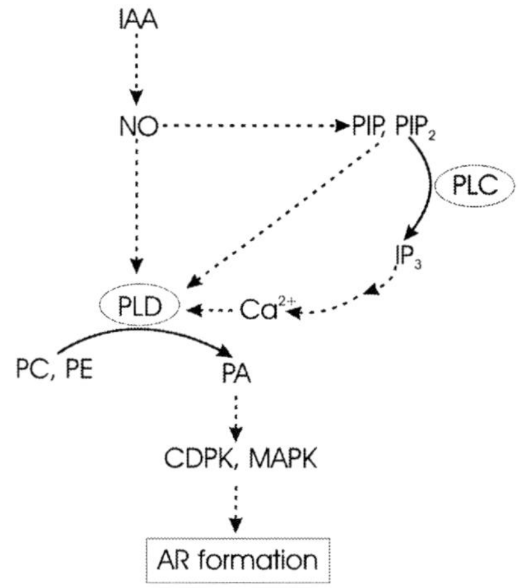

Fig. 4 Model of phospholipid signaling induced by auxin during adventitious root (AR) formation in cucumber explants. NO acts downstream of auxin (IAA) to trigger PA, PIP, and PIP_2 accumulation. PIP and PIP_2 can be hydrolyzed by PLC activity to generate IP_3, which induces the release of Ca^{2+} from internal stores. PLD is involved in PA formation induced by NO. PLD could be activated by (1) elevations in Ca^{2+}, (2) PIP and PIP_2 increases, and/or (3) other NO-mediated pathways. PA on its turn activates CDPK and MAPK signaling pathways leading to AR formation. *Solid arrows* indicate metabolic conversion, *dashed arrows* indicate activation (directly or indirectly)

Figure 4 illustrates a model of the phospholipid signaling pathway induced by auxin during AR formation in cucumber explants (seedlings without primary root system). NO acts downstream of auxin (IAA) to trigger the accumulation of PIP and PIP_2 by an unknown mechanism. PIP and PIP_2 can be hydrolyzed by PLC activity to generate IP_3. IP_3 production results in elevations in cytosolic Ca^{2+} concentration. PLD could be directly activated by NO via nitrosylation or nitration, and/or indirectly activated via elevations in either Ca^{2+} and/or PIP and PIP_2. PA could possibly activate CDPK and MAPK signaling pathways, as was previously shown (Farmer and Choi 1999; Lee et al. 2001).

A number of questions do arise. Firstly, how does NO induce PIP and PIP_2 accumulation? The levels of PIP and PIP_2 are regulated by activity of PLC, lipid kinases, and lipid phosphatases (see chapters, "The Emerging Roles of Phospholipase C in Plant Growth and Development," "Phosphatidylinositol 4-Phosphate is Required for Tip Growth in *Arabidopsis thaliana*," "PIP-Kinases as Key Regulators of Plant Function," "Signaling and the Polyphosphoinositide Phosphatases from Plants," "InsP3 in Plant Cells"). It would be interesting to analyze the activities of these enzymes upon auxin and NO treatments. Secondly, which PIP and PIP_2 isomers accumulate? An increase of phosphatidylinositol 3-phosphate (PI3P) has been reported upon auxin treatment in isolated membranes of *Arabidopsis* roots (Joo et al. 2005), but it should be noted that the identity of the PIP isomer was not checked there either. Future analysis will focus on the identification of the PIP and

PIP_2 isomers involved in auxin and NO responses in roots. Thirdly, which PLD isoform(s) is implicated in auxin- and NO-induced AR formation? A useful strategy to identify the PLD isoform(s) would be a reverse genetic screening of PLD mutants in *Arabidopsis*. PLDξ2 is a good candidate considering the recent findings reported by Li and Xue (2007) and Mancuso et al. (2007). Fourthly, are CDPK and MAPK participating as downstream messengers of PA? As already pointed out above, PA has been shown to induce CDPK and MAPK activities. It still remains to be shown whether this activation occurs during auxin- and NO-induced AR formation in cucumber explants. Further progress in addressing these and other questions will certainly improve our understanding of auxin, NO, and phospholipid signaling in plants.

8 Conclusions and Prospects

In the last years, important efforts were made to address the role of PA-generating enzymes in different physiological processes. At the same time, an effort has been directed at unraveling PA downstream targets in plants. The study of upstream regulation of the PA-generating enzymes is a less explored area. We postulated NO as an upstream regulator of PA signaling in plants and demonstrated such a role in three different physiological processes. Besides our reports, it has been shown that NO-stimulated vacuolar H^+-ATPase is blocked by 1-butanol in maize seedlings treated with NaCl (Zhang et al. 2006). However, activation of PLD upon NO treatments was not shown in this system.

How NO activates phospholipid-related enzymes is still unknown. A number of direct or indirect mechanisms by which NO activates PLC/DGK or PLD can be envisaged. NO could act directly on PLC, DGK, or PLD enzymes by nitrosylation of cysteines or nitration of tyrosines (Broillet 1999; Stamler et al. 2001; Schopfer et al. 2003). So far, there are no reports of in vivo or in vitro nitrosylation or nitration of these enzymes in either animals or plants. Nitrosylation of phospholipases is an interesting target of further study. As we already described in the introduction, the fact that local NO concentration and probably the presence of more nitrosative agents in the membrane makes nitrosylation a plausible regulatory mechanism of phospholipases. Motifs corresponding to NO-mediated [HKR]C [VILMFWC]x[DE] and GSNO-mediated [GSTCYNQ][KRHDE]C[DE] have been suggested (Wang et al. 2006; Broillet 1999). A PHI-Blast (Altschul et al. 1997) was performed using AtPLC1, AtPLDα1, and AtDGK1 sequences as queries and the above described motifs as patterns. No hits for PLC or DGK sequences were found for either motif but many plant PLDs appeared to have an NO-targeted motif. The motif position is conserved in α, β, γ, and δ C2-PLDs, which points to some kind of function, but is not found in C2-PLDε or in any PH/PX-PLD (Fig. 5). Whether or not these motifs correspond to NO-targeted sequences will be determined in future experiments. The fact that no PLC or DGK sequences contain either

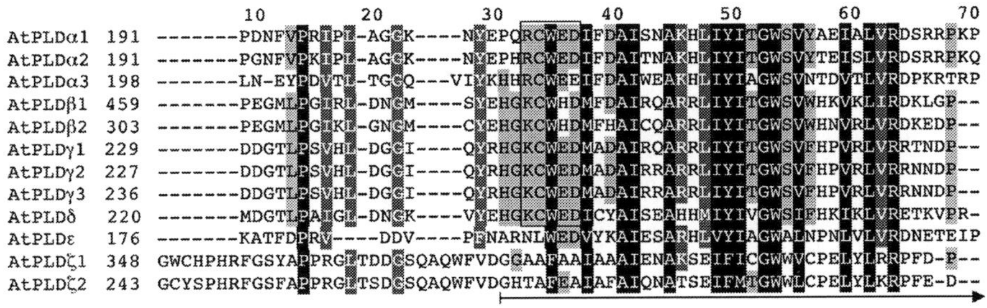

Fig. 5 Partial alignment of *Arabidopsis* PLD protein sequences. Sequences were aligned using T coffee (Notredame et al. 2000). The alignment was subsequently truncated at the positions indicated, in order to demonstrate similarity around the NO-targeted motif [HKR]CW[EH][DE], indicated by the *box*. The sequence corresponding to the first PLD catalytic domain, is *underlined*

of the nitrosylation motifs does not automatically mean that they cannot be subject to regulation by nitrosylation.

NO could also act indirectly on phospholipases and DGKs by regulating Ca^{2+} homeostasis. Biochemical and pharmacological experiments show that NO regulates the levels of cGMP via activation of the guanylate cyclase. On its turn, cGMP regulates cytosolic Ca^{2+} concentration via cADPR. IP_3 is another second messenger controlling the levels of cytosolic Ca^{2+}. Protein kinase inhibitors suppressed NO-evoked elevations of Ca^{2+} in *Vicia faba* guard cells (Sokolovski et al. 2005). In tobacco cells, Lamotte et al. (2006) demonstrated that NtOSAK (a member of the protein kinase SnRK2 family) activation precedes NO-mediated increases in cytosolic Ca^{2+} concentrations. Since PLC, PLD, and DGK are all Ca^{2+}-regulated proteins, it will be interesting to investigate whether these NO-mediated mechanisms affect the activity of the PA-generating enzymes, and/or whether PLC-derived IP_3 affects these NO-mediated mechanisms.

Another plausible indirect activation of phospholipases and DGKs by NO is via protein kinases. NO-dependent activation of MAPKs, CDPKs, and other protein kinases were reported in *Arabidopsis* (Clarke et al. 2000; Capone et al. 2004), cucumber (Pagnussat et al. 2004; Lanteri et al. 2006), tobacco (Klessig et al. 2000), *V. faba* (Sokolovski et al. 2005), and *Nicotiana plumbaginifolia* (Lamotte et al. 2006). Phosphorylation plays an important role in the regulation of animal PLDs. In plants, a PIP_2-dependent form of PLD associated to the membrane has been suggested to be regulated by phosphorylation/dephosphorylation (Novotna et al. 2003). More recently, a phosphoproteomic approach in elicitor-treated *Arabidopsis* cells showed that AtPLC2 becomes phosphorylated at a serine residue (Nuhse et al. 2007). Another scenario is that MAPK and CDPKs activation by NO could take place via generation of PA, and therefore be downstream of a phospholipase activation.

Recently, signaling has been accepted to take place within the confines of subcellular compartments, confines that are critical for specificity of targeting and perception of signals. The enzymes, the second messenger systems, and the effector molecules that propagate the signals are discretely colocalized. Even though NO is regarded as a diffusible signal molecule with a promiscuous sphere of influence, NO signaling is identified with a set of restricted and highly specific effects. NO activates the PLC/DGK pathway in plant defense, PLD during auxin treatments, and both PLC and PLD during stomatal closure. Because of the fact that PA is localized in membranes, it appears that NO/PA signaling is confined to membrane systems. NO activation of different phospholipase and/or kinase isoforms will depend on the type of isoform present at or near the location where NO is generated. Differential expression of large lipase and kinase gene families, both in time and location (e.g., hypocotyls, suspension cells, or guard cells), combined with differential organelle targetting, can explain further the compartmentalization and thereby the specificity of the response. Hence, although NO is readily diffusible, the interaction of NO with phospholipid signaling might generate the signal specificity required.

Acknowledgments This work was financially supported by Universidad Nacional de Mar del Plata, Consejo Nacional de Investigaciones Científicas y Técnicas (CONICET), and Agencia Nacional de Promoción Científica y Tecnológica (ANPCyT).

References

Altschul SF, Madden TL, Schaffer AA, Zhang J, Zhang Z, Miller W, Lipman DJ (1997) Gapped BLAST and PSI-BLAST: a new generation of protein database search programs. Nucleic Acids Res 25:3389–3402

An L, Liu Y, Zhang M, Chen T, Wang X (2005) Effects of nitric oxide on growth of maize seedling leaves in the presence or absence of ultraviolet-B radiation. J Plant Physiol 162:317–326

Andersson MX, Kourtchenko O, Dangl JL, Mackey D, Ellerstrom M (2006) Phospholipase-dependent signaling during the AvrRpm1- and AvrRpt2-induced disease resistance responses in *Arabidopsis thaliana*. Plant J 47:947–959

Bargmann BO, Laxalt AM, Riet BT, Schouten E, van Leeuwen W, Dekker HL, de Koster CG, Haring MA, Munnik T (2006) LePLDbeta1 activation and relocalization in suspension-cultured tomato cells treated with xylanase. Plant J 45:358–368

Baudouin E, Pieuchot L, Engler G, Pauly N, Puppo A (2006) Nitric oxide is formed in Medicago truncatula-Sinorhizobium meliloti functional nodules. Mol Plant Microbe Interact 19:970–975

Beligni MV, Lamattina L (2000) Nitric oxide stimulates seed germination and de-etiolation, and inhibits hypocotyl elongation, three light-inducible responses in plants. Planta 210:215–221

Beligni MV, Fath A, Bethke PC, Lamattina L, Jones R (2002) Nitric oxide acts as an antioxidant and delays programmed cell death in barley aleurone layers. Plant Physiol 129:1642–1650

Besson-Bard A, Pugin A, Wendehenne D (2008) New insights into nitric oxide signaling in plants. Annu Rev Plant Biol 59:21–39

Bright J, Desikan R, Hancock JT, Weir IS, Neill SJ (2006) ABA-induced NO generation and stomatal closure in *Arabidopsis* are dependent on H_2O_2 synthesis. Plant J 45:113–122

Broillet MC (1999) S-nitrosylation of proteins. Cell Mol Life Sci 55:1036–1042

Capone R, Tiwari BS, Levine A (2004) Rapid transmission of oxidative and nitrosative stress signals from roots to shoots in *Arabidopsis*. Plant Physiol Biochem 42:425–428

Chung JK, Sekiya F, Kang HS, Lee C, Han JS, Kim SR, Bae YS, Morris AJ, Rhee SG (1997) Synaptojanin inhibition of phospholipase D activity by hydrolysis of phosphatidylinositol 4,5-bisphosphate. J Biol Chem 272:15980–15985

Clarke A, Desikan R, Hurst RD, Hancock JT, Neill SJ (2000) NO way back: nitric oxide and programmed cell death in *Arabidosis thaliana* suspension cultures. Plant J 24:667–677

Correa-Aragunde N, Graziano M, Lamattina L (2004) Nitric oxide plays a central role in determining lateral root development in tomato. Planta 218:900–905
Davis KL, Martin E, Turko IV, Murad F (2001) Novel effects of nitric oxide. Annu Rev Pharmacol and Toxicol 41:203–236
de Jong CF, Laxalt AM, Bargmann BO, de Wit PJ, Joosten MH, Munnik T (2004) Phosphatidic acid accumulation is an early response in the Cf-4/Avr4 interaction. Plant J 39:1–12
de Pinto MC, Tommasie F, de Gara L (2002) Changes in the antioxidant systems as part of the signaling pathway responsible for the programmed cell death activated by nitric oxide and reactive oxygen species in tobacco bright-yellow 2 cells. Plant Physiol 130:698–708
de Pinto MC, Paradiso A, Leonetti P, De Gara L (2006) Hydrogen peroxide, nitric oxide and cytosolic ascorbate peroxidase at the crossroad between defence and cell death. Plant J 48:784–795
Delledonne M, Xia Y, Dixon RA, Lamb C (1998) Nitric oxide functions as a signal in plant disease resistance. Nature 394:585–588
Delledonne M, Zeier J, Marocco A, Lamb C (2001) Signal interactions between nitric oxide and reactive oxygen intermediates in the plant hypersensitive disease resistance response. Proc Natl Acad Sci USA 98:13454–13459
den Hartog M, Musgrave A, Munnik T (2001) Nod factor-induced phosphatidic acid and diacylglycerol pyrophosphate formation: a role for phospholipase C and D in root hair deformation. Plant J 25:55–65
den Hartog M, Verhoef N, Munnik T (2003) Nod factor and elicitors activate different phospholipid signaling pathways in suspension-cultured alfalfa cells. Plant Physiol 132:311–317
Desikan R, Graffiths R, Hancock J, Neill S (2002) A new role for an old enzyme: Nitrate reductase-mediated nitric oxide generation is required for abscisic acid-induced stomatal closure in *Arabidosis thaliana*. Proc Natl Acad Sci USA 99:16314–16318
Desikan R, Cheung MK, Bright J, Henson D, Hancock JT, Neill SJ (2004) ABA, hydrogen peroxide and nitric oxide signalling in stomatal guard cells. J Exp Bot 55:205–212
Distéfano AM, Garcia-Mata C, Lamattina L, Laxalt AM (2008) Nitric oxide-induced phosphatidic acid accumulation: a role for phospholipases C and D in stomatal closure. Plant Cell Environ 31:187–194
Drobak BK, Watkins PA (2000) Inositol(1,4,5)trisphosphate production in plant cells: an early response to salinity and hyperosmotic stress. FEBS Lett 481:240–244
Durner J, Klessig DF (1999) Nitric oxide as a signal in plants. Curr Opin Plant Biol 2:369–374
Durner J, Wendehenne D, Klessig DF (1998) Defense gene induction in tobacco by nitric oxide, cyclic GMP, and cyclic ADP-ribose. Proc Natl Acad Sci USA 95:10328–10333
Elias M, Potocky M, Cvrckova F, Zarsky V (2002) Molecular diversity of phospholipase D in angiosperms. BMC Genomics 3:2
Ettlinger C, Lehle L (1988) Auxin induces rapid changes in phosphatidylinositol metabolites. Nature 331:176–178
Fan L, Zheng S, Wang X (1997) Antisense suppression of phospholipase D(alpha) retards abscisic acid-and ethylene-promoted senscence of postharvest *Arabidopsis* leaves. Plant Cell 9:2183–2196
Farmer PK, Choi JH (1999) Calcium and phospholipid activation of a recombinant calcium-dependent protein kinase (DcCPK1) from carrot (Daucus carota L.). Biochim Biophys Acta 1434:6–17
Garcia-Mata C, Lamattina L (2001) Nitric oxide induces stomatal closure and enhances the adaptive plant responses against drought stress. Plant Physiol 126:1196–1204
Garcia-Mata C, Lamattina L (2002) Nitric oxide and abscisic acid cross talk in guard cells. Plant Physiol 128:790–792
Garcia-Mata C, Lamattina L (2007) Abscisic acid (ABA) inhibits light-induced stomatal opening through calcium- and nitric oxide-mediated signaling pathways. Nitric Oxide 17:143–151

Garcia-Mata C, Gay R, Sokolovski S, Hills A, Lamattina L, Blatt MR (2003) Nitric oxide regulates K^+ and Cl^- channels in guard cells through a subset of abscisic acid-evoked signaling pathways. Proc Natl Acad Sci USA 100:11116–11121

Gould KS, Lamotte O, Klinguer A, Pugin A, Wendehenne D (2003) Nitric oxide production in tobacco leaf cells: a generalized stress response? Plant Cell Environ 26:1851–1862

Gouvea CMCP, Souza JF, Magalhaes ACN, Martins IS (1997) NO-releasing substances that induce growth elongation in maize root segments. Plant Growth Regul 21:183–187

Grabowski L, Heim S, Wagner KG (1991) Rapid changes in the enzyme activities and metabolites of the phosphatidylinositol-cycle upon induction by growth substances of auxin-starved suspension cultured Catharanthus roseus cells. Plant Sci 75:33

Hari P, Raivonen M, Vesala T, Munger JW, Pilegaard K, Kulmala M (2003) Atmospheric science: ultraviolet light and leaf emission of NO(x). Nature 422:134

He J, Xu H, She X, Song X, Zhao W (2005) The role and the interrelationship of hydrogen peroxide and nitric oxide in the UV-B-induced stomatal closure in broad bean. Funct Plant Biol 32:237–247

Hirayama T, Ohto C, Mizoguchi T, Shinozaki K (1995) A gene encoding a phosphatidylinositol-specific phospholipase C is induced by dehydration and salt stress in *Arabidosis thaliana*. Proc Natl Acad Sci USA 92:3903–3907

Hu X, Neill SJ, Tang Z, Cai W (2005) Nitric oxide mediates gravitropic bending in soybean roots. Plant Physiol 137:663–670

Huang X, Stettmaier K, Michel C, Hutzler P, Mueller M, Jr D (2004) Nitric oxide is induced by wounding and influences jasmonic acid signaling in *Arabidopsis thaliana*. Planta 218:938–946

Huang A-X, She X-P, Huang C, Song T-S (2007) The dynamic distribution of NO and NADPH-diaphorase activity during IBA-induced adventitious root formation. Physiol Plant 130:240

Hunt L, Mills LN, Pical C, Leckie CP, Aitken FL, Kopka J, Mueller-Roeber B, McAinsh MR, Hetherington AM, Gray J (2003) Phospholipase C is required for the control of stomatal aperture by ABA. Plant J 34:47–55

Illes P, Schlicht M, Pavlovkin J, Lichtscheidl I, Baluska F, Ovecka M (2006) Aluminium toxicity in plants: internalization of aluminium into cells of the transition zone in *Arabidopsis* root apices related to changes in plasma membrane potential, endosomal behaviour, and nitric oxide production. J Exp Bot 57:4201–4213

Jacob T, Ritchie S, Assmann SM, Gilroy S (1999) Abscisic acid signal transduction in guard cells is mediated by phospholipase D activity. Proc Natl Acad Sci USA 96:12192–12197

Joo JH, Yoo HJ, Hwang I, Lee JS, Nam KH, Bae YS (2005) Auxin-induced reactive oxygen species production requires the activation of phosphatidylinositol 3-kinase. FEBS Lett 579:1243–1248

Katagiri T, Takahashi S, Shinozaki K (2001) Involvement of a novel *Arabidopsis* phospholipase D, AtPLDdelta, in dehydration-inducible accumulation of phosphatidic acid in stress signaling. Plant J 26:595–605

Katagiri T, Ishiyama K, Kato T, Tabata S, Kobayashi M, Shinozaki K (2005) An important role of phosphatidic acid in ABA signaling during germination in *Arabidopsis thaliana*. Plant J 43:107–117

Klessig DF, Durner J, Noad R, Navarre DA, Wendegenne D, Kumar D, Zhou JM, Shah J, Zhang S, Kachroo P, Trifa Y, Pontier D (2000) Nitric oxide and salicylic acid signaling in plant defense. Proc Natl Acad Sci USA 97:8849–8855

Kohler B, Hills A, Blatt MR (2003) Control of guard cell ion channels by hydrogen peroxide and abscisic acid indicates their action through alternate signaling pathways. Plant Physiol 131:385–388

Kolbert Z, Bartha B, Erdei L (2007) Exogenous auxin-induced NO synthesis is nitrate reductase-associated in *Arabidopsis thaliana* root primordia. J Plant Physiol 165:967–975

Kolla VA, Vavasseur A, Raghavendra AS (2007) Hydrogen peroxide production is an early event during bicarbonate induced stomatal closure in abaxial epidermis of Arabidopsis. Planta 225:1421–1429

Lamattina L, Garcia-Mata C, Graziano M, Pagnussat G (2003) Nitric oxide: the versatility of an extensive signal molecule. Annu Rev Plant Biol 54:109–136

Lamb C, Dixon R (1997) The oxidative burst in plant disease resistance. Annu Rev Plant Physiol Plant Mol Biol 48:251–275

Lamotte O, Gould K, Lecourieux D, Sequeira-Legrand A, Lebrun-Garcia A, Durner J, Pugin A, Wendehenne D (2004) Analysis of nitric oxide signaling functions in tobacco cells challenged by the elicitor cryptogein. Plant Physiol 135:516–529

Lamotte O, Courtois C, Dobrowolska G, Besson A, Pugin A, Wendehenne D (2006) Mechanisms of nitric-oxide-induced increase of free cytosolic Ca2+ concentration in Nicotiana plumbaginifolia cells. Free Radic Biol Med 40:1369–1376

Lanteri ML, Pagnussat GC, Lamattina L (2006) Calcium and calcium-dependent protein kinases are involved in nitric oxide- and auxin-induced adventitious root formation in cucumber. J Exp Bot 57:1341–1351

Lanteri ML, Laxalt AM, Lamattina L (2008) Nitric oxide triggers phosphatidic acid accumulation *via phospholipase D during auxin induced adventitious root formation in cucumber*. Plant Physiol 147:188–198

Laxalt AM, Raho N, ten Have AT, Lamattina L (2007) Nitric oxide is critical for inducing phosphatidic acid accumulation in xylanase-elicited tomato cells. J Biol Chem 282:21160–21168

Lee Y, Choi YB, Suh S, Lee J, Assmann SM, Joe CO, Kelleher JF, Crain R (1996) Abscisic acid-induced phosphoinositide turnover in guard cell protoplasts of Vicia faba. Plant Physiol 110:987–996

Lee S, Suh S, Kim S, Crain RC, Kwak JM, Nam H-G, Lee Y (1997) Systemic elevation of phosphatidic acid and lysophospholipid levels in wounded plants. Plant J 12:547–556

Lee S, Hirt H, Lee Y (2001) Phosphatidic acid activates a wound-activated MAPK in Glycine max. Plant J 26:479–486

Leshem YY, Haramaty E (1995) Plant ageing: the emission of NO and ethylene and effect of NO-releasing compounds on growth of pea (Pisum sativum) foliage. J Endoth Cell Res 3:66

Leshem YY, Pinchasov Y (2000) Non-invasive photoacoustic spectroscopic determination of relative endogenous nitric oxide and ethylene content stoichiometry during the ripening of strawberries Fragaria anannasa (Duch.) and avocados Persea americana (Mill.). J Exp Bot 51:1471–1473

Li L, Fleming N (1999) Aluminum fluoride inhibition of cabbage phospholipase D by a phosphate-mimicking mechanism. FEBS Lett 461:1–5

Li G, Xue H-W (2007) Arabidosis PLD{zeta}2 regulates vesicle trafficking and is required for auxin response. Plant Cell 19:281

Lindermayr C, Saalbach G, Durner J (2005) Proteomic identification of S-nitrosylated proteins in *Arabidosis*. Plant Physiol 137:921–930

Lombardo M, Graziano M, Polacco J, Lamattina L (2006) Nitric oxide functions as a positive regulator of root hair development. Plant Signal Behav 1:28–33

Lum HK, Butt YKC, Lo SCL (2002) Hydrogen peroxide induces a rapid production of nitric oxide in mung bean (Phaseolus aureus). Nitric oxide 6:205–213

Mackerness SAH, John CF, Jordan B, Thomas B (2001) Early signaling components in ultraviolet-B responses: distinct roles for different reactive oxygen species and nitric oxide. FEBS Lett 489:237–242

MacRobbie E (2006) Control of volume and turgor in stomatal guard cells. J Membr Biol 210:131

Mao LC, Huber DJ (2004) Induction of water-soaking and phospholipid catabolism in ripe watermelon fruit by ethylene. Biol Technol 30:284–290

Mancuso S, Marras AM, Mugnai S, Schlich M, Zársky V, Li G, Song L, Xue H, Baluska B (2007) Phospholipase Dζ2 drives vesicular secretion of auxin for its polar cell-cell transport in the transition zone of the root apex. Plant Signal Behav 2:240–244

Mishra G, Zhang W, Deng F, Zhao J, Wang X (2006) A bifurcating pathway directs abscisic acid effects on stomatal closure and opening in *Arabidosis*. Science 312:264–266

Moller M, Botti H, Batthyany C, Rubbo H, Radi R, Denicola A (2005) Direct measurement of nitric oxide and oxygen partitioning into liposomes and low density lipoprotein. J Biol Chem 280:8850

Mueller-Roeber B, Pical C (2002) Inositol phospholipid metabolism in *Arabidosis*. Characterized and putative isoforms of inositol phospholipid kinase and phosphoinositide-specific phospholipase C. Plant Physiol 130:22–46

Munnik T (2001) Phosphatidic acid: an emerging plant lipid second messenger. Trends Plant Sci 6:227–233

Munnik T, Meijer HJ, Ter Riet B, Hirt H, Frank W, Bartels D, Musgrave A (2000) Hyperosmotic stress stimulates phospholipase D activity and elevates the levels of phosphatidic acid and diacylglycerol pyrophosphate. Plant J 22:147–154

Neill SJ, Desikan R, Clarke A, Hancock JT (2002) Nitric oxide is a novel component of abscisic acid signaling in stomatal guard cells. Plant Physiol 128:13–16

Neill SJ, Desikan R, Hancock JT (2003) Nitric oxide signaling in plants. New Phytol 159:11–35

Notredame C, Higgins DG, Heringa J (2000) T-Coffee: a novel method for fast and accurate multiple sequence alignment. J Mol Biol 302:205–217

Novotna Z, Linek J, Hynek R, Martinec J, Potocky M, Valentova O (2003) Plant PIP2-dependent phospholipase D activity is regulated by phosphorylation. FEBS Lett 554:50

Nuhse TS, Bottrill AR, Jones AME, Peck SC (2007) Quantitative phosphoproteomic analysis of plasma membrane proteins reveals regulatory mechanisms of plant innate immune responses. Plant J 51:931

Ohashi Y, Oka A, Rodrigues-Pousada R, Possenti M, Ruberti I, Morelli G, Aoyama T (2003) Modulation of phospholipid signaling by GLABRA2 in root-hair pattern formation. Science 300:1427–1430

Osmont KS, Sibout R, Hardtke CS (2007) Hidden branches: developments in root system architecture. Annu Rev Plant Biol 58:93–113

Pagnussat GC, Simontacchi M, Puntarulo S, Lamattina L (2002) Nitric oxide is required for root organogenesis. Plant Physiol 129:954–956

Pagnussat GC, Lanteri ML, Lamattina L (2003) Nitric oxide and cyclic GMP are messengers in the indole acetic acid-induced adventitious rooting process. Plant Physiol 132:1241–1248

Pagnussat GC, Lanteri ML, Lombardo MC, Lamattina L (2004) Nitric oxide mediates the indole acetic acid induction activation of a mitogen-activated protein kinase cascade involved in adventitious root development. Plant Physiol 135:279–286

Pappan K, Qin W, Dyer JH, Zheng L, Wang X (1997) Molecular cloning and functional analysis of polyphosphoinositide-dependent Phospholipase D, PLDb from *Arabidopsis*. J Biol Chem 272:7055–7061

Park J, Gu Y, Lee Y, Yang Z, Lee Y (2004) Phosphatidic acid induces leaf cell death in *Arabidosis* by activating the rho-related small G protein GTPase-mediated pathway of reactive oxygen species generation. Plant Physiol 134:129–136

Parre E, Ghars MA, Leprince AS, Thiery L, Lefebvre D, Bordenave M, Richard L, Mazars C, Abdelly C, Savoure A (2007) Calcium signaling via phospholipase C is essential for proline accumulation upon ionic but not nonionic hyperosmotic stresses in *Arabidosis*. Plant Physiol 144:503–551

Pejchar P, Pleskot R, Schwarzerova K, Martinec J, Valentova O, Novotna Z (2008) Aluminum ions inhibit phospholipase D in a microtubule-dependent manner. Cell Biol Int 32:554–556

Qin C, Wang X (2002) The *Arabidosis* phospholipase D family. Characterization of a calcium-independent and phosphatidylcholine-selective PLD zeta 1 with distinct regulatory domains. Plant Physiol 128:1057–1068

Qin W, Pappan K, Wang X (1997) Molecular heterogenity of Phospholipase D (PLD). Cloning of PLDgamma and regulation of plant PLDgamma, -beta, and -alpha by polyphosphoinositides and calcium. J Biol Chem 272:28267–28273

Ramos-Diaz A, Brito-Argaez L, Munnik T, Hernandez-Sotomayor SM (2007) Aluminum inhibits phosphatidic acid formation by blocking the phospholipase C pathway. Planta 225:393–401

Rubbo H, Radi R, Anselmi D, Kirk M, Barnes S, Butler J, Eiserich JP, Freeman BA (2000) Nitric oxide reaction with lipid peroxyl radicals spares alpha -tocopherol during lipid peroxidation. Greater oxidant protection from the pair nitric oxide/alpha -tocopherol than alpha -tocopherol/ ascorbate. J Biol Chem 275:10812

Sang Y, Cui D, Wang X (2001a) Phospholipase D and phosphatidic acid-mediated generation of superoxide in *Arabidosis*. Plant Physiol 126:1449–1458

Sang Y, Zheng S, Li W, Huang B, Wang X (2001b) Regulation of plant water loss by manipulating the expression of phospholipase Dalpha. Plant J 28:135–144

Sarath G, Bethke PC, Jones R, Baird LM, Hou G, Mitchell RB (2006) Nitric oxide accelerates seed germination in warm-season grasses. Planta 223:1154–1164

Schopfer FJ, Baker PR, Freeman BA (2003) NO-dependent protein nitration: a cell signaling event or an oxidative inflammatory response? Trends Biochem Sci 28:646–654

Sokolovski S, Hills A, Gay R, García-Mata C, Lamattina L, Blatt MR (2005) Protein phosphorylation is a prerequisite for intracelluar Ca^{2+} release and ion channel control by nitric oxide and abscisic acid in guard cells. Plant J 43:520–529

Stamler JS (1994) Redox signaling: nitrosylation and related target interactions of nitric oxide. Cell Dev Biol 78:931–936

Stamler JS, Singel DJ, Loscalzo J (1992) Biochemistry of nitric oxide and its redox-activated forms. Science 258:1898–1902

Stamler JS, Lamas S, Fang FC (2001) Nitrosylation: the prototypic redox-based signaling mechanism. Cell 106:675–683

Suhita D, Raghavendra AS, Kwak JM, Vavasseur A (2004) Cytoplasmic alkalization precedes reactive oxygen species production during methyl jasmonate- and abscisic acid-induced stomatal closure. Plant Physiol 134:1536–1545

Takahashi S, Katagiri T, Hirayama T, Yamaguchi-Shinozaki K, Shinozaki K (2001) Hyperosmotic stress induces a rapid and transient increase in inositol 1,4,5-trisphosphate independent of abscisic acid in Arabidosis cell culture. Plant Cell Physiol 42:214–222

Testerink C, Munnik T (2005) Phosphatidic acid: a multifunctional stress signaling lipid in plants. Trends Plant Sci 10:368–375

Thiery L, Leprince AS, Lefebvre D, Ghars MA, Debarbieux E, Savoure A (2004) Phospholipase D is a negative regulator of proline biosynthesis in *Arabidosis* thaliana. J Biol Chem 279:14812–14818

Tian QY, Sun DH, Zhao MG, Zhang WH (2007) Inhibition of nitric oxide synthase (NOS) underlies aluminum-induced inhibition of root elongation in Hibiscus moscheutos. New Phytol 174:322–331

van der Luit AH, Piatti T, van Doorn A, Musgrave A, Felix G, Boller T, Munnik T (2000) Elicitation of suspension-cultured tomato cells triggers the formation of phosphatidic acid and diacylglycerol pyrophosphate. Plant Physiol 123:1507–1516

Villasuso AL, Molas M, Racagni G, Abdala G, Machado-Domenech E (2003) Gibberellin signal in barley aleurone: early activation of PLC by G protein mediates amylase secretion. Plant Growth Regul 41:197

Wang JW, Wu JU (2005) Nitric oxide is involved in methyl jasmonate-induced defense responses and secondary metabolism activities of taxus cells. Plant Cell Physiol 46:923

Wang YS, Yang ZM (2005) Nitric oxide reduces aluminum toxicity by preventing oxidative stress in the roots of Cassia tora L. Plant Cell Physiol 46:1915–1923

Wang C, Zien CA, Afitlhile M, Welti R, Hildebrand DF, Wang X (2000) Involvement of phospholipase D in wound-induced accumulation of jasmonic acid in *Arabidosis*. Plant Cell 12:2237–2246

Wang Y, Yun BW, Kwon E, Hong JK, Yoon J, Loake GJ (2006) S-nitrosylation: an emerging redox-based post-translational modification in plants. J Exp Bot 57:1777–1784

Wang CR, Yang AF, Yue GD, Gao Q, Yin HY, Zhang JR (2008) Enhanced expression of phospholipase C 1 (ZmPLC1) improves drought tolerance in transgenic maize. Planta 227:1127–1140

Yamaguchi T, Tanabe S, Minami E, Shibuya N (2004) Activation of phospholipase D induced by hdrogen peroxide in suspension-cultured rice cells. Plant Cell Physiol 45:1261–1270

Yamaguchi T, Minami E, Ueki J, Shibuya N (2005) Elicitor-induced activation of phospholipases plays an important role for the induction of defense responses in suspension-cultured rice cells. Plant Cell Physiol 46:579–587

Zbell B, Walter-Back C (1989) Signal transduction of auxin on isolated plant cell membranes: indications for a rapid polyphosphoinositide response stimulated by indoleacetic acid. J Plant Physiol 133:353–360

Zbell BA, Walter-Back C, Bucher H (1989) Evidence of an auxin-mediated phosphoinositide turnover and an inositol (1,4,5)trisphosphate effect on isolated membranes of Daucus carota L. J Cell Biochem 40:331–340

Zhang X, Zhang L, Dong F, Gao J, Galbraith DW, Song C-P (2001) Hydrogen peroxide is involved in abscicic acid-induced stomatal closure in Vicia faba. Plant Physiol 126:1438–1448

Zhang W, Wang C, Qin C, Wood T, Olafsdottir G, Welti R, Wang X (2003) The oleate-stimulated phospholipase D, PLD{delta}, and phosphatidic acid decrease H2O2-induced cell death in *Arabidosis*. Plant Cell 15:2285

Zhang W, Qin C, Zhao J, Wang X (2004) Phospholipase D alpha 1-derived phosphatidic acid interacts with ABI1 phosphatase 2C and regulates abscisic acid signaling. Proc Natl Acad Sci USA 101:9508–9513

Zhang Y, Wang L, Liu Y, Zhang Q, Wei Q, Zhang W (2006) Nitric oxide enhances salt tolerance in maize seedlings through increasing activities of proton-pump and Na^+/H^+ antiport in the tonoplast. Planta 224:545–555

Zhang X, Takemiya A, Kinoshita T, Shimazaki K-I (2007) Nitric oxide inhibits blue light-specific stomatal opening via abscisic acid signaling pathways in Vicia guard cells. Plant Cell Physiol 48:715

Zhao J, Guo Y, Kosaihira A, Sakai K (2004a) Rapid accumulation and metabolism of polyphosphoinositol and its possible role in phytoalexin biosynthesis in yeast elicitor-treated Cupressus lusitanica cell cultures. Planta 219:121–131

Zhao L, Zhang F, Guo J, Yang Y, Li B, Zhang L (2004b) Nitric oxide functions as a signal in salt resistance in the calluses from two ecotypes of reed. Plant Physiol 134:849–857

Zhao DY, Tian QY, Li LH, Zhang WH (2007a) Nitric oxide is involved in nitrate-induced inhibition of root elongation in Zea mays. Ann Bot 100:497–503

Zhao J, Fujita K, Sakai K (2007b) Reactive oxygen species, nitric oxide, and their interactions play different roles in Cupressus lusitanica cell death and phytoalexin biosynthesis. New Phytol 175:215–229

3-Phosphoinositide-Dependent Protein Kinase is a Switchboard from Signaling Lipids to Protein Phosphorylation Cascades

Christine Zalejski and László Bögre

Abstract While signal perception relies on local assembly of receptor complexes from proteins that have slow diffusion kinetics, signal propagation within cells often depend on more freely diffusible second messenger molecules, such as Ca^{2+}, reactive oxygen species, nitric oxide within the cytoplasm, or signaling phospholipids that move only in two dimensions within lipid bilayers. These signaling systems are typically composed of the signal-dependent production of second messenger molecules with spatial and temporal dynamics and the availability of sensors that bind these molecules and decode the information. Lipid binding can activate enzymes or can recruit proteins to membranes via distinct lipid-binding domains, where the local increase in their concentration promotes interactions and downstream signaling. One such characterized downstream signaling component is the 3′-phosphoinositide-dependent kinase-1 (PDK1), that in plants, through its lipid-binding PH domain interacts and is activated by the phospholipids: phosphatidic acid and $PI(4,5)P_2$. PDK1 is a master kinase that supervises a number of downstream protein kinases belonging to the protein kinase A, cGMP-activated kinase, and protein kinase C family (AGC kinases). These kinases typically possess a C-terminal hydrophobic motif that serves as a docking site for PDK1 to enable the phosphorylation of these kinases by PDK1 at their activation loop. These downstream kinases than can regulate a number of cellular processes, such as the localisation of auxin efflux carriers, the PINs by the pinoid kinase (PID), ROS signaling (OX1), regulation of cell death in pathogen response (ADI3), or regulation of growth and protein translation, the ribosomal S6 kinase (S6K).

L. Bögre (✉)
Royal Holloway, University of London, School of Biological Sciences, Egham Hill, Egham TW20 0EX, UK
e-mail: l.bogre@rhul.ac.uk

T. Munnik (ed.), *Lipid Signaling in Plants*, Plant Cell Monographs 16,
DOI 10.1007/978-3-642-03873-0_16,

1 Introduction

Downstream effectors to signaling lipids are little understood in plants, but it appears that many of the paradigms developed in animal systems cannot be directly adapted (Munnik and Testerink 2009). Important targets to signaling lipids are protein kinases that contain various lipid-binding domains and whereby lipid-binding can lead to activation, corecruitment, and specific localisation of signaling components. Futile hunt for plant homologs of animal lipid-regulated protein kinases, such as the animal protein kinase C, led to the conclusion that not only distinct lipid signal generation mechanisms but also distinct downstream effector mechanisms have evolved in plants. One notable exception, however, is the 3-phosphoinositide-dependent protein kinase (PDK1). There are ample differences in the way PDK1 is regulated by phospholipids in yeast, plant, and animal systems. PDK1 have distinct targets with plant-specific biological functions, but the principal mechanism of how PDK1 operates as a master switch from signaling lipids to protein phosphorylation cascades is conserved. Within this chapter, we will start with what we know about PDK regulation in animals and draw parallels with plant PDK1 regulatory mechanisms.

1.1 AGC Kinase Family

The 3-phosphoinositide-dependent protein kinase (PDK1) was originally identified in mammalian cells as an activator of protein kinase B (PKB), an enzyme that mediates many intracellular actions of insulin and growth factors (Alessi 2001; Alessi et al. 1997; Woodgett 2005). Subsequently, it was realized that PDK1 is conserved and also present in higher plants (Deak et al. 1999). On the basis of sequence conservation of the catalytic domain, PDK1 belongs to the AGC kinase family named after the cAMP-dependent protein kinase A, cGMP-dependent protein kinase G, and phospholipid-dependent protein kinase C (Hanks and Hunter 1995). Computational tools have been developed to sensitively detect and analyze amino acid sequence signatures and so to classify protein kinases to specific groups across kingdoms; yeasts, animals, and plants (Martin et al. 2009). In this way, eukaryotic protein kinases are grouped into 12 families, one of which is the AGC kinase family. Systematic search based on sequence conservation has also identified genes coding for protein kinases within the AGC kinase family in *Arabidopsis thaliana* (Bogre et al. 2003; Galvan-Ampudia and Offringa 2007). Sequence similarity analysis based on the full-length sequence of *Arabidopsis* AGC kinases, together with sequence motifs specific to the plant AGC kinases allowed us to establish six plant-specific subgroups, AGCVIIIa, AGCVIIIb, AGCVII (NDR), AGC other (IRE), AGCVI (S6K), and PDK1. Subsequently, the sequence relatedness of members of the AGCVIII group was reevaluated based on sequence similarities only within the kinase domain, conservation in intron positions, and phylogenic conservations across species within the plant kingdom (Galvan-Ampudia and Offringa 2007). On the basis of this analysis, PID, WAG1, and WAG2 together with AGC3-4 (PID2) form a clearly separate group, and PHOT1 and PHOT2 are more related to this PID subgroup than to the AGCVIIIb subgroup (Galvan-Ampudia and Offringa 2007). The sequence conservation well correlates with the related functions of the members within this subgroup, which is a directional growth. The reordering of the grouping of *Arabidopsis* AGC kinases also lead to suggestions to change the numbering system given to members with no attached functions, as indicated in Table 1 (Bogre et al. 2003; Galvan-Ampudia and Offringa 2007).

The *A. thaliana* and rice (*Oryza sativa*) homolog of mammalian PDK1 displays 40% overall identity with human PDK1 (Deak et al. 1999). There are two closely related *PDK1* genes in *Arabidopsis*, *PDK1-1* (At5g04510) and *PDK1-2* (At3g10540), sharing 93% amino acid identities. *Arabidopsis* PDK1 can rescue lethality in *Saccharomyces cerevisiae* caused by disruption of the genes encoding yeast PDK1 homologs (Pkh1 and Pkh2), showing the conserved function of this gene. It was also demonstrated that *Arabidopsis* PDK1 is able to activate human protein kinase B in vivo (Deak et al. 1999).

2 Structural Features of PDK1

PDK1 possesses a kinase domain, an N-terminal catalytic domain with a PDK1-Interacting Fragment (PIF) binding pocket and a C-terminal domain with a T-loop and a Pleckstrin Homology (PH) domain (Fig. 1).

2.1 PIF Pocket

How PDK1 interacts with its substrates, the AGC kinases, is well understood in animals (Biondi 2004). PDK1 contains a specific region for substrate docking, called PDK1-interacting fragment (PIF)-binding pocket (Biondi et al. 2000). This hydrophobic region at the N-terminal part of the kinase domain interacts with the C-terminal hydrophobic motif (HM) present on AGC kinases. It has been shown that the *Arabidopsis* PDK1 can efficiently phosphorylate a peptide substrate only when it contains the PIF close to a PDK1 phosphorylation site, called PIFtide, showing that the docking interaction between PDK1 and its substrate is conserved between animals and plants (Anthony et al. 2004). Correspondingly, the PIF-binding pocket is conserved in the *Arabidopsis* PDK1, and the *Arabidopsis* AGCVIII kinases, excluding WAG and PHOT all have a C-terminal hydrophobic motif (Phe–Xaa–Xaa–Phe), and are therefore potential binding partners and substrates for PDK1 (Bogre et al. 2003). PDK1–AGC kinase interactions have been experimentally

Table 1 The Arabidopsis AGC protein kinases family (AGI number and gene name)

AGC VIII	At5g55910	AGC1-1/PK64/D6PK
	At2g36350	AGC1-9
	At5g03640	AGC1-8
	At3g27580	PK7
	At3g44610	AGC1-12
	At4g26610	AGC1-2
	At5g47750	PK5
	At2g44830	AGC1-3
	At5g40030	AGC1-4
	At3g12690	AGC1-5
	At1g16440	AGC1-6
	At1g79250	AGC1-7
	At3g52890	KIPK
	At2g34650	PID
	At2g26700	AGC3-4/AGC1-10/PID2
	At1g53700	WAG1/PK3
	At3g14370	WAG2/AGC1-11
	At3g45780	PHOT1
	At5g58140	PHOT2
	At3g25250	AGC2-1/OXI1
	At4g13000	AGC2-2
	At1g51170	AGC2-3
	At3g20830	AGC2-4
AGC VII	At4g14350	NDR-1
	At1g03920	NDR-2
	At3g23310	NDR-3
	At2g19400	NDR-4
	At2g20470	NDR-5
	At4g33080	NDR-6
	At1g30640	NDR-7
	At5g09890	NDR-8
AGC VI	At3g08730	S6K-1
	At3g08720	S6K-2
PDK1	At5g04510	PDK1-1
	At3g10540	PDK1-2
AGC Other	At5g62310	IRE-1
	At3g17850	IRE-2
	At1g48490	IRE-3
	At1g45160	IRE-4

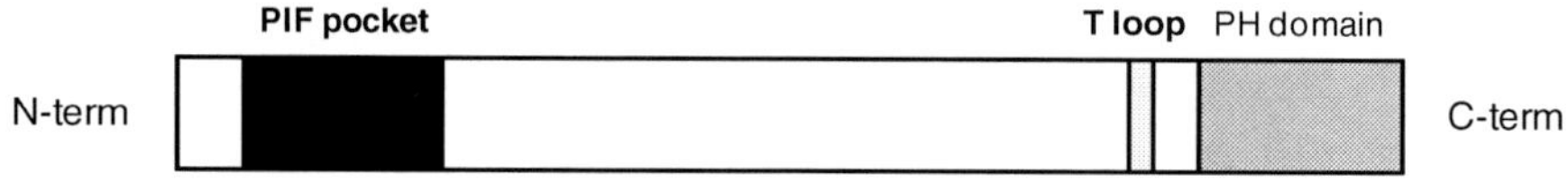

Fig. 1 PDK1 domain structure

demonstrated for a number of plant AGC kinases (Anthony et al. 2004; Zegzouti et al. 2006a, 2006b).

It was shown for the animal PDK1 that the binding of HM to the PIF-binding pocket stabilizes the active conformation of PDK1 and promotes the T-loop phosphorylation of AGC kinases by PDK1 (Biondi et al. 2000). Some AGC kinases, e.g., S6K and SGK, have a hydrophobic motif that contains conserved phosphorylation sites, that has to be prior phosphorylated for effective PDK1 binding (Biondi et al. 2001). S6K and SGK homologs with phosphorylatable HM motifs are also present in plants. Recently, it was shown that TOR kinase in two distinct complexes (TOR complex 1 and TOR complex 2) is responsible for the activation of a number of AGC kinases through hydrophobic domain phosphorylation (Jacinto and Lorberg 2008). TOR can also interact and activate the plant S6K (Mahfouz et al. 2006).

The hydrophobic motif phosphorylation can convert the AGC kinases into substrates that can be activated by PDK1 (Biondi et al. 2001; Silber et al. 2004). Thus, PDK1 has a fascinating mechanism for sensing protein conformation (Biondi 2004). PDK1 detects inactive conformations of AGC substrates because inactive conformations have disrupted PIF pockets and their HMs are available to interact with the PDK1 PIF pocket. Interaction of phosphorylated HMs with PDK1 PIF-binding pocket promotes activation of PDK1, enabling the phosphorylation of substrates at the activation loop, which, in turn, prompts the binding of the phosphorylated HM to their own PIF pockets and helps stabilize the active conformation of PDK1 substrates. PDK1 does not detect active conformations of AGC kinase substrates because their HMs are not available for interaction with PDK1 PIF-binding pocket.

The role of the PIF pocket in the regulation of the intrinsic activity of PDK1 is further supported by the generation of PIF pocket mutants, which can stabilize active conformations of PDK1 (Biondi et al. 2000). It was shown that mutation of the hydrophobic motif on *Arabidopsis* AGC kinases can also disrupt the PDK1–AGC kinase interaction, and block the activation of AGC kinases by PDK1, again showing that this mechanism is conserved (Anthony et al. 2004).

2.2 T-Loop

All AGC kinases are activated by phosphorylation of a threonine and a serine that lies in a region of kinase catalytic domain termed the T-loop. In vivo, the animal PDK1 is capable of autophosphorylation at position Ser-241 by an intermolecular reaction and is, thus, constitutively phosphorylated at Ser-241 (Wick et al. 2003). The structural analysis of the PDK1 kinase domain has revealed that, similar to what has been observed in other kinases, the phosphorylated Ser-241 residue forms the key interactions by coordinating and aligning important catalytic motifs such as the α C-helix of the N-terminal lobe. The α C-helix plays a key role in all kinases as it contributes crucial residues to coordinating ATP. In PDK1, it positions the conserved Glu-130 residue such that it coordinates the conserved Lys-111, which interacts with the α-phosphate of ATP. This hydrogen bonding network is conserved in most protein kinases and is required for phosphoryl transfer (Komander et al. 2005). As opposed to animal PDK1 with constitutive basal activity, the plant

PDK1 can be further activated above the base level by phospholipids, phosphatidic acid, but the mechanism for this is not clear (Anthony et al. 2004).

PDK1 carries out the T-loop phosphorylation of a number of AGC kinases, including the ribosomal S6 kinase, PKB, various isoforms of PKC, and serum and glucocorticoid-inducible kinase in animals (Biondi and Nebreda 2003), and this role of PDK1 to phosphorylate and activate a number of plant AGC kinases is conserved in plants (Anthony et al. 2004; Bogre et al. 2003; Zegzouti et al. 2006a, 2006b). Other AGC kinases, such as PKA, can self-activate by autophosphorylation at the T-loop, and this can also occur for some of the plant AGC kinases, as it was shown for AGC1-1 (Anthony et al. 2004).

2.3 PH Domain

In animals, the PH domain of PDK1 interacts with PI(3,4)P_2 and PI(3,4,5)P_3 and also, but with a lower affinity, with PI3P or PI(4,5)P_2 and this lipid interaction is thought to be important for the recruitment or retention of PDK1 within the plasma membrane (Komander et al. 2004). Animal PDK1 can also interact with Ins (1,3,4,5,6)P5 and InsP6, which are present at much higher micromolar levels in the cytosol, and this could serve to anchor a portion of cellular PDK1 in the cytosol, where it could activate its cytosolic substrates, such as p70 S6-kinase and p90 ribosomal S6 kinase, that do not interact with phosphoinositides (Komander et al. 2004). The PH domain of *Arabidopsis* and rice PDK1 lack, in comparison to yeast and animal PDK1, two conserved amino acid residues that are required for high-affinity binding to PI(3,4,5)P_3 (Deak et al. 1999). Correspondingly, the Arabidopsis PDK1 was shown to have altered lipid binding specificity (Table 2), its interaction with lipids is promiscuous, can interact with PI3P, PA, and also PI(4,5)P_2 (Anthony et al. 2004; Deak et al. 1999). Moreover, PDK1 is the only plant AGC kinase to possess a lipid-binding pleckstrin homology domain (PH). Therefore, it is not clear whether the lipid binding of PDK1 could serve as a mechanism to become cor-ecruited to specific cellular locations with its substrates, the AGC kinases. Interestingly, it was found that several *Arabidopsis* AGC kinases can bind phosphatidic acid through a domain inserted within the T-loop of the kinase part, and this region is important for the peripheral localisation of the AGC kinase, e.g., PID (Zegzouti

Table 2 Lipid binding and activation of PDK1 in *Arabidopsis*

	Interactor of PDK1	Activator of PDK1 activity	References
PA	+	+	Deak et al. (1999); Anthony et al. (2004)
PI3P	+	–	Deak et al. (1999); Anthony et al. (2004)
PI(3,4,5)P3	+	Not tested	Deak et al. (1999)
PI(3,4)P2	+	Not tested	Deak et al. (1999)
PI(4,5)P2	+	+	Deak et al. (1999); Anthony et al. (2004)
PI4P	+	–	Deak et al. (1999)

Table 3 Functions of AGC kinases in plants

Targets	Physiological role	References
ADI3	Cell death	Devarenne et al. (2006)
AGC2-1/OXI1	Root hair growth, oxidative stress	Anthony et al. (2004)
	Pathogen defence	Anthony et al. (2006)
PINOID (PID)	Polar auxin transport	Sauer et al. (2006); Zegzouti et al. (2006); Robert and Offringa (2008)
S6K1 and S6K2	Growth, osmotic stress	Mahfouz et al. (2006); Otterhag et al. (2006)
AGCVIII kinases	?	Zegzouti et al. (2006)
WAG1, WAG2, PID2	Root waving, auxin biosynthesis	Santner and Watson (2006); Cheng et al. (2007;2008)
AGC1-1/PK64/D6PK	Auxin transport	Zourelidou et al. (2009)
PHOT1; PHOT2	Photoperception	Tokutomi et al. (2008)

In bold: AGC kinases where PDK1 interaction has been experimentally demonstrated

et al. 2006b). Alternatively, lipid binding might activate the enzymatic activity of PDK1, as it was shown that exogenously applied PA increased very rapidly (less than 5 min) the activity of *Arabidopsis* PDK1 (Anthony et al. 2004).

Genome-wide analyses of PH domains in yeast suggests that membrane targeting is important for only 30% of yeast PH domains, indicating that the PH fold appears to represent a structural module adaptable to several different binding functions, interacting with phosphoinositides in some cases and with protein targets in others. So far, it is uncertain as to how lipids regulate PDK1 activity and modulate its localisation, activation, stability, or a combination of these (Storz and Toker 2002).

3 PDK1 is a Switchboard to Downstream AGC Kinases Regulating Distinct Cellular Responses

In animal cells, PDK1 is a central integrator for signaling events downstream of receptors that stimulate PI-3 kinase and in the crossroads of the maintenance of balance between cell proliferation, differentiation, and apoptosis. It had been demonstrated that PDK1 able to supervise a plethora of cellular functions through the phosphorylation of different targets belonging to the AGC kinase family of proteins kinases, of which include protein kinase B (PKB) isoforms, serum and glucocorticoid-inducible kinase (SGK), protein kinase C (PKC) isoforms, P90 S6 kinase (S6K), and P70 ribosomal (Alessi 2001).

In plants, PDK1 acts downstream of phospholipids like PA and PI4,5P_2 and also activates AGC kinases involved in diverse processes, such as root hair growth, pathogen response, reactive oxygen species (ROS) signaling, plant cell death, and auxin regulation (Bogre et al. 2003; Galvan-Ampudia and Offringa 2007).

3.1 PDK1 in Oxidative Stress and Pathogen Signaling

Plant defence relies on the detection of pathogen-associated molecular patterns (PAMPs) and pathogenic virulence factors. PAMPs trigger innate immunity, while the recognition of virulence factors often leads to hypersensitive reaction. Lipid signaling, and specifically the generation of PI(4,5)P_2 and PA, is involved in both these pathogen resistance pathways (Laxalt and Munnik 2002). Anthony et al. (2004) demonstrated that PDK1 activity is directly stimulated by PA. PDK1 was linked to pathogen defence mechanisms because it was found to be specifically activated by xylanase (a fungal cell wall degrading enzyme that is recognized as PAMP). Using butanol for blocking PLD pathways and PLC inhibitors, they demonstrated that PDK1 activity increases significantly only when xylanase triggers the PLD pathway and is independent of PA generation via the PLC pathway. It was shown that *AGC2-1/OXIDATIVE SIGNAL-INDUCIBLE1 (OXI1)* can physically interact with PDK1 in yeast two-hybrid assay and in vivo. This interaction was shown to be dependent on the C-terminal PIF motif, and that PDK1 is required for the activation of OXI1/AGC2-1 by direct phosphorylation in the catalytic domain of OXI1 (Anthony et al. 2004). They went on to show that PA activates AGC2-1/OX1 in a PDK1-dependent manner. Cellular localization of GFP-AGC2-1 fusion protein is highly dynamic, confined to cell periphery and to nuclei. Loss-of-function mutants of AGC2-1/OX1 display defects in directional cell growth, such as root hair growth. Both PLDε and PLDζ have been found to regulate directional plant cell growth (Hong et al. 2009; Ohashi et al. 2003), as well as the PIP5K3, which generates PI(4,5)P2 (Kusano et al. 2008), and the downstream effectors of PI3P (Lee et al. 2008). Some of the directional growth process may rely on ROS generation (Takeda et al. 2008), auxin biosynthesis, and transport (Ikeda et al. 2009; Jones et al. 2009). Consistent with the role of ROS in cell growth regulation, AGC2-1/OXI1 expression and activity is strongly induced by ROS (Rentel et al. 2004). Further downstream signaling pathways to AGC2-1/OXI1 was searched by yeast two-hybrid interaction screen, and was found that the Ser/Thr protein kinase, PTI1-2, interacts with AGC2-1/OX1 (Anthony et al. 2006). The tomato homolog is Pti kinase, which has been implicated in the hypersensitive response, a localized programmed cell death that occurs at the site of pathogen infection. Similar to AGC2-1/OXI1, PTI1-2 is activated by the elicitor xylanase but additionally it is also activated by flagellin. With RNA interference knock down of PDK1 and AGC2-1/OXI1, the activation of PTI1-2 by these elicitor signals was strongly compromised, showing that PDK1 and AGC2-1/OXI1 are upstream regulators. PA also converges on PTI1-2 activations via the PDK1-AGC2-1/OXI1 axis. In plants, ROS are thought to be produced in a similar way to mammalian cells, where NADPH oxidase is being recruited to the plasmamembrane during the inflammatory response by signaling pathway involving a PA-dependent protein kinase (Waite et al. 1997). Further emerging signaling pathways downstream of PA, PDK1, and AGC2-1/OXI1 are stress-activated MAPKs. It is known that AGC2-1/OX1 is required for full activation of two MAPKs, MPK3 and MPK6, in response to

H_2O_2 (Rentel et al. 2004). The formation of PA via wounding has also been coupled to the activation of a MAPK cascade; PA was shown to activate MAPK indirectly by an unidentified protein kinase (Lee et al. 2001). It was shown that the activation of MPK6 by xylanase and PA depends on PDK1, as in the case of RNAi knock down of PDK1 where this activation is almost completely abolished (Anthony et al. 2006). Thus, the sequence of events in elicitor-induced defence response includes the activation of an unknown PLD isoform via an elicitor receptor, which generates a rapid accumulation of PA (van der Luit et al. 2000). This, in turn, activates PDK1 and subsequently MPK6 and AGC2-1/OXI1 and PTI1-2 (Fig. 2).

The tomato Adi3, the closest homolog of AGC1-3 in *Arabidopsis*, is involved in pathogen defence (Devarenne et al. 2006). Adi3 interacts with a central hypersensitive response generating signaling component, the Pto kinase, dependent on the presence of the avirulence factor, AvrPto. Pdk1 and Pto simultaneously phosphorylate Adi3, but with different outcomes. ADI3 also interacts with PDK1 via the C-terminal hydrophobic PIF motif. PDK1 phosphorylation enhances Adi3 activity but is not absolutely required for its activation. Use of a chemical inhibitor of PDK1, OSU-03012, implicates PDK1 and ADI3 in suppression of cell death, an

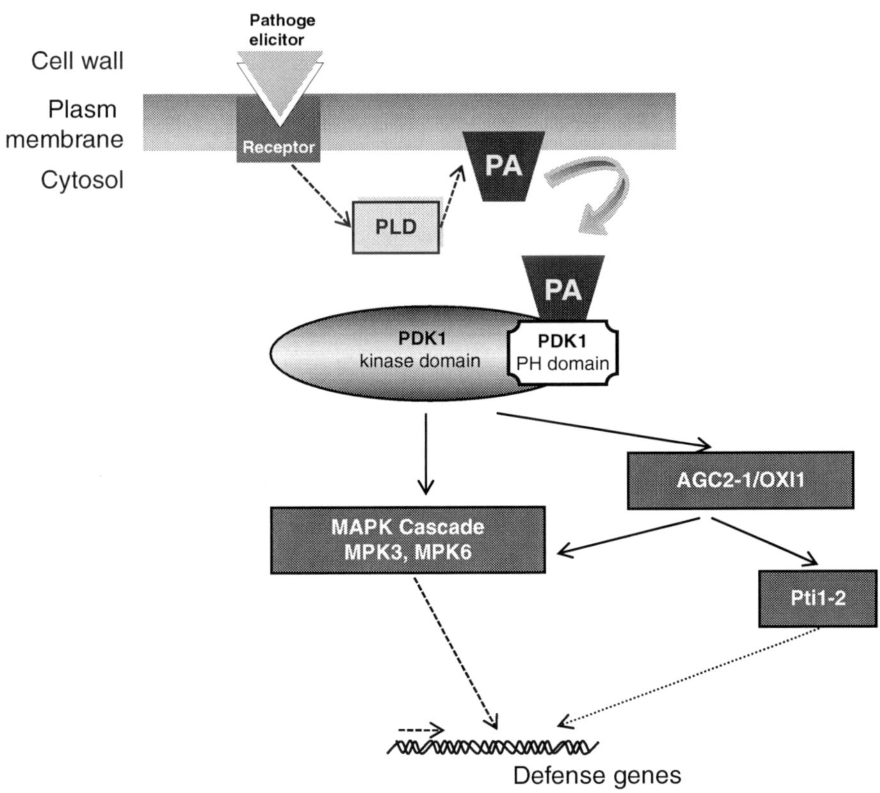

Fig. 2 Pathogen elicitor signaling pathways involving PDK1

analogous function to the mammalian PDK1–protein kinase B module (Manning and Cantley 2007), while the negative regulation of Adi3 by Pto/AVRPto leads to host cell death. Interestingly, the host cell death downstream of Adi3 involves the tomato MAPKKKα and further MAPK signaling components (Devarenne et al. 2006; Pedley and Martin 2004).

3.2 Regulating the Extent of Growth, PDK1–TOR–S6K Constitute a Signaling Module

TOR (target of rapamicin) is a conserved integrator of several growth inputs, including nutrients, energy levels, stress, and growth factors, and TOR channels these signals to downstream effectors to positively or negatively influence cellular growth and proliferation in yeasts, animals, and plants (Diaz-Troya et al. 2008; Ma and Blenis 2009; Wullschleger et al. 2006). The ribosomal protein S6 kinase (S6K) is a major known downstream effector of TOR to regulate growth. The canonical substrate for S6K is the ribosomal protein S6 (RPS6), one of the components of the small 40S ribosome subunit. Surprisingly, the knock out of S6K1 and S6K2 or mutation of S6K phosphorylation target sites on RPS6 resulted in a drastic reduction of size in mice cells, but protein synthesis was not downregulated, indicating that S6K phosphorylation on RPS6 might regulate cell size checkpoint control, whereas protein translation would probably depend on other S6K and TOR targets (Meyuhas 2008; Pende et al. 2004; Ruvinsky et al. 2005). Consistent with the role of TOR–S6K signaling module in cell size regulation, the fission yeast TOR homologs have been shown to connect growth and cell proliferation by restraining the mitotic onset through the regulation of Cdc2 activity (Petersen and Nurse 2007). Activation of TOR signaling pathway by insulin can also delay progression through G2 to M phase in cultured *Drosophila* cells (Wu et al. 2007), while in a genome-wide RNAi survey for genes with function to alter cell cycle and cell size parameters in *Drosophila*, the silencing of S6K1 led to reduced cell size with a concomitant decrease in cells with G2 content (Bettencourt-Dias et al. 2004). In budding yeast, the animal S6K counterpart is Sch9 that involves in regulating cell size as well as in nutrient signaling pathways and ageing (Jorgensen et al. 2004; Steffen et al. 2008; Urban et al. 2007). Under favourable growth conditions, the TOR kinase activates Sch9 leading to the expression of genes related to ribosomal function; under stress, Sch9 regulates the activation of stress-responsive genes (Pascual-Ahuir and Proft 2007; Roosen et al. 2005). Nevertheless, Sch9 regulation of the switch between stress and growth responses depends on TOR activity (Smets et al. 2008).

In *Arabidopsis*, S6 kinases (S6K1 and S6K2) are encoded by two structurally similar genes that map in proximity. The *S6K2* gene, when introduced into human cells, can be activated by insulin, suggesting a conserved regulatory mechanism (Turck et al. 1998). Furthermore, AtS6K2 is dependent on signaling pathways that are sensitive to phosphoinositide-3-kinase inhibitors (Turck et al. 2004). AtS6K1 and AtS6K2 associate with the adaptor protein RAPTOR (Mahfouz et al. 2006) and are activated by PDK1 (Mahfouz et al. 2006; Otterhag et al. 2006). It was shown that S6K2 can phosphorylate the ribosomal protein S6 and regulate selective recruitment of ribosomal mRNAs to polysomes (Turck et al. 2004). S6 phosphorylation is positively regulated by the phytohormones auxin and cytokinin (Turck et al. 2004), while osmotic stress inactivates S6K that is counteracted by RAPTOR1 overexpression (Mahfouz et al. 2006). Correspondingly, heat and oxidative stress were shown to inhibit S6 phosphorylation, but cold stress is an inducer of RPS6 phosphorylation in maize (Williams et al. 2003).

Plant growth is regulated in the meristems by temporal and spatial coordination of cell division and differentiation. AtTOR is expressed in the embryo and endosperm and at or near meristems (Menand et al. 2002). Mutation in TOR leads to embryo arrest during the transition from intensive proliferation with little growth to proliferation coupled with growth, while the endosperm undergoes initial rounds of proliferation, but fails to cellularizse (Menand et al. 2002). S6K1 was also shown to be highly expressed in metabolically active tissues, though in root it is expressed higher in the elongation zone compared with the meristem (Zhang et al. 1994).

3.3 Regulating the Directionality of Growth, All Around Auxin

PINOID (PID) is a pivotal determinant for the asymmetrical localisation of PIN auxin efflux carrier proteins, and thus regulation of polar auxin transport (Friml et al. 2004; Michniewicz et al. 2007). It was shown that *Arabidopsis* PDK1 interacts with PID, and that transphosphorylation by PDK1 increases PID activity and autophosphorylation (Zegzouti et al. 2006a). This activation is rapid and requires the PIF domain and the phosphorylation of a serine residue on the activation loop by PDK1. PID coimmunoprecipitates with PDK1 from *Arabidopsis* cells and PID activity is dependent on PDK1 expression (Zegzouti et al. 2006a). Furthermore, PID-dependent polar targeting of PINs is tightly regulated by combined actions of PDK1, phospholipids (PA), and binding of a calcium-binding protein, TCH3 (Benjamins et al. 2003; Robert and Offringa 2008; Zegzouti et al. 2006b). Correspondingly, auxin transport was shown to be regulated by PLDζ (Li and Xue 2007). The BTP and TAZ domain proteins, NPY have been genetically identified to regulate PID and other PID-related AGC kinases, WAG1, WAG2, and PID2, possibly as a scaffolding molecule (Cheng et al. 2007, 2008; Robert and Offringa 2008; Robert et al. 2008).

Two PID-related AGC kinases, WAG1 and WAG2, lack both the PIF domain and the activation loop $p + 1$ phenylalanine, and are not phosphorylated by PDK1, despite the fact that both proteins can efficiently interact with PDK1 in pull-down assays (Zegzouti et al. 2006b). Seedlings carrying knockout mutations in WAG genes exhibit constitutive root waving in *Arabidopsis* (Santner and Watson 2006). Given the relationship between waving and curling, and the involvement of auxin transport in both processes, they speculate that alterations in auxin transport account for the distinctive phenotype of the *wag* mutants. Interestingly, the PID group of AGC kinases might not only regulate auxin transport, but also auxin biosynthesis through the YUC family of auxin biosynthesis enzymes (Cheng et al. 2008).

Interestingly, it is not only the PID group of AGC kinases that might have a role to regulate auxin transport and PIN localisation. It was shown that mutants in the D6 protein kinase (D6PKs) that was previously designated as AGC1-1 has defects in lateral root initiation, root gravitropism, and axillary shoot development, phenotypes that correlate with a reduction in auxin transport. Interestingly, D6PK localizes to the basal (lower) membrane of *Arabidopsis* root cells, where it colocalizes with PIN1, PIN2, and PIN4. *D6PK* and *PIN1* interact genetically, and D6PK phosphorylates PIN proteins in vitro and in vivo (Zourelidou et al. 2009).

Phototropin (PHOT1 and PHOT2) are blue light-regulated protein kinase with a photoreceptive LOV1 and LOV2 domains at the N-terminal half and AGC kinase at the C-terminal. PHOTs mediate phototropism, chloroplast movement, stomata opening, and leaf expansion (Tokutomi et al. 2008). Docking simulation of the LOV2 domain with PHOT kinase provided useful information regarding the molecular mechanism underlying the photoregulation of PHOT kinase. PHOT does not seem to be regulated by PDK1, but by intramolecular interaction. However, downstream targets of PHOTs might overlap with other AGC kinases regulating directional plant growth, and might also depend on regulation of auxin transport and auxin biosynthesis. It was shown that PIN polarity can be rapidly regulated by light (Laxmi et al. 2008). On the basis of phylogenic sequence analysis, it was proposed that in plants, the AGCVIII kinase gene family has expanded from the ancestral phototropin gene, while the PID family first appeared in land plants, moss, and possibly coevolved with PIN proteins as a module to regulate auxin transport

Table 4 Prediction of cellular localisation of AGC kinases based on GO-term over-representation in genes showing correlated transcriptional regulation

GO-term cell comp	AGI	Gene name
Cell surface	At5g03640	AGC1-8
	At5g62310	IRE-1
Cell wall	At1g16440	AGC1-6
Chloroplast	At2g19400	NDR-4
	At3g45780	PHOT1
Citrate lyase complex	At5g40030	AGC1-4
Cytoskeleton	At1g45160	IRE-4
	At1g51170	AGC2-3
	At1g79250	AGC1-7
	At3g10540	PDK1-2
	At2g44830	AGC1-3
	At5g62310	IRE-1
Membrane	At3g14370	WAG2/AGC1-11
	At3g44610	AGC1-12
	At5g04510	PDK1-1
Nucleus	At2g26700	AGC3-4/AGC1-10
	At2g34650	PID
	At2g36350	AGC1-9
	At3g52890	KIPK
	At5g47750	PK5

and thus to support tissue organization and multicellular development (Galvan-Ampudia and Offringa 2007).

4 Functional Prediction for AGC Kinases Based on Gene Expression Correlations of Microarray Datasets

Many of the members of the AGC kinase gene family have not yet been experimentally studied, and their function is therefore elusive. High throughput data, such as gene expression microarrays, have provided us with opportunities to study functional relationships among genes. The classical approach to gene function prediction using microarray data is to correlate gene expression profiles based on the assumption that genes with similar expression profiles may have similar cellular

Table 5 Prediction of molecular function of AGC kinases based on GO-term over-representation in genes showing correlated transcriptional regulation

GO-term mol func	AGI	Gene name
Adenosylmethionine decarboxylase activity	At3g08730	S6K1
Antioxidant activity	At1g16440	AGC1-6
	At1g53700	WAG1/PK3
	At3g14370	WAG2/AGC1-11
DNA-dependent ATPase activity	At5g04510	PDK1-1
GDP-dissociation inhibitor activity	At3g10540	PDK1-2
	At3g12690	AGC1-5
GTPase regulator activity	At3g10540	PDK1-2
	At3g12690	AGC1-5
Hydrolase activity	At1g30640	NDR-7
Kinase activity	At1g51170	AGC2-3
	At3g23310	NDR-3
	At4g14350	NDR-1
	At4g33080	NDR-6
	At5g09890	NDR-8
	At5g40030	AGC1-4
	At5g58140	PHOT2
Lyase activity	At3g08730	S6K1
	At5g58140	PHOT2
Lysine *N*-methyltransferase activity	At2g44830	AGC1-3
Metal ion binding	At3g08720	S6K2
Microtubule motor activity	At2g44830	AGC1-3
Nucleic acid binding	At2g34650	PID
	At2g36350	AGC1-9
	At3g52890	KIPK
	At5g47750	PK5
Photoreceptor activity	At3g45780	PHOT1
Structural constituent of cell wall	At5g03640	AGC1-8
	At5g62310	IRE-1
Transporter activity	At5g55910	AGC1-1/PK64

localisation and molecular function. This approach is popularly known as Guilt by Association (Quackenbush 2003). We have performed global expression correlation of the AGC kinase genes with each of the represented 21,692 *Arabidopsis* genes on a large set (over 1,000) microarray experiments as we described before (Menges et al. 2008). Overrepresentation of specific gene ontology (GO) categories of cellular localisation and molecular function with correlated gene expression patterns for each AGC kinase gene was analyzed using the BiNGO algorithm (Maere et al. 2005). With these tools, we have uncovered several known molecular function, such as the photoreceptor activity for PHOT1, and have uncovered interesting predictions for other members of the AGC kinase family for cellular localization (Table 4) and molecular function (Table 5). AGC1-1/D6PK expression correlates with genes having transporter activity that might relate to the proposed regulatory function of auxin transport (Zourelidou et al. 2009). PDK1-2 appears to be coregulated with small GTPases and cytoskeletal protein, perhaps reflecting a regulatory role in cell shape and growth control. These types of analyses should help directing future experiments for functional analysis of this important class of protein kinases.

References

Alessi DR (2001) Discovery of PDK1, one of the missing links in insulin signal transduction. Colworth Medal Lecture. Biochem Soc Trans 29:1–14

Alessi DR, James SR, Downes CP, Holmes AB, Gaffney PR, Reese CB, Cohen P (1997) Characterization of a 3-phosphoinositide-dependent protein kinase which phosphorylates and activates protein kinase Balpha. Curr Biol 7:261–269

Anthony RG et al (2004) A protein kinase target of a PDK1 signalling pathway is involved in root hair growth in Arabidopsis. EMBO J 23:572–581

Anthony RG, Khan S, Costa J, Pais MS, Bogre L (2006) The Arabidopsis protein kinase PTI1–2 is activated by convergent phosphatidic acid and oxidative stress signaling pathways downstream of PDK1 and OXI1. J Biol Chem 281:37536–37546

Benjamins R, Ampudia CS, Hooykaas PJ, Offringa R (2003) PINOID-mediated signaling involves calcium-binding proteins. Plant Physiol 132:1623–1630

Bettencourt-Dias M et al (2004) Genome-wide survey of protein kinases required for cell cycle progression. Nature 432:980–987

Biondi RM (2004) Phosphoinositide-dependent protein kinase 1, a sensor of protein conformation. Trends Biochem Sci 29:136–142

Biondi RM, Nebreda AR (2003) Signalling specificity of Ser/Thr protein kinases through docking-site-mediated interactions. Biochem J 372:1–13

Biondi RM, Cheung PC, Casamayor A, Deak M, Currie RA, Alessi DR (2000) Identification of a pocket in the PDK1 kinase domain that interacts with PIF and the C-terminal residues of PKA. EMBO J 19:979–988

Biondi RM, Kieloch A, Currie RA, Deak M, Alessi DR (2001) The PIF-binding pocket in PDK1 is essential for activation of S6K and SGK, but not PKB. EMBO J 20:4380–4390

Bogre L, Okresz L, Henriques R, Anthony RG (2003) Growth signalling pathways in Arabidopsis and the AGC protein kinases. Trends Plant Sci 8:424–431

Cheng Y, Qin G, Dai X, Zhao Y (2007) NPY1, a BTB-NPH3-like protein, plays a critical role in auxin-regulated organogenesis in Arabidopsis. Proc Natl Acad Sci USA 104:18825–18829

Cheng Y, Qin G, Dai X, Zhao Y (2008) NPY genes and AGC kinases define two key steps in auxin-mediated organogenesis in Arabidopsis. Proc Natl Acad Sci USA 105:21017–21022

Deak M, Casamayor A, Currie RA, Downes CP, Alessi DR (1999) Characterisation of a plant 3-phosphoinositide-dependent protein kinase-1 homologue which contains a pleckstrin homology domain. FEBS Lett 451:220–226

Devarenne TP, Ekengren SK, Pedley KF, Martin GB (2006) Adi3 is a Pdk1-interacting AGC kinase that negatively regulates plant cell death. EMBO J 25:255–265

Diaz-Troya S, Perez-Perez ME, Florencio FJ, Crespo JL (2008) The role of TOR in autophagy regulation from yeast to plants and mammals. Autophagy 4:851–865

Friml J et al (2004) A PINOID-dependent binary switch in apical-basal PIN polar targeting directs auxin efflux. Science 306:862–865

Galvan-Ampudia CS, Offringa R (2007) Plant evolution: AGC kinases tell the auxin tale. Trends Plant Sci 12:541–547

Hanks SK, Hunter T (1995) Protein kinases 6. The eukaryotic protein kinase superfamily: kinase (catalytic) domain structure and classification. FASEB J 9:576–596

Hong Y, Devaiah SP, Bahn SC, Thamasandra BN, Li M, Welti R, Wang X (2009) Phospholipase D epsilon and phosphatidic acid enhance Arabidopsis nitrogen signaling and growth. Plant J 58:376–387

Ikeda Y, Men S, Fischer U, Stepanova AN, Alonso JM, Ljung K, Grebe M (2009) Local auxin biosynthesis modulates gradient-directed planar polarity in Arabidopsis. Nat Cell Biol 11: 731–738

Jacinto E, Lorberg A (2008) TOR regulation of AGC kinases in yeast and mammals. Biochem J 410:19–37

Jones AR, Kramer EM, Knox K, Swarup R, Bennett MJ, Lazarus CM, Leyser HM, Grierson CS (2009) Auxin transport through non-hair cells sustains root-hair development. Nat Cell Biol 11:78–84

Jorgensen P, Rupes I, Sharom JR, Schneper L, Broach JR, Tyers M (2004) A dynamic transcriptional network communicates growth potential to ribosome synthesis and critical cell size. Genes Dev 18:2491–2505

Komander D, Fairservice A, Deak M, Kular GS, Prescott AR, Peter Downes C, Safrany ST, Alessi DR, van Aalten DM (2004) Structural insights into the regulation of PDK1 by phosphoinositides and inositol phosphates. EMBO J 23:3918–3928

Komander D, Kular G, Deak M, Alessi DR, van Aalten DM (2005) Role of T-loop phosphorylation in PDK1 activation, stability, and substrate binding. J Biol Chem 280:18797–18802

Kusano H, Testerink C, Vermeer JE, Tsuge T, Shimada H, Oka A, Munnik T, Aoyama T (2008) The Arabidopsis phosphatidylinositol phosphate 5-kinase PIP5K3 is a key regulator of root hair tip growth. Plant Cell 20:367–380

Laxalt AM, Munnik T (2002) Phospholipid signalling in plant defence. Curr Opin Plant Biol 5:332–338

Laxmi A, Pan J, Morsy M, Chen R (2008) Light plays an essential role in intracellular distribution of auxin efflux carrier PIN2 in Arabidopsis thaliana. PLoS One 3:e1510

Lee S, Hirt H, Lee Y (2001) Phosphatidic acid activates a wound-activated MAPK in Glycine max. Plant J 26:479–486

Lee Y, Bak G, Choi Y, Chuang WI, Cho HT (2008) Roles of phosphatidylinositol 3-kinase in root hair growth. Plant Physiol 147:624–635

Li G, Xue HW (2007) Arabidopsis PLDzeta2 regulates vesicle trafficking and is required for auxin response. Plant Cell 19:281–295

Ma XM, Blenis J (2009) Molecular mechanisms of mTOR-mediated translational control. Nat Rev Mol Cell Biol 10:307–318

Maere S, Heymans K, Kuiper M (2005) BiNGO: a Cytoscape plugin to assess overrepresentation of gene ontology categories in biological networks. Bioinformatics 21:3448–3449

Mahfouz MM, Kim S, Delauney AJ, Verma DP (2006) Arabidopsis TARGET OF RAPAMYCIN interacts with RAPTOR, which regulates the activity of S6 kinase in response to osmotic stress signals. Plant Cell 18:477–490

Manning BD, Cantley LC (2007) AKT/PKB signaling: navigating downstream. Cell 129: 1261–1274

Martin DM, Miranda-Saavedra D, Barton GJ (2009) Kinomer v. 1.0: a database of systematically classified eukaryotic protein kinases. Nucleic Acids Res 37:D244–D250

Menand B, Desnos T, Nussaume L, Berger F, Bouchez D, Meyer C, Robaglia C (2002) Expression and disruption of the Arabidopsis TOR (target of rapamycin) gene. Proc Natl Acad Sci USA 99:6422–6427

Menges M, Doczi R, Okresz L, Morandini P, Mizzi L, Soloviev M, Murray JA, Bogre L (2008) Comprehensive gene expression atlas for the Arabidopsis MAP kinase signalling pathways. New Phytol 179:643–662

Meyuhas O (2008) Physiological roles of ribosomal protein S6: one of its kind. Int Rev Cell Mol Biol 268:1–37

Michniewicz M et al (2007) Antagonistic regulation of PIN phosphorylation by PP2A and PINOID directs auxin flux. Cell 130:1044–1056

Munnik T, Testerink C (2009) Plant phospholipid signaling: "in a nutshell". J Lipid Res 50(Suppl): S260–S265

Ohashi Y, Oka A, Rodrigues-Pousada R, Possenti M, Ruberti I, Morelli G, Aoyama T (2003) Modulation of phospholipid signaling by GLABRA2 in root-hair pattern formation. Science 300:1427–1430

Otterhag L, Gustavsson N, Alsterfjord M, Pical C, Lehrach H, Gobom J, Sommarin M (2006) *Arabidopsis* PDK1: identification of sites important for activity and downstream phosphorylation of S6 kinase. Biochimie 88:11–21

Pascual-Ahuir A, Proft M (2007) The Sch9 kinase is a chromatin-associated transcriptional activator of osmostress-responsive genes. EMBO J 26:3098–3108

Pedley KF, Martin GB (2004) Identification of MAPKs and their possible MAPK kinase activators involved in the Pto-mediated defense response of tomato. J Biol Chem 279:49229–49235

Pende M et al (2004) S6K1-/-/S6K2-/- mice exhibit perinatal lethality and rapamycin-sensitive 5′-Terminal Oligopyrimidine mRNA translation and reveal a mitogen-activated protein kinase-dependent S6 kinase pathway. Mol Cell Biol 24:3112–3124

Petersen J, Nurse P (2007) TOR signalling regulates mitotic commitment through the stress MAP kinase pathway and the Polo and Cdc2 kinases. Nat Cell Biol 9:1263–1272

Quackenbush J (2003) Genomics. Microarrays–guilt by association. Science 302:240–241

Rentel MC et al (2004) OXI1 kinase is necessary for oxidative burst-mediated signalling in Arabidopsis. Nature 427:858–861

Robert HS, Offringa R (2008) Regulation of auxin transport polarity by AGC kinases. Curr Opin Plant Biol 11:495–502

Robert HS, Quint A, Brand D, Vivian-Smith A, Offringa R (2009) BTB AND TAZ DOMAIN scaffold proteins perform a crucial function in Arabidopsis development. Plant J 58:109–121

Roosen J et al (2005) PKA and Sch9 control a molecular switch important for the proper adaptation to nutrient availability. Mol Microbiol 55:862–880

Ruvinsky I, Sharon N, Lerer T, Cohen H, Stolovich-Rain M, Nir T, Dor Y, Zisman P, Meyuhas O (2005) Ribosomal protein S6 phosphorylation is a determinant of cell size and glucose homeostasis. Genes Dev 19:2199–2211

Santner AA, Watson JC (2006) The WAG1 and WAG2 protein kinases negatively regulate root waving in Arabidopsis. Plant J 45:752–764

Silber J, Antal TL, Gammeltoft S, Rasmussen TE (2004) Phosphoinositide-dependent kinase-1 orthologues from five eukaryotes are activated by the hydrophobic motif in AGC kinases. Biochem Biophys Res Commun 321:823–827

Smets B, De Snijder P, Engelen K, Joossens E, Ghillebert R, Thevissen K, Marchal K, Winderickx J (2008) Genome-wide expression analysis reveals TORC1-dependent and -independent functions of Sch9. FEMS Yeast Res 8:1276–1288

Steffen KK et al (2008) Yeast life span extension by depletion of 60S Ribosomal subunits is mediated by Gcn4. Cell 133:292–302

Storz P, Toker A (2002) 3′-phosphoinositide-dependent kinase-1 (PDK-1) in PI 3-kinase signaling. Front Biosci 7:d886–d902

Takeda S, Gapper C, Kaya H, Bell E, Kuchitsu K, Dolan L (2008) Local positive feedback regulation determines cell shape in root hair cells. Science 319:1241–1244

Tokutomi S, Matsuoka D, Zikihara K (2008) Molecular structure and regulation of phototropin kinase by blue light. Biochim Biophys Acta 1784:133–142

Turck F, Kozma SC, Thomas G, Nagy F (1998) A heat-sensitive *Arabidopsis thaliana* kinase substitutes for human p70s6k function *in vivo*. Mol Cell Biol 18:2038–2044

Turck F, Zilbermann F, Kozma SC, Thomas G, Nagy F (2004) Phytohormones participate in an S6 Kinase signal transduction pathway in *Arabidopsis*. Plant Physiol 134:1527–1535

Urban J et al (2007) Sch9 is a major target of TORC1 in Saccharomyces cerevisiae. Mol Cell 26:663–674

van der Luit AH, Piatti T, van Doorn A, Musgrave A, Felix G, Boller T, Munnik T (2000) Elicitation of suspension-cultured tomato cells triggers the formation of phosphatidic acid and diacylglycerol pyrophosphate. Plant Physiol 123:1507–1516

Waite KA, Wallin R, Qualliotine-Mann D, McPhail LC (1997) Phosphatidic acid-mediated phosphorylation of the NADPH oxidase component p47-phox. Evidence that phosphatidic acid may activate a novel protein kinase. J Biol Chem 272:15569–15578

Wick MJ, Ramos FJ, Chen H, Quon MJ, Dong LQ, Liu F (2003) Mouse 3-phosphoinositide-dependent protein kinase-1 undergoes dimerization and trans-phosphorylation in the activation loop. J Biol Chem 278:42913–42919

Williams AJ, Werner-Fraczek J, Chang IF, Bailey-Serres J (2003) Regulated phosphorylation of 40S ribosomal protein S6 in root tips of maize. Plant Physiol 132:2086–2097

Woodgett JR (2005) Recent advances in the protein kinase B signaling pathway. Curr Opin Cell Biol 17:150–157

Wu MY, Cully M, Andersen D, Leevers SJ (2007) Insulin delays the progression of Drosophila cells through G2/M by activating the dTOR/dRaptor complex. EMBO J 26:371–379

Wullschleger S, Loewith R, Hall MN (2006) TOR signaling in growth and metabolism. Cell 124:471–484

Zegzouti H, Anthony RG, Jahchan N, Bogre L, Christensen SK (2006a) Phosphorylation and activation of PINOID by the phospholipid signaling kinase 3-phosphoinositide-dependent protein kinase 1 (PDK1) in Arabidopsis. Proc Natl Acad Sci USA 103:6404–6409

Zegzouti H, Li W, Lorenz TC, Xie M, Payne CT, Smith K, Glenny S, Payne GS, Christensen SK (2006b) Structural and functional insights into the regulation of Arabidopsis AGC VIIIa kinases. J Biol Chem 281:35520–35530

Zhang SH, Lawton MA, Hunter T, Lamb CJ (1994) atpk1, a novel ribosomal protein kinase gene from Arabidopsis. I. Isolation, characterization, and expression. J Biol Chem 269: 17586–17592

Zourelidou M, Muller I, Willige BC, Nill C, Jikumaru Y, Li H, Schwechheimer C (2009) The polarly localized D6 PROTEIN KINASE is required for efficient auxin transport in Arabidopsis thaliana. Development 136:627–636

Part VI
Additional Lipid Signals

Diacylglycerol Pyrophosphate, A Novel Plant Signaling Lipid

Emmanuelle Jeannette, Sophie Paradis, and Christine Zalejski

Abstract Diacylglycerol pyrophosphate (DGPP) is a phosphorylated form of phosphatidic acid (PA) found in plants and yeast but not in mammals. DGPP is a minor lipid that accumulates transiently under various abiotic stresses and during biotic interactions. DGPP formation may be a way of attenuating PA content, but DGPP itself might also be a signaling lipid. DGPP is the product of the phosphorylation of PA catalyzed by PA kinase (PAK). Unfortunately, studies describing the role of PAK are limited as gene encoding PAK has not been identified yet. DGPP is dephosphorylated by lipid phosphate phosphatase (LPP) activity to produce PA. LPPs with DGPP phosphatase activity are found in a wide variety of organisms including bacteria, yeast, plants, and mammals. In Arabidopsis, four genes encoding LPPs have been identified, and a role for DGPP in abscisic acid signaling is becoming apparent.

1 Introduction

Understanding signaling pathways that are activated in response to abiotic stresses and biotic interactions is an important challenge in plant biology. Recently, the key role of phospholipids in plant growth, development, and in response to environmental stimuli has been well documented (Meijer and Munnik 2003; Wang 2004; Xue et al. 2007). Phospholipids and phospholipid-derived compounds involved in plant intracellular communication include lysophosphatidic acid (LPA), phosphatidic acid (PA), polyphosphoinositides, inositol hexakisphosphate, and sphingosine-1-phosphate (Lemtiri-Chlieh et al. 2000; Mueller-Roeber and Pical 2002; Lemtiri-Chlieh et al. 2003; Spiegel and Milstien 2003; Ryu 2004; Wang et al.

E. Jeannette (✉)
Physiologie Cellulaire et Moléculaire des Plantes, UR5 Université Pierre et Marie Curie, 3 rue Galilée, F-94200 Ivry-sur-Seine, France
e-mail: emmanuelle.jeannette@upmc.fr

T. Munnik (ed.), *Lipid Signaling in Plants*, Plant Cell Monographs 16,
DOI 10.1007/978-3-642-03873-0_17,

2006 and see chapters, "Phosphatidylinositol 4-Phosphate is Required for Tip Growth in *Arabidopsis thaliana*," "PIP-Kinases as Key Regulators of Plant Function," "Plant Phosphatidylinositol 3-Kinase," "Diacylglycerol Kinase," "Phosphatidic Acid – An Electrostatic/Hydrogen-Bond Switch?," "Nitric Oxide and Phosphatidic Acid Signaling in Plants," "Sphingolipid Signaling in Plants" of this book). Of these, PA is a minor lipid, representing 1–2% of the total phospholipid content of cells that has been shown to mediate many signaling pathways and developmental processes (Wang et al. 2006). Indeed, PA is rapidly generated in response to various stresses including pathogen infection, drought, salinity, wounding, and cold (Testerink and Munnik 2005). Diacylglycerol pyrophosphate (DGPP) is a phosphorylated form of PA found in plants and yeast but not in mammals (Oshiro et al. 2003; van Schooten et al. 2006). The formation of DGPP could be a way of attenuating the PA content of cells, but the molecule itself may also be a second messenger. This chapter focuses on current data on DGPP and the enzymes that produce and metabolize it. A few examples implicating DGPP in plant signaling and hypotheses about its mechanisms of action are also given.

2 Identification

DGPP, previously named pyrophosphatidic acid, is a glycerophospholipid consisting of a pyrophosphate group attached to diacylglycerol (DAG) (Fig. 1). It was first identified by mass spectrometry and ^{31}P-NMR analysis as an in vitro product of a lipid kinase activity in the microsomal fraction of suspension-cultured *Catharanthus roseus* cells that were able to phosphorylate PA, hence called PA kinase (PAK) (Wissing and Behrbohm 1993a). The first evidence of the in vivo occurrence of DGPP was obtained when lipids from the green alga *Chlamydomonas eugametos* were analyzed after labeling with ^{32}P (Munnik et al. 1996). When cells were treated with the G-protein activator mastoparan, the formation of PA plus an unknown phospholipid was induced. By successive deacylation and hydrolysis, this novel phospholipid was shown to be DGPP (Munnik et al. 1996). Meanwhile, the presence of DGPP in various plant species is well established (Table 1). DGPP has been found in leaves, roots, petals, microspores, and pollen tubes of higher plants. Moreover, DGPP has been found in the calli of 16 different plant species (Wissing and Behrbohm 1993b) and in suspension-cultured cells of tomato, potato

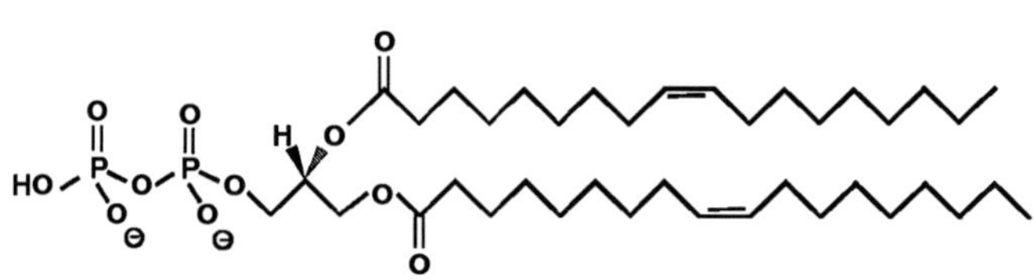

Fig. 1 Diacylglycerol pyrophosphate (DGPP)

Table 1 Identification of DGPP in different species

Plant species	Organ or tissue	References
Adiantum capillaris	Callus	Wissing and Behrbohm (1993b)
Agrostis tenuis	Callus	Wissing and Behrbohm (1993b)
Arabidopsis thaliana	Suspension cells	Pical et al. (1999); Zalejski et al. (2005)
	Seed, seedling	Katagiri et al. (2005)
Atropa belladonna	Callus	Wissing and Behrbohm (1993b)
Beta vulgaris	Callus	Wissing and Behrbohm (1993b)
Brassica napus	Microspore	Munnik et al. (1996)
Catharanthus roseus	Suspension cells, leaf	Wissing and Behrbohm (1993b)
	Flower, stem, root	Wissing and Behrbohm (1993b)
Chenopodium ficifoltum	Leaf	Munnik et al. (1996)
Chenopodium rubrum	Callus	Wissing and Behrbohm (1993b)
Chlamydomonas eugametos		Munnik et al. (1996)
Chlamydomonas moewusii		Munnik et al. (2000); Meijer and Munnik (2003)
Craterostigma plantagineum	Leaf	Munnik et al. (2000)
Datura sanguinaria	Callus	Wissing and Behrbohm (1993b)
Daucus carota	Callus	Wissing and Behrbohm (1993b)
Dianthus caryophyllus	Petal	Munnik et al. (1996)
Ecchinofossulocactus crispatus	Callus	Wissing and Behrbohm (1993b)
Helianthus annuus	Callus	Wissing and Behrbohm (1993b)
Hordeum vulgare	Seed	Villasuso et al. (2003)
Lycopersicon esculentum	Suspension cells, callus	Munnik et al. (2000); van der Luit et al. (2000); Wissing and Behrbohm (1993b)
Malus domestica	Callus	Wissing and Behrbohm (1993b)
Marchantia polymorpha	Callus	Wissing and Behrbohm (1993b)
Medicago sativa	Suspension cells	Munnik et al. (2000); den Hartog et al. (2003)
Nicotiana tabacum	Suspension cells, callus	Wissing and Behrbohm (1993b); den Hartog et al (2001)
	Pollen tube	Zonia and Munnik (2004)
Phaseolus vulgaris	Callus	Wissing and Behrbohm (1993b)
Pinus pinea	Callus	Wissing and Behrbohm (1993b)
Polygonum persicaria	Leaf	Munnik et al. (1996)
Polygonum tinetorum	Callus	Wissing and Behrbohm (1993b)
Solanum tuberosum	Suspension cells	Munnik et al. (1996)
Vicia sativa	Suspension cells	den Hartog et al. (2001)
	Seedling, root	Munnik et al. (1996)
Other organisms		
Phytophtora infestans		Latijnhouvwers et al. (2002)
Saccharomyces cerevisiae		Wu et al. (1996)
Trypanosoma cruzi		Marchesini et al. (1998)

(Munnik et al. 1996), and *Arabidopsis thaliana* (Pical et al. 1999). The subcellular localization of DGPP, however, remains unknown.

DGPP has also been found in the yeast *Saccharomyces cerevisiae* (Wu et al. 1996; Han et al. 2004), in the oomycete *Phytophtora infestans* (Latijnhouwers et al. 2002), and in *Trypanosoma cruzi* (Marchesini et al. 1998; Santander et al. 2002).

Data pertaining to the natural occurrence of DGPP are not comprehensive as DGPP has not been measured in most organisms. However, despite several attempts, DGPP or PAK activity has never been found in mammalian cells (van Schooten et al. 2006). Therefore, DGPP seems to be a lipid specific to plants and lower eukaryotes.

3 DGPP is a Minor Lipid that Accumulates Under Stress Conditions

In plants and yeast, DGPP is a minor lipid that is almost undetectable under standard conditions of culture. For example, in *A. thaliana* suspension cells, DGPP accounts for less than 0.1% of all the phospholipids (Zalejski et al. 2005). Given the small absolute amounts involved, ^{33}P or ^{32}P radioactive labeling followed by one- or two-dimensional thin-layer chromatography (TLC) is a convenient method to quantify DGPP content (for examples see Munnik et al. 1996; Meijer et al. 2001; Zalejski et al. 2005). However, nonradioactive methods based on phosphorus mineralization after TLC separation (Han et al. 2004) or high performance liquid chromatography (Wu et al. 1996; Toke et al. 1998a) have also been used to quantify DGPP. Altogether, these methods have been used to show that DGPP is formed when plants are challenged with abiotic stresses or biotic interactions (Table 2).

Drought, osmotic stress, and salinity trigger the formation of DGPP in various plants (Pical et al. 1999; Munnik et al. 2000). For example, hyperosmotic stress induces an increase in DGPP in *Chlamydomonas moewusii* (Munnik et al. 2000; Meijer et al. 2001), the resurrection plant *Craterostigma plantagineum*, and in tomato, alfalfa, and *A. thaliana* suspension cells (Pical et al. 1999; Munnik et al. 2000).

Table 2 DGPP formation in response to different stresses

Stress	Species	Reference
Hyperosmotic	*Chlamydomonas moewusii*	Munnik et al. (2000); Meijer et al. (2001)
	Craterostigma plantagineum	Munnik et al. (2000)
	Lycopersicon esculentum suspension cells	Munnik et al. (2000)
	Medicago sativa suspension cells	Munnik et al. (2000)
	Arabidopsis thaliana suspension cells	Pical et al. (1999)
Elicitors	*Lycopersicon esculentum* suspension cells	van der Luit et al. (2000)
	Nicotiana tabacum suspension cells expressing the *Cladosporium fulvum* Cf-4^{+} resistance gene	de Jong et al. (2004)
Nod factors	*Vicia sativa* seedlings	den Hartog et al. (2001)
	Medicago sativa suspension cells	den Hartog et al. (2003)
Drought	*Craterostigma plantagineum*	Munnik et al. (2000)
ABA	*Arabidopsis thaliana* suspension cells	Zalejski et al. (2005)
	Arabidopsis thaliana seeds	Katagiri et al. (2005)
	Hordeum vulgare aleurone cells	Villasuso et al. (2003)

Moreover, application of ABA, which mediates many stress responses and is involved in several processes of development (Finkelstein and Rock 2002), leads to an increase of DGPP in cell suspensions, aleurone layer cells, and seeds (Villasuso et al. 2003; Katagiri et al. 2005; Zalejski et al. 2005). Addition of plant elicitors (van der Luit et al. 2000; de Jong et al. 2004) or nodulation factors (den Hartog et al. 2001, 2003) causes an increase in DGPP demonstrating that DGPP participates in the cell's responses to pathogenic and symbiotic interactions. The G-protein activator mastoparan induces DGPP formation in *Chlamydomonas* (Munnik et al. 1996) and *Vicia sativa* (den Hartog et al. 2001). As heterotrimeric G-proteins are components of many abiotic and biotic signaling pathways, these data might suggest that DGPP accumulation is a factor in a wide variety of cellular responses (Assmann 2002). Overall, the variety of situations in which DGPP levels change suggests that this lipid could play an important role in the capacity of plant cells to respond to environmental changes. Interestingly, DGPP could also play a role in development, especially in seed germination. In *S. cerevisiae*, the amounts of DGPP and PA decrease to undetectable levels in response to zinc depletion (Han et al. 2004; Carman and Han 2007). PA and DGPP accumulate in ABA-treated seeds (Katagiri et al. 2005) and application of both GA and ABA modulate the DGPP content of isolated barley aleurone layers (Villasuso et al. 2003). In addition, it has been reported that DGPP is formed in the oomycete *P. infestans* when it undergoes differentiation (Latijnhouwers et al. 2002).

Whatever initiates an increase in the amount of DGPP, its production is transient and is always associated with an increase in the amount of PA. DGPP formation would thus seems to be a general feature of PA signaling in plants. However, it is often difficult to quantitatively correlate changes in DGPP levels with changes in PA levels. Indeed, the subcellular origin of the two pools of lipids is often unknown and the methods used to determine them may vary, even within the same study. As a consequence, an overview of the relative changes in PA and DGPP over time can be obtained, but it is not always possible to accurately evaluate what proportion of PA is converted to DGPP and vice versa.

Depending on the stimulus, PA can be the product of PLD activity (Hallouin et al. 2002), or the combined action of PLC and diacylglycerol kinase (DGK) activity (Pical et al. 1999; van der Luit et al. 2000; de Jong et al. 2004) or from both routes (van der Luit et al. 2000; den Hartog et al. 2001); see also chapters, "The Emerging Roles of Phospholipase C in Plant Growth and Development," "Plant Phospholipase D," "Diacylglycerol Kinase," "Phosphatidic Acid – An Electrostatic/Hydrogen-Bond Switch?," "Nitric Oxide and Phosphatidic Acid Signaling in Plants"). Although LPA acylation also produces PA (Ohlrogge and Browse 1995), this reaction is not likely to involve the plant's PA response to environmental stimuli, as there is, to our knowledge, no evidence for this.

The metabolism of DGPP is due to the combined action of PAK and LPP with DGPP phosphatase activity. Therefore, the increase in the level of DGPP might result from the stimulation of PAK activity and/or the inhibition of DGPP phosphatase activity. So understanding the role of these enzymes is crucial in understanding how DGPP is formed.

4 DGPP Metabolism

The understanding of DGPP metabolism in plants is still fragmentary (van Schooten et al. 2006). Nonetheless, detailed studies on DGPP phosphatases in yeast by Dr. G. Carman's laboratory (Rutger's NJ, USA) have provided valuable data to serve as a basis for further analysis (Oshiro et al. 2003).

4.1 PA Kinase

DGPP is the product of the phosphorylation of PA in a reaction catalyzed by PAK (Wissing and Behrbohm 1993b). As mentioned above, the PA substrate may originate from the combined action of PLC and DGK activities or from PLD, depending on the physiological situation studied. PAK activity was first discovered in vitro, in a microsomal membrane fraction isolated from suspension-cultured *C. roseus* cells in Dr. J. Wissing's lab. Most subsequent PAK activity data came from his studies (Wissing and Behrbohm 1993b; Wissing et al. 1994). PAK was especially enriched in the plasma membrane, similar to DGK activity and demonstrated to be an integral membrane protein (Wissing and Behrbohm 1993b). Interestingly, PAK activity has been measured in detergent insoluble membrane isolated from the plasma membrane of tobacco BY-2 cells and *Medicago truncatula* roots (F. Furt and S. Mongrand, personal communication).

When PAK was partially purified from *C. roseus*, a 39 kDa protein band with a specific activity of about 10 μmol min^{-1} mg^{-1} was obtained in SDS-PAGE (Wissing and Behrbohm 1993b). In vitro, PA has been shown to be the only lipid substrate of PAK despite of its molecular species. In contrast, the phosphorylation of PA can be achieved with either ATP or GTP. The optimum pH of the enzyme activity was 6.1 and the isoelectric point of the protein is 4.8 (Wissing et al. 1994). PAK activity requires Mg^{2+}, whereas addition of Ca^{2+} or Mn^{2+} inhibits the activity. In Arabidopsis suspension cells, in vivo DGPP formation was shown to require calcium (Zalejski et al. 2006). However, it remains unclear whether calcium is necessary for PAK activity or for the upstream PLD activity, generating PA. PAK activity was found in all *C. roseus* organs (Wissing and Behrbohm 1993b), but always at a lower level than in cell cultures. Data on PAK in *Trypanosoma* have confirmed the results obtained in *C. roseus* with respect to the enzyme kinetics (Marchesini et al. 1998). PAK can also catalyze the reverse reaction and produce PA and ATP from DGPP and ADP (Wissing et al. 1994).

Unfortunately, no genetic tools are currently available for studying the role of PAK because the gene encoding PAK has not been identified in either plants or yeast. In *S. cerevisiae*, the level of PAK activity is very low (Wu et al. 1996) and the protein has not been identified directly.

4.2 DGPP Phosphatase Activity

DGPP degradation occurs in two dephosphorylation steps catalyzed by lipid phosphate phosphatases (LPPs). LPPs first remove the β-phosphate from DGPP to form PA and inorganic phosphate, and PA is then dephosphorylated to produce DAG and inorganic phosphate (Wu et al. 1996). LPPs with DGPP phosphatase activity belong to the larger LPP family but have some specific features (see also chapter, "Phosphatidic Acid Phosphatases in Seed Plants"). This phosphatidate phosphatase activity can be distinguished from the conventional PA phosphatase activity by its enzymatic characteristics (Oshiro et al. 2003). The activity of conventional PA phosphatases is inhibited by *N*-ethylmaleimide (NEM) and, depending on the PA phosphatase in question, is sensitive to inhibition by Mg^{2+} (Carman 1997; Carman and Han 2006). In contrast, the PA phosphatase activity of LPPs that are also able to dephosphorylate DGPP is Mg^{2+} independent and insensitive to NEM (Wu et al. 1996; Toke et al. 1998a, b). These specialized LPPs have been cloned from yeast, bacteria, plants, and mammals (Table 3). Therefore, the presence of DGPP phosphatases in a wide range of organisms would indicate that the function of these enzymes is important in cell biology.

In *Saccharomyces*, *DPP1* and *LPP1* encode proteins with DGPP phosphatase activity (Toke et al. 1998a, b). DPP1 is more abundant than LPP1 and is responsible for almost all the DGPP phosphatase activity measured (Oshiro et al. 2003). *DPP1* encodes a protein of 34 kDa (289 aa) and is predicted to have six transmembrane-spanning regions (Toke et al. 1998a, b). The phosphatase sequence is a novel one and has motifs in three domains characteristic of the family of LPPs having DGPP phosphatase activity, namely: KXXXXXXRP (domain 1), PSGH (domain 2) and SRXXXXXHXXXD (domain 3). DPP1 is located in the tonoplast with its catalytic site facing the cytosol (Han et al. 2004). The expression of *DPP1* is induced during the stationary phase of growth and when cells are supplemented with inositol or when zinc is limiting (Oshiro et al. 2000; Han et al. 2001; Carman and Han 2007). In *E. coli*, *pgpB* encodes a 28 kDa DGPP phosphatase (Dillon et al. 1996). In mammals, *PAP2* has been identified in rat liver (Dillon et al. 1997) and two additional genes encoding LPPs with DGPP phosphatase activity were recently cloned, DGPP Phosphatase Like (*DPPL*) 1 and *DPPL2* (Takeuchi et al. 2007). Both have a higher affinity for DGPP than for LPA or PA, and DPPL2 is sensitive to NEM (Takeuchi et al. 2007).

In Arabidopsis, four genes encoding LPPs with DGPP phosphatase activity were identified based on their homology to the yeast *DPP1* gene (Pierrugues et al. 2001; Katagiri et al. 2005). The expression of *AtLPP1* is induced by UV radiation, mastoparan, and the plant elicitor harpin (Pierrugues et al. 2001). In vitro LPP1 has a higher affinity for DGPP than for PA and LPP2 has the same affinity for both PA and DGPP. A proteomic study has shown that LPP3 is present in the plasma membrane of Arabidopsis suspension cells (Marmagne et al. 2004). In an experiment where *GFP* was fused to *LPP2* cDNA, the expressed fusion protein was observed at the periphery of seed cells (Katagiri et al. 2005).

Table 3 LPPs with DGPP phosphatase activity. Sphingosine-1 phosphate (S1P), ceramide-1 phosphate (C1P), plasma membrane (PM)

	GGene	Protein MW (kDa)	In vitro substrate (Regulation of activity)	Regulation of expression	Subcellular location	Tissue/organ location
S. cerevisiae Oshiro et al. (2003); Toke et al. (1998a; 1998b)	*DPP1*	34	DGPP > PA > LPA (Mn^{2+}, Zn^{2+}, CDP-DAG)	Growth phase, inositol, zinc	Tonoplast	
	LPP1		PA > DGPP > LPA (Mn^{2+}, Zn^{2+}, Ca^{2+})			
E. coli						
Dillon et al. (1996)	*pgpB*	28				
Mammals						
Dillon et al. (1997)	*PAP2*		DGPP > PA (LPA, S1P, C1P)			
Takeuchi et al. (2007)	*DPPL1*	±29	DGPP > LPA, PA (NEM-sensitive)			All tissues
	DPPL2	±31				Endothelial cells
Arabidopsis thaliana						
	AtLPP1		DGPP > PA	UV, harpin, Mastoparan		Leaf, root
Pierrugues et al. (2001); Katagiri et al. (2005)	*AtLPP2*		DGPP = PA		PM	All tissues
Marmagne et al. (2004)	*AtLPP3*				PM	
	AtLPP4					

5 DGPP and Plant Signaling

DGPP formation is related to signaling pathways because (1) DGPP and PA levels increase transiently under stress conditions, (2) DGPP formation attenuates PA content, a factor in many signaling pathways and developmental processes (Wang et al. 2006), and (3) the expression of *LPP* genes is induced under stress conditions. While DGPP formation could be simply a means of inactivating PA, conversely DGPP could serve as a precursor of a specific pool of PA. However, a key question is whether DGPP could be a signaling lipid in its own right. With regards to this, studies of ABA signal transduction strongly suggest that DGPP, like PA, is a lipid messenger in cell signaling.

5.1 The Example of Abscisic Acid Signaling

ABA is a phytohormone that mediates a variety of stress responses and several aspects of plant development (Finkelstein and Rock 2002). The importance of lipid signaling in ABA responses is well established especially in the regulation of stomatal movements (Mishra et al. 2006). Several lipids and their related enzymes have been shown to participate in guard cell physiology (Jacob et al. 1999; Staxen et al. 1999; Lemtiri-Chlieh et al. 2000; Jung et al. 2002; Coursol et al. 2003; Hunt et al. 2003; Lemtiri-Chlieh et al. 2003). Of these lipids, PA is known to play a crucial role (Mishra et al. 2006) since a knockout mutant of PLDα1 is affected in stomatal closure and the protein phosphatase ABI1 has been identified as a PA target (Zhang et al. 2004). In our laboratory, we studied the ABA transduction pathway that regulates gene expression in *A. thaliana* suspension cells. We demonstrated that part of the signaling cascade is located at the plasma membrane and that DGPP is likely to be an important signaling partner (Jeannette et al. 1999; Ghelis et al. 2000; Hallouin et al. 2002; Zalejski et al. 2005, 2006). Application of ABA stimulates a transient increase in PA, followed by an increase in DGPP. Interestingly, application of dioleoyl-DGPP, but not dioleoyl-PA, mimicked ABA in stimulating anion currents and changes in gene expression (Zalejski et al. 2005, 2006). Moreover, we showed that DGPP formation was calcium-dependent (Zalejski et al. 2006). Transcriptomic analysis has confirmed the effectiveness of DGPP in altering gene expression and has strengthened the hypothesis that DGPP is a second messenger in ABA and other signaling pathways, as the expression of ABA-independent genes was triggered by ABA (S. Paradis and E. Jeannette, unpublished data).

In *A. thaliana* suspension cells, ABA is perceived by a plasma membrane receptor (Jeannette et al. 1999) and calcium influx occurs which probably stimulates PLD activity (Ghelis et al. 2000; Hallouin et al. 2002). The hydrolysis of phosphatidyl choline (PC) produces PA. The level of DGPP increases in turn due to the stimulation of PAK activity and/or the inhibition of LPP activity (Zalejski et al.

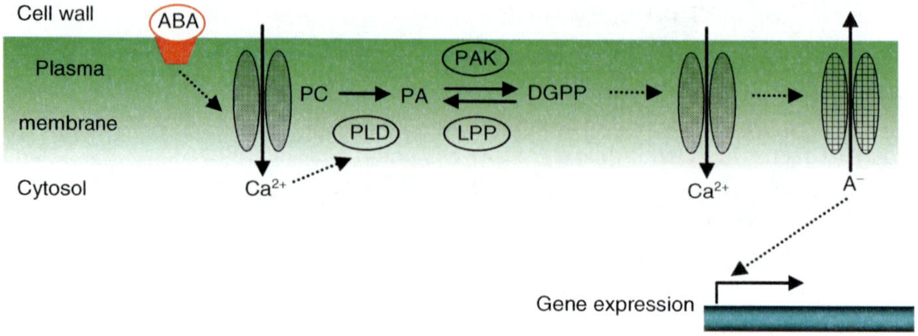

Fig. 2 DGPP mediates ABA in *Arabidopsis thaliana* suspension cells

2005). Finally, DGPP stimulates the activity of calcium-dependent anion (A^-) channels (Hallouin et al. 2002; Zalejski et al. 2005). All these events, from perception of ABA to anion efflux, are part of a signaling pathway leading to the regulation of gene expression (Zalejski et al. 2006) (Fig. 2).

5.1.1 Possible Mechanisms of DGPP Action

Proposing a role for DGPP in cell signaling can only be tentative in the absence of any identification of DGPP targets. However, hypotheses based on the effect of the high negative charge of DGPP might be considered with more certainty. Depending on the pH, the pyrophosphate moiety of DGPP has two or three negative charges making it a highly polar molecule, so DGPP may be able to interact with positively charged compounds. Although not demonstrated in vivo, it has been suggested, again tentatively, that zinc may be chelated by DGPP (Han et al. 2001). It has also been proposed that DGPP can bind polyamines in *Trypanosoma* (Marchesini et al. 1998). Electrostatic interactions could also be involved in the docking of cytosolic protein targets of DGPP to the plasma membrane, similar to the mechanism proposed for PA (see chapter, "Phosphatidic Acid – An Electrostatic/Hydrogen-Bond Switch?"). Additionally, the formation of DGPP could change the membrane's biophysical properties, especially its polarity and curvature. It has been shown that generation of PA, LPA, and the minor lipid bisphosphatidic acid modifies membrane curvature, thus potentially making PA and LPA essential in the fission of biomembranes (Powell and Hui 1996; Kooijman et al. 2003). In this respect, studies of an artificial membrane system would be useful in predicting the biophysical consequences of the presence of DGPP.

Interestingly, although DGPP has not been identified in mammalian cells, exogenous application of DGPP has several pharmacological effects. In macrophages, DGPP triggers specific inflammatory responses and, especially, the activation of phospholipase A_2, which is responsible for the release of arachidonic acid-derived mediators (Balboa et al. 1999; Balsinde et al. 2000). Furthermore, DGPP application antagonizes LPA receptors in several cell lines (Fischer et al. 2001). Consequently, the medical use of DGPP in the care of patients with thrombosis diseases is under consideration (Spector 2003).

6 Conclusion and Future Developments

Much work is still needed to have a better understanding of the role of DGPP in plant physiology. DGPP is obviously a minor lipid that accumulates transiently during plant stress. However, several aspects of DGPP are completely unknown and many questions need to be addressed. For example, during standard growth conditions, in which tissues and in which cell membrane is DGPP located? Conversely, in response to different stresses, where within the plant and cell do the recorded increases in DGPP occur? Because of the tight metabolic link between DGPP and PA, it would be interesting to consider DGPP's function in specialized cells, in which PA has been shown to play a crucial role. For example, studies on guard cells, aleurone cells, and pollen tubes would be relevant. For all these purposes, the identification of a gene encoding PAK activity would be invaluable. Attention should also be focussed on identifying the molecular DGPP species that accumulate under stress. Indeed, this could lead to the detection and isolation of proteins that are targets of DGPP – an unusual lipid with intriguing properties.

Acknowledgments We acknowledge Prof. Emile Miginiac and Dr Bruno Sotta for critical reading of the manuscript. We thank Dr Régis Maldiney for his help in drawing the figures. This work was supported by grants from *Ministère de la Recherche et de l'Enseignement Supérieur-France* to SP and CZ.

References

Assmann SM (2002) Heterotrimeric and unconventional GTP binding proteins in plant cell signaling. Plant Cell 14(Suppl):355–373

Balboa MA, Balsinde J, Dillon DA, Carman GM, Dennis EA (1999) Proinflammatory macrophage-activating properties of the novel phospholipid diacylglycerol pyrophosphate. J Biol Chem 274:522–526

Balsinde J, Balboa MA, Dennis EA (2000) Group IV cytosolic phospholipase A2 activation by diacylglycerol pyrophosphate in murine P388D1 macrophages. Ann N Y Acad Sci 905:11–15

Carman GM (1997) Phosphatidate phosphatases and diacylglycerol pyrophosphate phosphatases in Saccharomyces cerevisiae and Escherichia coli. Biochim Biophys Acta 1348:45–55

Carman GM, Han GS (2006) Roles of phosphatidate phosphatase enzymes in lipid metabolism. Trends Biochem Sci 31:694–699

Carman GM, Han GS (2007) Regulation of phospholipid synthesis in Saccharomyces cerevisiae by zinc depletion. Biochim Biophys Acta 1771:322–330

Coursol S, Fan L, Le Stunff H, Spiegel S, Gilroy S, Assmann SM (2003) Sphingolipid signaling in Arabidopsis guard cells involves heterotrimeric G proteins. Nature 423:651–654

de Jong CF, Laxalt AM, Bargmann BO, de Wit PJ, Joosten MH, Munnik T (2004) Phosphatidic acid accumulation is an early response in the Cf-4/Avr4 interaction. Plant J 39:1–12 den Hartog M, Musgrave A, Munnik T (2001) Nod factor-induced phosphatidic acid and diacylglycerol pyrophosphate formation: a role for phospholipase C and D in root hair deformation. Plant J 25:55–65

den Hartog M, Verhoef N, Munnik T (2003) Nod factor and elicitors activate different phospholipid signaling pathways in suspension-cultured alfalfa cells. Plant Physiol 132:311–317

Dillon DA, Wu WI, Riedel B, Wissing JB, Dowhan W, Carman GM (1996) The Escherichia coli pgpB gene encodes for a diacylglycerol pyrophosphate phosphatase activity. J Biol Chem 29:30548–30553

Dillon DA, Chen X, Zeimetz GM, Wu WI, Waggoner DW, Dewald J, Brindley DN, Carman GM (1997) Mammalian Mg2 + -independent phosphatidate phosphatase (PAP2) displays diacylglycerol pyrophosphate phosphatase activity. J Biol Chem 272:10361–10366

Finkelstein RR, Rock CD (2002) Abscisic acid biosynthesis and response. In: Somerville CR, Meyerowitz EM (eds) The Arabidopsis book. American Society of Plant Biologists, Rockville, MD, USA

Fischer DJ, Nusser N, Virag T, Yokoyama K, Wang D, Baker DL, Bautista D, Parrill AL, Tigyi G (2001) Short-chain phosphatidates are subtype-selective antagonists of lysophosphatidic acid receptors. Mol Pharmacol 60:776–784

Ghelis T, Dellis O, Jeannette E, Bardat F, Miginiac E, Sotta B (2000) Abscisic acid plasmalemma perception triggers a calcium influx essential for RAB18 gene expression in Arabidopsis thaliana suspension cells. FEBS Lett 483:67–70

Hallouin M, Ghelis T, Brault M, Bardat F, Cornel D, Miginiac E, Rona JP, Sotta B, Jeannette E (2002) Plasmalemma abscisic acid perception leads to RAB18 expression via phospholipase D activation in Arabidopsis suspension cells. Plant Physiol 130:265–272

Han GS, Johnston CN, Chen X, Athenstaedt K, Daum G, Carman GM (2001) Regulation of the Saccharomyces cerevisiae DPP1-encoded diacylglycerol pyrophosphate phosphatase by zinc. J Biol Chem 276:10126–10133

Han GS, Johnston CN, Carman GM (2004) Vacuole membrane topography of the DPP1-encoded diacylglycerol pyrophosphate phosphatase catalytic site from Saccharomyces cerevisiae. J Biol Chem 279:5338–5345

Jeannette E, Rona JP, Bardat F, Cornel D, Sotta B, Miginiac E (1999) Induction of RAB18 gene expression and activation of K + outward rectifying channels depend on an extracellular perception of ABA in Arabidopsis thaliana suspension cells. Plant J 18:13–22

Jung JY, Kim YW, Kwak JM, Hwang JU, Young J, Schroeder JI, Hwang I, Lee Y (2002) Phosphatidylinositol 3- and 4-phosphate are required for normal stomatal movements. Plant Cell 14:2399–2412

Katagiri T, Ishiyama K, Kato T, Tabata S, Kobayashi M, Shinozaki K (2005) An important role of phosphatidic acid in ABA signaling during germination in Arabidopsis thaliana. Plant J 43:107–117

Kooijman EE, Chupin V, de Kruijff B, Burger KN (2003) Modulation of membrane curvature by phosphatidic acid and lysophosphatidic acid. Traffic 4:162–174

Latijnhouwers M, Munnik T, Govers F (2002) Phospholipase D in Phytophthora infestans and its role in zoospore encystment. Mol Plant Microbe Interact 15:939–946

Lemtiri-Chlieh F, MacRobbie E, Brearley CA (2000) Inositol hexakisphosphate is a physiological signal regulating the K + -inward rectifying conductance in guard cells. Proc Natl Acad Sci USA 97:8687–8692

Lemtiri-Chlieh F, MacRobbie E, Webb AA, Manison NF, Brownlee C, Skepper JN, Chen J, Prestwich GD, Brearley CA (2003) Inositol hexakisphosphate mobilizes an endomembrane store of calcium in guard cells. Proc Natl Acad Sci USA 100:10091–10095

Marchesini N, Santander V, Machado-Domenech E (1998) Diacylglycerol pyrophosphate: a novel metabolite in the Trypanosoma cruzi phosphatidic acid metabolism. FEBS Lett 436:377–381

Marmagne A, Rouet MA, Ferro M, Rolland N, Alcon C, Joyard J, Garin J, Barbier-Brygoo H, Ephritikhine G (2004) Identification of new intrinsic proteins in Arabidopsis plasma membrane proteome. Mol Cell Proteomics 3:675–691

Meijer HJ, Munnik T (2003) Phospholipid-based signaling in plants. Annu Rev Plant Biol 54:265–306
Meijer HJ, Arisz SA, Van Himbergen JA, Musgrave A, Munnik T (2001) Hyperosmotic stress rapidly generates lyso-phosphatidic acid in Chlamydomonas. Plant J 25:541–548
Mishra G, Zang W, Deng F, Zhao J, Wang X (2006) A bifurcating pathway directs abscisic acid effects on stomatal closure and opening in Arabidopsis. Science 312:264–266
Hunt L, Pical C, Leckie CP, Aitken FL, Kopka J, Mueller-Roeber B, McAinsh MR, Hetherington AM, Gray JE (2003) Phospholipase C is required for the control of stomatal aperture by ABA. Plant J 34:47–55
Mueller-Roeber B, Pical C (2002) Inositol phospholipid metabolism in Arabidopsis. Characterized and putative isoforms of inositol phospholipid kinase and phosphoinositide-specific phospholipase C. Plant Physiol 130:22–46
Munnik T, de Vrije T, Irvine RF, Musgrave A (1996) Identification of diacylglycerol pyrophosphate as a novel metabolic product of phosphatidic acid during G-protein activation in plants. J Biol Chem 271:15708–15715
Munnik T, Meijer HJ, Ter Riet B, Hirt H, Frank W, Bartels D, Musgrave A (2000) Hyperosmotic stress stimulates phospholipase D activity and elevates the levels of phosphatidic acid and diacylglycerol pyrophosphate. Plant J 22:147–154
Ohlrogge J, Browse J (1995) Lipid biosynthesis. Plant Cell 7:957–970
Oshiro J, Rangaswamy S, Chen X, Han GS, Quinn JE, Carman GM (2000) Regulation of the DPP1-encoded diacylglycerol pyrophosphate (DGPP) phosphatase by inositol and growth phase. Inhibition of DGPP phosphatase activity by CDP-diacylglycerol and activation of phosphatidylserine synthase activity by DGPP. J Biol Chem 275:40887–40896
Oshiro J, Han GS, Carman GM (2003) Diacylglycerol pyrophosphate phosphatase in Saccharomyces cerevisiae. Biochim Biophys Acta 1635:1–9
Pical C, Westergren T, Dove SK, Larsson C, Sommarin M (1999) Salinity and hyperosmotic stress induce rapid increases in phosphatidylinositol 4, 5-bisphosphate, diacylglycerol pyrophosphate, and phosphatidylcholine in Arabidopsis thaliana cells. J Biol Chem 274: 38232–38240
Pierrugues O, Brutesco C, Oshiro J, Gouy M, Deveaux Y, Carman GM, Thuriaux P, Kazmaier M (2001) Lipid phosphate phosphatases in Arabidopsis. Regulation of the AtLPP1 gene in response to stress. J Biol Chem 276:20300–20308
Powell GL, Hui SW (1996) Tetraoleoylpyrophosphatidic acid: a four acyl-chain lipid which forms a hexagonal II phase with high curvature. Biophys J 70:1402–1406
Jacob T, Assmann SM, Gilroy S (1999) Abscisic acid signal transduction in guard cells is mediated by phospholipase D activity. Proc Natl Acad Sci U S A 96:12192–12197
Ryu SB (2004) Phospholipid-derived signaling mediated by phospholipase A in plants. Trends Plant Sci 9:229–235
Santander V, Bollo M, Machado-Domenech E (2002) Lipid kinases and Ca(2+) signaling in Trypanosoma cruzi stimulated by a synthetic peptide. Biochem Biophys Res Commun 293:314–320
Spector AA (2003) Plaque rupture, lysophosphatidic acid, and thrombosis. Circulation 108: 641–643
Spiegel S, Milstien S (2003) Sphingosine-1-phosphate: an enigmatic signaling lipid. Nat Rev Mol Cell Biol 4:397–407
Staxen I, Pical C, Montgomery LT, Gray JE, Hetherington AM, McAinsh MR (1999) Abscisic acid induces oscillations in guard-cell cytosolic free calcium that involve phosphoinositide-specific phospholipase C. Proc Natl Acad Sci USA 96:1779–1784
Takeuchi M, Harigai M, Momohara S, Ball E, Abe J, Furuichi K, Kamatani N (2007) Cloning and characterization of DPPL1 and DPPL2, representatives of a novel type of mammalian phosphatidate phosphatase. Gene 399:174–180
Testerink C, Munnik T (2005) Phosphatidic acid: a multifunctional stress signaling lipid in plants. Trends Plant Sci 10:368–375

Toke DA, Bennett WL, Dillon DA, Wu WI, Chen X, Ostrander DB, Oshiro J, Cremesti A, Voelker DR, Fischl AS, Carman GM (1998a) Isolation and characterization of the Saccharomyces cerevisiae DPP1 gene encoding diacylglycerol pyrophosphate phosphatase. J Biol Chem 273:3278–3284

Toke DA, Bennett WL, Oshiro J, Wu WI, Voelker DR, Carman GM (1998b) Isolation and characterization of the Saccharomyces cerevisiae LPP1 gene encoding a Mg2 + -independent phosphatidate phosphatase. J Biol Chem 273:14331–14338

van der Luit AH, Piatti T, van Doorn A, Musgrave A, Felix G, Boller T, Munnik T (2000) Elicitation of suspension-cultured tomato cells triggers the formation of phosphatidic acid and diacylglycerol pyrophosphate. Plant Physiol 123:1507–1516

van Schooten B, Testerink C, Munnik T (2006) Signaling diacylglycerol pyrophosphate, a new phosphatidic acid metabolite. Biochim Biophys Acta 1761:151–159

Villasuso AL, Molas ML, Racagni G, Abdala G, Machado-Domenech E (2003) Gibberellin signal in barley aleurone : early activation of PLC by G protein mediates amylase secretion. Plant Growth Regul 41:197–205

Wang X (2004) Lipid signaling. Curr Opin Plant Biol 7:329–336

Wang X, Devaiah SP, Zhang W, Welti R (2006) Signaling functions of phosphatidic acid. Prog Lipid Res 45:250–278

Wissing JB, Behrbohm H (1993a) Diacyglycerol pyrophosphate, a novel phospholipid compound. FEBS Lett 315:95–99

Wissing JB, Behrbohm H (1993b) Phosphatidate kinase, a novel enzyme in phospholipid metabolism purification, subcellular localization, and occurrence in the plant kingdom. Plant Physiol 102:1243–1249

Wissing JB, Kornak B, Funke A, Riedel B (1994) Phosphatidate kinase, a novel enzyme in phospholipid metabolism. Characterization of the enzyme from suspension-cultured Catharanthus roseus cells. Plant Physiol 105:903–909

Wu WI, Liu Y, Riedel B, Wissing JB, Fischl AS, Carman GM (1996) Purification and characterization of diacylglycerol pyrophosphate phosphatase from Saccharomyces cerevisiae. J Biol Chem 271:1868–1876

Xue H, Chen X, Li G (2007) Involvement of phospholipid signaling in plant growth and hormone effects. Curr Opin Plant Biol 10:483–489

Zalejski C, Zhang Z, Quettier AL, Maldiney R, Bonnet M, Brault M, Demandre C, Miginiac E, Rona JP, Sotta B, Jeannette E (2005) Diacylglycerol pyrophosphate is a second messenger of abscisic acid signaling in Arabidopsis thaliana suspension cells. Plant J 42:145–152

Zalejski C, Paradis S, Maldiney R, Habricot Y, Miginiac E, Rona JP, Jeannette E (2006) Induction of abscisic acid-regulated gene expression by diacylglycerol pyrophosphate involves Ca2+ and anion currents in Arabidopsis suspension cells. Plant Physiol 141:1555–1562

Zhang W, Qin C, Zhao J, Wang X (2004) Phospholipase D alpha 1-derived phosphatidic acid interacts with ABI1 phosphatase 2C and regulates abscisic acid signaling. Proc Natl Acad Sci USA 101:9508–9513

Zonia L, Munnik T (2004) Osmotically induced cell swelling versus cell shrinking elicits specific changes in phospholipid signals in tobacco pollen tubes. Plant Physiol 134:813–823

Oxylipin Signaling and Plant Growth

Alina Mosblech, Ivo Feussner, and Ingo Heilmann

Abstract Oxylipins are derived from the oxidation of polyunsaturated fatty acids. Further conversion of the resulting fatty acid hydroperoxides gives rise to a multitude of oxylipin classes, including hydroxy-, oxo-, or keto fatty acids, volatile aldehydes, and the phytohormone, jasmonic acid (JA). Oxylipins may be structurally further diversified by esterification, i.e., to plastidial glycolipids, Arabidopsides, or conjugation to amino acids. Oxylipin research so far has focused mainly on the investigation of jasmonates and their roles in wound signaling and plant development. In contrast, the physiological roles of other oxylipins are by far less well understood, in part because enzymes responsible for their formation are not well characterized. This chapter aims at giving an overview of plant oxylipin signaling, highlighting recent discoveries of new roles for different oxylipins in the regulation of developmental or adaptational processes.

1 Oxylipins are a Diverse Class of Signaling Molecules Derived from Fatty Acids

Plant oxylipins are a diverse class of lipid metabolites that are derived from the oxidation of polyunsaturated fatty acids (PUFAs). Oxylipins formed in plants include fatty acid hydroperoxides, hydroxy-, oxo-, or keto-fatty acids, divinyl ethers, volatile aldehydes, or the plant hormone, jasmonic acid (JA) (Grechkin 1998). The first committed step of oxylipin biosynthesis is the formation of fatty acid hydroperoxides (Blee 1998), which can occur by enzymatic processes or by chemical (auto)oxidation (Esterbauer et al. 1991; Spiteller et al. 2001). Figure 1

I. Feussner and I. Heilmann (✉)
Department of Plant Biochemistry, Albrecht-von-Haller-Institute for Plant Sciences, Georg-August-University Göttingen, Justus-von-Liebig-Weg 11, 37077, Göttingen, Germany
e-mail: ifeussn@uni-goettingen.de, iheilma@uni-goettingen.de

T. Munnik (ed.), *Lipid Signaling in Plants*, Plant Cell Monographs 16,
DOI 10.1007/978-3-642-03873-0_18, © Springer-Verlag Berlin Heidelberg 2010

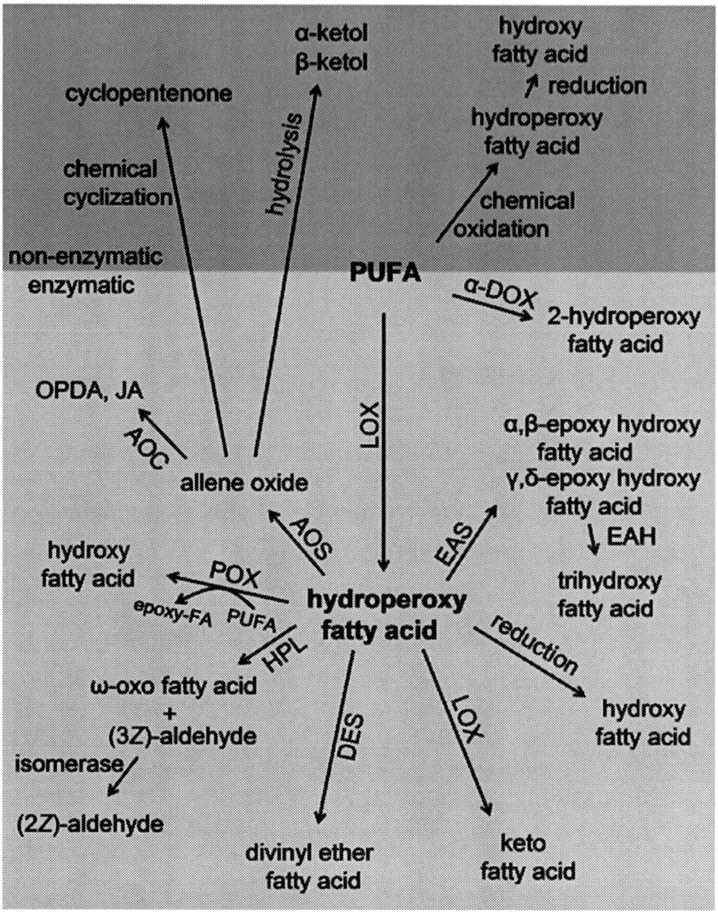

Fig. 1 *Overview of oxylipin biosynthesis.* The formation of oxylipins starts with the conversion of polyunsaturated fatty acids (PUFAs) containing a (1Z,4Z)-pentadiene system, such as linoleic acid or α-linolenic acid. Initial conversion of PUFAs by lipoxygenase (LOX) or 2-dioxygenase (α-DOX) generates fatty acid hydroperoxides that are substrates for alternative metabolic pathways defined by the key enzymes indicated. *AOS*, allene oxide synthase; *AOC*, allene oxide cyclase; *DES*, divinylether synthase; *EAS*, epoxy alcohol synthase; *EAH*, epoxy alcohol hydrolase; *epoxy-FA*, epoxy fatty acid; *HPL*, hydroperoxide lyase; *POX*, peroxidase. PUFAs can also be non-enzymatically converted into fatty acid hydroperoxides and hydroxy fatty acids. Allene oxides formed by *AOS* can spontaneously cyclise to form cyclopentenones, or hydrolyze into α-ketols and β-ketols. Nonenzymatic and enzymatic reactions are indicated by the *dark* and *light gray* background, respectively

gives an overview of oxylipin metabolism in plants. Reactive hydroperoxides of the abundant fatty acids linoleic acid (18:2, LA), α-linolenic acid (18:3, α-LeA), or roughanic acid (16:3) are formed predominantly by lipoxygenases (LOXs) (Feussner and Wasternack 2002) or can also be formed by α-dioxygenase (α-DOX) (Hamberg et al. 2005). Subsequent conversion of hydroperoxides can occur by various alternative pathways, including those initiated by allene oxide synthase (AOS), divinyl ether synthase (DES), hydroperoxide lyases (HPL), peroxygenases (POX), or epoxy alcohol synthase (EAS), as indicated in Fig. 1. The resulting oxygenated derivatives include the phytohormone, JA, as well as oxylipins with characteristic reactive epoxide, α,β-unsaturated carbonyl, or aldehyde functionalities.

Oxylipins occur not only in a free form, but can also be bound to phospholipids, glycolipids, neutral lipids, sulfate, glutathione, carbohydrates, or amino acids in the form of esters or conjugates. The most prominent example is JA, which can occur as a methyl ester (MeJA), be bound in glycosyl esters, can form amide-conjugates with various amino acids (Sembdner et al. 1994), or is hydroxylated at the ω or ($\omega-1$) end of its side chain that is then further sulfated (Wasternack 2007). The conjugate of JA and isoleucine (JA-Ile) has been implicated as a biologically active form of JA which interacts with a JA-receptor (Chini et al. 2007; Thines et al. 2007). Oxylipins may, in addition, occur esterified to complex lipids (Feussner et al. 1995; Feussner et al. 1997). This was first discovered for phospholipids and neutral lipids and later on analyzed in detail for 12-oxo phytodienoic acid (OPDA) and dinor-OPDA (dn-OPDA) (Stelmach et al. 2001; Hisamatsu et al. 2003, 2005; Andersson et al. 2006; Buseman et al. 2006; Kourtchenko et al. 2007; Nakajyo et al. 2007), such as the plastidial galactolipids, mono- and digalactosyldiacylglycerol (MGDG and DGDG). Glycerolipids containing OPDA and/or dn-OPDA instead of fatty acyl moieties are collectively called Arabidopsides. For a more detailed view of the chemical characteristics of various oxylipins, the reader is referred to more specialized reviews on the topic (Grechkin 1998; Blee 2002; Liavonchanka and Feussner 2006).

2 Physiological Roles of Oxylipins

Some oxylipins are signaling compounds with profound relevance for plant function. The following paragraphs provide a summary of the most prominent examples.

2.1 *Responses to Wounding and Infection with Pathogens*

Metabolites that are either precursors, derivatives, or metabolites of JA may be collectively called octadecanoids, in analogy to eicosanoids from the mammalian field. They mediate plant responses to wounding and pathogen attack (Leon et al. 2001; Browse 2005; Schilmiller and Howe 2005; Wasternack et al. 2006; Wasternack 2007). The primary observation that initiated this research field was that wounding, by chewing insects or mechanical damage, results in rapid local accumulation of JA at the site of wounding. Infestation with herbivorous insects or necrotrophic pathogens results in the production of various oxylipin compounds that act as regulators of defence signaling (Blee 2002; Browse 2005). JA and octadecanoids together with other oxylipins generate an integrated "signal signature," specific for the particular stress that the plant is exposed to. The finding that mechanical wounding or wounding caused by chewing insects elicits similar but distinct signaling events (Reymond et al. 2000) may indicate that a particular "signature" is triggered by the recognition of elicitors specific for the herbivore (Kessler and Baldwin 2002).

The initiation of downstream aspects of the wounding response, including the induction of gene expression, involves the enzymatic modification of JA. These later reactions include the formation of the volatile MeJA and conjugation of JA to amino acids, such as Ile. Recent evidence indicates that OPDA signaling involves downstream cascades differing from those of JA-derived signals such as JA-Ile (Weiler et al. 1999; Kramell et al. 2000; Taki et al. 2005), and OPDA has, thus, been demonstrated to act as a signaling factor mediating defence responses independently of JA. Other 13-LOX-derived compounds, such as 13-hydroxy α-LeA (13-HOT), HPL-derived C6 aldehydes, or electrophilic ketodienes that derive either from LA (KODs) or α-LeA (KOTs) may also be regulators of plant defence gene expression (Bate and Rothstein 1998; Weichert et al. 1999; Vollenweider et al. 2000; Almeras et al. 2003). In addition to the induction of defence responses against herbivory, JA and LOX-generated fatty acid 9- and 13-hydroperoxides, as well as KODs and KOTs have been shown to interact with other signaling cascades involved in the mediation of resistance against necrotrophic pathogens as part of the so-called hypersensitive-response (HR) (Penninckx et al. 1996; Rusterucci et al. 1999; Vollenweider et al. 2000; Glazebrook 2005), which is employed to limit the spreading of microorganisms through infected plant tissue.

Insights into the physiological roles of oxylipins in plant signaling have been gained by the analysis of various mutants. Both the CORONATINE INSENSITIVE 1 (*coi1*) and JA-resistant 1 (*jar1*) mutants are defective in their responses to jasmonates. *coi1* mutants are insensitive to octadecanoid signals, and COI1 is required for all known JA-dependent physiological responses (Feys et al. 1994; Xie et al. 1998; Devoto et al. 2005). COI1 is an F-box protein involved in the JA–Ile-dependent degradation of transcriptional repressors (Chini et al. 2007; Thines et al. 2007). The JAR1 gene codes for an enzyme that conjugates JA predominantly to the hydrophobic amino acid Ile (Staswick and Tiryaki 2004). A reported strong reduction of Arabidopside levels in *jar1* mutant plants suggests that JA–amino acid conjugates significantly contribute to the regulation of Arabidopside synthesis. Octadecanoid signaling is interlinked with that of Arabidopside species, and wounding and pathogen challenge have been found to induce the accumulation of Arabidopsides (Stelmach et al. 2001; Andersson et al. 2006; Buseman et al. 2006; Böttcher and Weiler 2007; Kourtchenko et al. 2007). While it has been suggested that Arabidopsides serve as a pool for the release of OPDA, which is then converted to JA (Stelmach et al. 2001; Andersson et al. 2006; Buseman et al. 2006), it has been demonstrated that Arabidopside E has antimicrobial effects by itself and inhibits the growth of *P. syringae* cultures (Andersson et al. 2006).

The identification of a number of genes encoding α-DOXs in different plant species allows the investigation of physiological roles of this enzymatic activity. On the basis of the studies on tobacco and *Arabidopsis*, α-DOX1's involvement in plant defence against microbial infection has been proposed (De Leon et al. 2002; Hamberg et al. 2003; Mueller 2004). The characterisation of α-DOX1 mutants revealed a correlation between α-DOX1 induction and the manifestation of HR. Consistent with this hypothesis, high levels of 2-hydroxy α-LeA accumulated in HR-responding leaves (Hamberg et al. 2005). Other possible roles of α-DOX1 include responses to insect attack and salt stress (Hamberg et al. 2005).

2.2 Inter-Organismic Signaling

A number of oxylipins are volatile and have been proposed to function in the interaction between plants and other inhabitants of their biosphere. Such volatiles are collectively called "green leaf volatiles" (GLVs). For instance, MeJA has been discussed as a potent signal in the communication between individual plants (Farmer and Ryan 1990). *Arabidopsis* plants overexpressing a JA methyl transferase have increased levels of MeJA and exhibit enhanced resistance to pathogen infection, supporting a role for MeJA in pathogen-defence signaling (Seo et al. 2001). Similarly, some GLVs have been shown to convey resistance against fungal pathogens (Kishimoto et al. 2006). An interesting situation is provided by the specific induction of volatile organic compounds, including many oxylipins, by *Heliothis virescens* larvae feeding on tobacco, corn, or cotton (*Gossypium hirsutum*) plants. The resulting mixture (bouquet) of volatiles specifically attracts the parasitic wasp *Cardiochiles nigriceps* (De Moraes et al. 1998), and a number of related studies indicate that the emission of volatiles can specifically attract natural enemies of particular herbivores (Van Poecke et al. 2001; D'Alessandro and Turlings 2005; Fatouros et al. 2005; Shiojiri et al. 2006b). Besides the attraction of predators, plants perceiving volatile signals from damaged neighbors may gain a benefit by readying their defences (Engelberth et al. 2004). The HPL-derived oxylipin, (3*Z*)-hexene-1-ol, is a frequent component of wounding-induced mixtures of volatiles that are emitted as parts of direct and indirect defence mechanisms (Kessler and Baldwin 2002). Another HPL-derived compound, (3*Z*)-hexenal, the precursor of (3*Z*)-hexene-1-ol, is active directly against sucking insects (Vancanneyt et al. 2001). The production of GLVs has been harnessed to improve plant resistance against insect-herbivory (Shiojiri et al. 2006a).

Another function for JA in interspecies-communication has been reported for the establishment of symbiotic interactions between plants and fungi, for instance, during the formation of arbuscular mycorrhizal symbiosis. In arbuscular mycorrhiza, a fungus of the phylum *Glomeromycota* enters a symbiotic association with a vascular plant, for instance *M. truncatula* (Schüssler et al. 2001), which provides the fungus with carbohydrates, while the fungus improves the accessibility of micronutrients for the plant. The establishment of arbuscular mycorrhiza in

M. truncatula (Stumpe et al. 2005) and barley plants (Hause et al. 2002) has been correlated with increases in endogenous JA. Partial suppression of one of two *AOS* genes in *M. truncatula* decreased the rate of arbuscule formation (Isayenkov et al. 2004; Isayenkov et al. 2005).

2.3 Developmental Signaling

Despite their obvious relevance for plant function, the involvement of oxylipins in developmental processes has so far not received as much attention as oxylipin-functions in plant defence responses. One of the earliest physiological effects of JA reported was a generic inhibition of root growth (Staswick et al. 1992). Mutants affected in JA- and COI1-dependent signal transduction exhibit reduced sensitivity to JA and develop normal roots in its presence (Browse 2005; Wasternack et al. 2006). In contrast, mutants characterized by constitutive overexpression of JA-responsive genes due to constitutively elevated JA levels, for instance *cev1*, *cet1*, *joe2*, or *cex1*, exhibit reduced root length and stunted growth, similar to plants treated with exogenous JA (Ellis et al. 2002a, 2002b). Besides JA, Arabidopsides A, B, and D also seem to exert inhibitory effects on root growth (Hisamatsu et al. 2005). Another Arabidopside, a lyso-MGDG from morning glory (*Ipomoea tricolor*), enhanced stomatal opening of dayflower (*Commelina communis*) in darkness at micromolar concentrations (Ohashi et al. 2005). Noncyclic oxylipins generated from LOX-derived fatty acid hydroperoxides by different pathways of further conversion also have distinct effects on root growth, suggesting that functional specialization of different oxylipins has been accompanied by the diversification of their signaling pathways. Hydroxy, keto, and keto-hydroxy oxylipins induce waving of roots accompanied by the inhibition of lateral root-formation (Vellosillo et al. 2007). Divinyl ethers and short-chain ω-oxo acids effected a general arrest of root growth combined with a loss of apical dominance. The detailed characterisation of the root-waving response to treatment with 9-HOT, a 9-LOX derivative, suggests a role of this oxylipin in the formation of lateral roots (Vellosillo et al. 2007).

Besides altered root growth, another key characteristic of mutants deficient in octadecanoid formation is defective flower development. Evidence of the JA-dependency for flower development has come from *Arabidopsis* mutants affected in JA biosynthesis, such as *dad1*, *fad3–fad7–fad8*, *dde2-2*, *dde1*, *opr3*, and *aim1* as well as mutants compromised in JA-signaling, such as *coi1*. All these mutants are male-sterile due to delayed anther development and incomplete anther dehiscence, or are impaired in filament elongation (Turner et al. 2002; Delker et al. 2006). However, it should be pointed out that mutations in the same genes in tomato cause a defect in the development of the female reproductive organs and in particular the egg cells (Xie et al. 1998; Li et al. 2004). The levels of JA–Ile have been shown to change during tomato flower development, suggesting a regulatory role in the control of reproductive development (Hause et al. 2000). While some aldehydes and ketols were initially detected as constituents of leaves (Theodoulou et al. 2005; Stumpe and Feussner 2006; Stumpe et al. 2006), the recently isolated (12*Z*,15*Z*)-9-hydroxy-10-oxo-12,15-octadecadienoic acid was shown to have strong flower-inducing activity (Suzuki et al. 2003).

A modified form of JA, 12-hydroxy-JA, has been shown to induce tuber-formation in Solanaceae, and was aptly named tuberonic acid (Koda et al. 1988). Tuberonic acid can form glucosides likely representing transport forms of the compound. While initially detected in Solanacean species, tuberonic acid has also been detected in other plants, including *Arabidopsis* (Gidda et al. 2003; Gidda and Varin 2006).

Touch-mediated growth, such as tendril-coiling, has been shown to involve octadecanoid signals (Falkenstein et al. 1991; Weiler 1997). Dramatic increases of OPDA with touch indicate a role for OPDA, rather than JA, in the mediation of the touch response (Stelmach et al. 1998).

Another developmental process controlled by oxylipins is that of programmed cell death or senescence. Senescing leaves of *Arabidopsis* contain increased levels of JA and exhibit enhanced expression of genes encoding JA-biosynthetic enzymes (He and Gan 2001; He et al. 2002). In addition to JA, Arabidopside A has been reported to have senescence-promoting effects on barley leaf tissue (Hisamatsu et al. 2006). There is substantial overlap of gene-expression patterns associated with pathogen-responses and with senescence (Weaver et al. 1998; Quirino et al. 1999), indicating some elements regulating senescence or HR-associated programmed cell death are shared. Gene-expression analyses during leaf senescence have revealed an involvement of JA in combination with other signaling pathways, such as SA or ethylene, in the regulation of the senescence process.

3 Oxylipins are Part of a Signaling-Network Involving Other Phytohormone Pathways

The view of signaling events of eukaryotic cells has changed from the notion of linear pathways to that of integrated networks of interdependent factors (Schwartz and Baron 1999; Klipp and Liebermeister 2006). As interdependencies within signaling networks are exceedingly complex, interactions can be described on a smaller scale as cross-talk between pathways. Cross-talk between signaling pathways must be assumed if input from two or more signals synergistically or antagonistically regulates specific plant stress responses or development.

Evidence for cross-talk between JA and SA-signals is provided by studies on the roles of these signaling molecules in pathogen defence. Infection of *Arabidopsis* plants by the biotrophic leaf-pathogen *Pseudomonas syringae* pv. *tomato* (*Pst*) DC3000 triggers simultaneous increases in the levels of SA and JA (van Wees et al. 2000). While in wild-type plants JA levels decrease in response to SA accumulation, JA levels remain elevated in *nahG* plants, in which SA fails to accumulate (Prithiviraj et al. 2005), concomitant with enhanced expression of the JA-responsive genes *LOX2*, *PDF1.2*, and *VSP* in response to infection by the biotrophic pathogen. The reciprocal situation is presented by *coi1*-mutant plants, which are deficient in JA-signaling and exhibit elevated SA levels, activation of *PR* genes, and increased resistance to bacterial pathogens (Kloek et al. 2001). The data indicate that in wild-type plants, JA exerts an inhibitory effect on SA-signaling. The defence response manifested, thus, is a consequence of the cross-talk between SA and JA, which in turn represents the output of specific recognition of the particular pathogen (van Wees et al. 2000).

JA and SA have differential effects on the downstream formation of oxylipins. Exogenous application of JA induces 13-LOX activity and an increase in 13-HPOD, which may then be a substrate for either HPL or AOS, leading to amplification of JA as signal. Alternative conversion by reduction (cf. Fig. 1), indicated by the accumulation of 13-HOT, is activated after SA treatment (Weichert et al. 1999). It is interesting to note that increases in endogenous JA induced by sorbitol treatment (Stenzel et al. 2003), do not result in the accumulation of oxylipin metabolites other than those arising from AOS-activity (Weichert et al. 2000).

While the interactions of JA and SA are complex by themselves, it is clear that other signaling pathways also contribute to the function of the signaling network in plants. There are extensive interactions among various defence response pathways activated in pathogen-infected plants. Antagonism between the JA/ethylene-dependent and the SA-regulated pathways has been clearly established, even though the factors regulating such relationships are still unresolved (Petersen et al. 2000; Kunkel and Brooks 2002; Andreasson et al. 2005; Veronese et al. 2006).

Transient increases in cytosolic Ca^{2+} levels have been demonstrated to occur in wounded *Arabidopsis* plants (Knight et al. 1993), and application of exogenous JA to plant cells results in transient influx of Ca^{2+} from the extracellular space (Sun et al. 2006). Several recent studies have supported a role for Ca^{2+} in defence signaling and have proposed a role for the Ca^{2+}-channel, TPC1, in plant defence responses (Bonaventure et al. 2007a, 2007b). The effects of the pharmacological agent, (2,5-di-tert-butyl)-1,4-hydroquinone, on Ca^{2+} levels during wound signaling led Leon and coworkers to suggest that Ca^{2+} may be released from internal stores rather than their originating from the surrounding medium, and that the phosphoinositide-derived second messenger, inositol 1,4,5-trisphosphate ($InsP_3$), may act as an intermediate in wound signaling (Leon et al. 1998). It has recently been demonstrated that in *Arabidopsis*, timing and kinetics of wounding-induced changes in phosphoinositides and $InsP_3$ coincide with increases in JA (Mosblech et al. 2008). Analyses of JA and $InsP_3$ levels in plants impaired in JA production or $InsP_3$ accumulation suggest that $InsP_3$ acts downstream of JA in the same signaling pathway. Monitoring of wounding-induced changes in transcript levels of downstream target genes indicates an involvement of phosphoinositide signals in the induction of gene expression in response to wounding. Caterpillar feeding experiments suggest a role for phosphoinositide-signals in the mediation of plant defences to herbivory (Mosblech et al. 2008). $InsP_3$ is one of the best-characterized effectors of Ca^{2+} release from internal stores in animal cells (Berridge 2005), and Ca^{2+} release by $InsP_3$ has also been demonstrated in plants (Alexandre and Lassalles 1990; see chapter, "InsP3 in Plant Cells").

A possible alternative role for inositolpolyphosphates in the mediation of plant stress responses is suggested by the recent identification of an inositolhexakisphosphate ($InsP_6$)-cofactor in the binding site of the auxin-receptor protein TIR1 (Tan et al. 2007). Amino acid residues coordinating $InsP_6$ in the TIR1 structure are conserved in other F-box proteins of *Arabidopsis* related to TIR1, including COI1. $InsP_6$ can be formed from $InsP_3$ by action of inositolpolyphosphate-kinases as summarized in chapter, "Inositol Polyphosphates and Kinases" (Stevenson-Paulik et al. 2005), and it is possible that $InsP_3$ serves as a precursor of $InsP_6$ that is required for optimal receptor function in auxin or JA perception.

4 Conclusions

Cellular functions for oxylipins emerge to be as diverse as the appearances of oxylipins themselves. The biosynthesis of oxylipins is highly dynamic and occurs in both a constitutive mode and as a consequence of various stresses. Oxylipin signals are involved in numerous signaling processes and are, thus, integral components of the plant signaling network. While the biosynthesis, perception, and physiological role of, e.g., OPDA or JA are well-defined, the relevance of other oxylipins for plant function is only recently becoming apparent. The latest technological advances in the field of oxylipin analysis open the field for nontargeted metabolomic approaches, which may provide further insights into the contributions of little-known oxylipin species in plant function and development.

Acknowledgements We thank Dr. Cornelia Göbel for critical reading of this manuscript. We are gratefully acknowledging financial support through the German Research Foundation (DFG) to I. H. and I. F.

References

Alexandre J, Lassalles JP (1990) Effect of d-myo-inositol 1, 4, 5-trisphosphate on the electrical properties of the red beet vacuole membrane. Plant Physiol 93:837–840

Almeras E, Stolz S, Vollenweider S, Reymond P, Mene-Saffrane L, Farmer EE (2003) Reactive electrophile species activate defense gene expression in Arabidopsis. Plant J 34:205–216

Andersson MX, Hamberg M, Kourtchenko O, Brunnstrom A, McPhail KL, Gerwick WH, Gobel C, Feussner I, Ellerstrom M (2006) Oxylipin profiling of the hypersensitive response in *Arabidopsis thaliana*. Formation of a novel oxo-phytodienoic acid-containing galactolipid, arabidopside E. J Biol Chem 281:31528–31537

Andreasson E, Jenkins T, Brodersen P, Thorgrimsen S, Petersen NH, Zhu S, Qiu JL, Micheelsen P, Rocher A, Petersen M, Newman MA, Bjorn Nielsen H, Hirt H, Somssich I, Mattsson O, Mundy J (2005) The MAP kinase substrate MKS1 is a regulator of plant defense responses. EMBO J 24:2579–2589

Bate NJ, Rothstein SJ (1998) C6-volatiles derived from the lipoxygenase pathway. Plant J 16:561–569

Berridge MJ (2005) Unlocking the secrets of cell signaling. Annu Rev Physiol 67:1–21

Blee E (1998) Phytooxylipins and plant defense reactions. Prog Lipid Res 37:33–72

Blee E (2002) Impact of phyto-oxylipins in plant defense. Trends Plant Sci 7:315–322

Bonaventure G, Gfeller A, Proebsting WM, Hortensteiner S, Chetelat A, Martinoia E, Farmer EE (2007a) A gain-of-function allele of TPC1 activates oxylipin biogenesis after leaf wounding in Arabidopsis. Plant J 49:889–898

Bonaventure G, Gfeller A, Rodriguez VM, Armand F, Farmer EE (2007b) The fou2 gain-of-function allele and the wild-type allele of two pore channel 1 contribute to different extents or by different mechanisms to defense gene expression in Arabidopsis. Plant Cell Physiol 48:1775–1789

Böttcher C, Weiler EW (2007) Cyclo-Oxylipin-galactolipids in plants: occurrence and dynamics. Planta 226:629–637

Browse J (2005) Jasmonate: an oxylipin signal with many roles in plants. Vitam Horm 72:431–456

Buseman CM, Tamura P, Sparks AA, Baughman EJ, Maatta S, Zhao J, Roth MR, Esch SW, Shah J, Williams TD, Welti R (2006) Wounding stimulates the accumulation of glycerolipids containing oxophytodienoic acid and dinor-oxophytodienoic acid in Arabidopsis leaves. Plant Physiol 142:28–39

Chini A, Fonseca S, Fernandez G, Adie B, Chico JM, Lorenzo O, Garcia-Casado G, Lopez-Vidriero I, Lozano FM, Ponce MR, Micol JL, Solano R (2007) The JAZ family of repressors is the missing link in jasmonate signalling. Nature 448:666–671

D'Alessandro M, Turlings TC (2005) In situ modification of herbivore-induced plant odors: a novel approach to study the attractiveness of volatile organic compounds to parasitic wasps. Chem Senses 30:739–753

De Leon IP, Sanz A, Hamberg M, Castresana C (2002) Involvement of the Arabidopsis alpha-DOX1 fatty acid dioxygenase in protection against oxidative stress and cell death. Plant J 29:61–62

De Moraes CM, Lewis WJ, Pare PW, Alborn HT, Tumlinson JH (1998) Herbivore-infested plants selectively attract parasitoids. Nature 393:570–573

Delker C, Stenzel I, Hause B, Miersch O, Feussner I, Wasternack C (2006) Jasmonate biosynthesis in Arabidopsis thaliana–enzymes, products, regulation. Plant Biol 8:297–306

Devoto A, Ellis C, Magusin A, Chang HS, Chilcott C, Zhu T, Turner JG (2005) Expression profiling reveals COI1 to be a key regulator of genes involved in wound- and methyl jasmonate-induced secondary metabolism, defence, and hormone interactions. Plant Mol Biol 58:497–513

Ellis C, Karafyllidis I, Turner JG (2002a) Constitutive activation of jasmonate signalling in an Arabidopsis mutant correlates with enhanced resistance to Erysiphe cichoracearum, Pseudomonas syringae, and Myzus persicae. Mol Plant Microbe Interact 15:1025–1030

Ellis C, Karafyllidis I, Wasternack C, Turner JG (2002b) The Arabidopsis mutant cev1 links cell wall signalling to jasmonate and ethylene responses. Plant Cell 14:1557–1566

Engelberth J, Alborn HT, Schmelz EA, Tumlinson JH (2004) Airborne signals prime plants against insect herbivore attack. Proc Natl Acad Sci USA 101:1781–1785

Esterbauer H, Schaur RJ, Zollner H (1991) Chemistry and biochemistry of 4-hydroxynonenal, malonaldehyde and related aldehydes. Free Radic Biol Med 11:81–128

Falkenstein E, Groth B, Mithöfer A, Weiler EW (1991) Methyl jasmonate and -linolenic acid are potent inducers of tendril coiling. Planta 185:316–322

Farmer EE, Ryan CA (1990) Interplant communication: airborne methyl jasmonate induces synthesis of proteinase inhibitors in plant leaves. Proc Natl Acad Sci USA 87:7713–7716

Fatouros NE, van Loon JJ, Hordijk KA, Smid HM, Dicke M (2005) Herbivore-induced plant volatiles mediate in-flight host discrimination by parasitoids. J Chem Ecol 31:2033–2047

Feussner I, Wasternack C (2002) The lipoxygenase pathway. Annu Rev Plant Biol 53:275–297

Feussner I, Wasternack C, Kindl H, Kuhn H (1995) Lipoxygenase-catalyzed oxygenation of storage lipids is implicated in lipid mobilization during germination. Proc Natl Acad Sci USA 92:11849–11853

Feussner I, Balkenhohl TJ, Porzel A, Kühn H, Wasternack C (1997) Structural elucidation of oxygenated storage lipids in cucumber cotyledons.Implication of lipid body lipoxygenase in lipid mobilization during germination. J Biol Chem 272:21635–21641

Feys B, Benedetti CE, Penfold CN, Turner JG (1994) Arabidopsis mutants selected for resistance to the phytotoxin coronatine are male sterile, insensitive to methyl jasmonate, and resistant to a bacterial pathogen. Plant Cell 6:751–759

Gidda SK, Varin L (2006) Biochemical and molecular characterization of flavonoid 7-sulfotransferase from Arabidopsis thaliana. Plant Physiol Biochem 44:628–636

Gidda SK, Miersch O, Levitin A, Schmidt J, Wasternack C, Varin L (2003) Biochemical and molecular characterization of a hydroxyjasmonate sulfotransferase from Arabidopsis thaliana. J Biol Chem 278:17895–17900

Glazebrook J (2005) Contrasting mechanisms of defense against biotrophic and necrotrophic pathogens. Annu Rev Phytopathol 43:205–227

Grechkin A (1998) Recent developments in biochemistry of the plant lipoxygenase pathway. Prog Lipid Res 37:317–352

Hamberg M, Sanz A, Rodriguez MJ, Calvo AP, Castresana C (2003) Activation of the fatty acid alpha-dioxygenase pathway during bacterial infection of tobacco leaves. Formation of oxylipins protecting against cell death. J Biol Chem 278:51796–51805

Hamberg M, Ponce de Leon I, Rodriguez MJ, Castresana C (2005) Alpha-dioxygenases. Biochem Biophys Res Commun 338:169–174

Hause B, Stenzel I, Miersch O, Maucher H, Kramell R, Ziegler J, Wasternack C (2000) Tissue-specific oxylipin signature of tomato flowers: allene oxide cyclase is highly expressed in distinct flower organs and vascular bundles. Plant J 24:113–126

Hause B, Maier W, Miersch O, Kramell R, Strack D (2002) Induction of jasmonate biosynthesis in arbuscular mycorrhizal barley roots. Plant Physiol 130:1213–1220

He Y, Gan S (2001) Identical promoter elements are involved in regulation of the OPR1 gene by senescence and jasmonic acid in Arabidopsis. Plant Mol Biol 47:595–605

He Y, Fukushige H, Hildebrand DF, Gan S (2002) Evidence supporting a role of jasmonic acid in Arabidopsis leaf senescence. Plant Physiol 128:876–884

Hisamatsu Y, Goto N, Hasegawa K, Shigemori H (2003) Arabidopsides A and B, two new oxylipins from Arabidopsis thaliana. Tetrahedron Lett 44:5553–5556

Hisamatsu Y, Goto N, Sekiguchi M, Hasegawa K, Shigemori H (2005) Oxylipins arabidopsides C and D from Arabidopsis thaliana. J Nat Prod 68:600–603

Hisamatsu Y, Goto N, Hasegawa K, Shigemori H (2006) Senescence-promoting effect of arabidopside A. Z Naturforsch [C] 61:363–366

Isayenkov S, Fester T, Hause B (2004) Rapid determination of fungal colonization and arbuscule formation in roots of Medicago truncatula using real-time (RT) PCR. J Plant Physiol 161:1379–1383

Isayenkov S, Mrosk C, Stenzel I, Strack D, Hause B (2005) Suppression of allene oxide cyclase in hairy roots of Medicago truncatula reduces jasmonate levels and the degree of mycorrhization with Glomus intraradices. Plant Physiol 139:1401–1410

Kessler A, Baldwin IT (2002) Plant responses to insect herbivory: the emerging molecular analysis. Annu Rev Plant Biol 53:299–328

Kishimoto K, Matsui K, Ozawa R, Takabayashi J (2006) Components of C6-aldehyde-induced resistance in Arabidopsis thaliana against a necrotrophic fungal pathogen, *Botrytis cinerea*. Plant Sci 170:715–723

Klipp E, Liebermeister W (2006) Mathematical modeling of intracellular signalling pathways. BMC Neurosci 7(Suppl 1):S10

Kloek AP, Verbsky ML, Sharma SB, Schoelz JE, Vogel J, Klessig DF, Kunkel BN (2001) Resistance to Pseudomonas syringae conferred by an Arabidopsis thaliana coronatine-insensitive (coi1) mutation occurs through two distinct mechanisms. Plant J 26:509–522

Knight MR, Read ND, Campbell AK, Trewavas AJ (1993) Imaging calcium dynamics in living plants using semi-synthetic recombinant aequorins. J Cell Biol 121:83–90

Koda Y, Omer EA, Yoshihara T, Shibata H, Sakamura S, Okazawa Y (1988) Isolation of a specific tuber-inducing substance from potato leaves. Plant Cell Physiol 29:1047–1051

Kourtchenko O, Andersson MX, Hamberg M, Brunnstrom A, Gobel C, McPhail KL, Gerwick WH, Feussner I, Ellerstrom M (2007) Oxo-phytodienoic acid-containing galactolipids in Arabidopsis: jasmonate signalling dependence. Plant Physiol 145:1658–1669

Kramell R, Miersch O, Atzorn R, Parthier B, Wasternack C (2000) Octadecanoid-derived alteration of gene expression and the "oxylipin signature" in stressed barley leaves. Implications for different signalling pathways. Plant Physiol 123:177–188

Kunkel BN, Brooks DM (2002) Cross talk between signalling pathways in pathogen defense. Curr Opin Plant Biol 5:325–331

Leon J, Rojo E, Titarenko E, Sanchez-Serrano JJ (1998) Jasmonic acid-dependent and -independent wound signal transduction pathways are differentially regulated by Ca^{2+}/calmodulin in *Arabidopsis thaliana*. Mol Gen Genet 258:412–419

Leon J, Rojo E, Sanchez-Serrano JJ (2001) Wound signalling in plants. J Exp Bot 52:1–9

Li L, Zhao Y, McCaig BC, Wingerd BA, Wang J, Whalon ME, Pichersky E, Howe GA (2004) The tomato homolog of CORONATINE-INSENSITIVE1 is required for the maternal control of seed maturation, jasmonate-signaled defense responses, and glandular trichome development. Plant Cell 16:126–143

Liavonchanka A, Feussner I (2006) Lipoxygenases: occurrence, functions and catalysis. J Plant Physiol 163:348–357

Mosblech A, König S, Stenzel I, Grzeganek P, Feussner I, Heilmann I (2008) Phosphoinositide and inositolpolyphosphate-signalling in defense responses of *Arabidopsis thaliana* challenged by mechanical wounding. Mol Plant 1:249–261

Mueller MJ (2004) Archetype signals in plants: the phytoprostanes. Curr Opin Plant Biol 7:441–448

Nakajyo H, Hisamatsu Y, Sekiguchi M, Goto N, Hasegawa K, Shoigemori H (2007) Arabidopside F, a new oxylipin from Arabidopsis thaliana. ChemInform 38. doi: 10.1002/chin.200716205

Ohashi T, Ito Y, Okada M, Sakagami Y (2005) Isolation and stomatal opening activity of two oxylipins from Ipomoea tricolor. Bioorg Med Chem Lett 15:263–265

Penninckx IA, Eggermont K, Terras FR, Thomma BP, De Samblanx GW, Buchala A, Metraux JP, Manners JM, Broekaert WF (1996) Pathogen-induced systemic activation of a plant defensin gene in Arabidopsis follows a salicylic acid-independent pathway. Plant Cell 8:2309–2323

Petersen M, Brodersen P, Naested H, Andreasson E, Lindhart U, Johansen B, Nielsen HB, Lacy M, Austin MJ, Parker JE, Sharma SB, Klessig DF, Martienssen R, Mattsson O, Jensen AB, Mundy J (2000) Arabidopsis map kinase 4 negatively regulates systemic acquired resistance. Cell 103:1111–1120

Prithiviraj B, Bais HP, Weir T, Suresh B, Najarro EH, Dayakar BV, Schweizer HP, Vivanco JM (2005) Down regulation of virulence factors of Pseudomonas aeruginosa by salicylic acid attenuates its virulence on Arabidopsis thaliana and Caenorhabditis elegans. Infect Immun 73:5319–5328

Quirino BF, Normanly J, Amasino RM (1999) Diverse range of gene activity during Arabidopsis thaliana leaf senescence includes pathogen-independent induction of defense-related genes. Plant Mol Biol 40:267–278

Reymond P, Weber H, Damond M, Farmer EE (2000) Differential gene expression in response to mechanical wounding and insect feeding in Arabidopsis. Plant Cell 12:707–720

Rusterucci C, Montillet JL, Agnel JP, Battesti C, Alonso B, Knoll A, Bessoule JJ, Etienne P, Suty L, Blein JP, Triantaphylides C (1999) Involvement of lipoxygenase-dependent production of fatty acid hydroperoxides in the development of the hypersensitive cell death induced by cryptogein on tobacco leaves. J Biol Chem 274:36446–36455

Schilmiller AL, Howe GA (2005) Systemic signalling in the wound response. Curr Opin Plant Biol 8:369–377

Schüssler A, Schwarzott D, Walker C (2001) A new fungal phylum, the *Glomeromycota*: phylogeny and evolution. Mycol Res 105:1413–1421

Schwartz MA, Baron V (1999) Interactions between mitogenic stimuli, or, a thousand and one connections. Curr Opin Cell Biol 11:197–202

Sembdner G, Atzorn R, Schneider G (1994) Plant hormone conjugation. Plant Mol Biol 26:1459–1481

Seo JA, Proctor RH, Plattner RD (2001) Characterization of four clustered and coregulated genes associated with fumonisin biosynthesis in Fusarium verticillioides. Fungal Genet Biol 34:155–165

Shiojiri K, Kishimoto K, Ozawa R, Kugimiya S, Urashimo S, Arimura G, Horiuchi J, Nishioka T, Matsui K, Takabayashi J (2006a) Changing green leaf volatile biosynthesis in plants: an approach for improving plant resistance against both herbivores and pathogens. Proc Natl Acad Sci USA 103:16672–16676

Shiojiri K, Ozawa R, Matsui K, Kishimoto K, Kugimiya S, Takabayashi J (2006b) Role of the lipoxygenase/lyase pathway of host-food plants in the host searching behavior of two parasitoid species, Cotesia glomerata and Cotesia plutellae. J Chem Ecol 32:969–979

Spiteller P, Kern W, Reiner J, Spiteller G (2001) Aldehydic lipid peroxidation products derived from linoleic acid. Biochim Biophys Acta 1531:188–208

Staswick PE, Tiryaki I (2004) The oxylipin signal jasmonic acid is activated by an enzyme that conjugates it to isoleucine in Arabidopsis. Plant Cell 16:2117–2127

Staswick PE, Su W, Howell SH (1992) Methyl jasmonate inhibition of root growth and induction of a leaf protein are decreased in an Arabidopsis thaliana mutant. Proc Natl Acad Sci USA 89:6837–6840

Stelmach BA, Muller A, Hennig P, Laudert D, Andert L, Weiler EW (1998) Quantitation of the octadecanoid 12-oxo-phytodienoic acid, a signalling compound in plant mechanotransduction. Phytochemistry 47:539–546

Stelmach BA, Muller A, Hennig P, Gebhardt S, Schubert-Zsilavecz M, Weiler EW (2001) A novel class of oxylipins, sn1-O-(12-oxophytodienoyl)-sn2-O-(hexadecatrienoyl)-monogalactosyl Diglyceride, from Arabidopsis thaliana. J Biol Chem 276:12832–12838

Stenzel I, Hause B, Miersch O, Kurz T, Maucher H, Weichert H, Ziegler J, Feussner I, Wasternack C (2003) Jasmonate biosynthesis and the allene oxide cyclase family of Arabidopsis thaliana. Plant Mol Biol 51:895–911

Stevenson-Paulik J, Bastidas RJ, Chiou ST, Frye RA, York JD (2005) Generation of phytate-free seeds in Arabidopsis through disruption of inositol polyphosphate kinases. Proc Natl Acad Sci USA 102:12612–12617

Stumpe M, Feussner I (2006) Formation of oxylipins by CYP74 enzymes. Phytochem Rev 5:347–357

Stumpe M, Carsjens JG, Stenzel I, Gobel C, Lang I, Pawlowski K, Hause B, Feussner I (2005) Lipid metabolism in arbuscular mycorrhizal roots of Medicago truncatula. Phytochemistry 66:781–791

Stumpe M, Gobel C, Demchenko K, Hoffmann M, Klosgen RB, Pawlowski K, Feussner I (2006) Identification of an allene oxide synthase (CYP74C) that leads to formation of alpha-ketols from 9-hydroperoxides of linoleic and linolenic acid in below-ground organs of potato. Plant J 47:883–896

Sun QP, Guo Y, Sun Y, Sun DY, Wang XJ (2006) Influx of extracellular Ca^{2+} involved in jasmonic-acid-induced elevation of $[Ca^{2+}]_{cyt}$ and JR1 expression in *Arabidopsis thaliana*. J Plant Res 119:343–350

Suzuki M, Yamaguchi S, Iida T, Hashimoto I, Teranishi H, Mizoguchi M, Yano F, Todoroki Y, Watanabe N, Yokoyama M (2003) Endogenous alpha-ketol linolenic acid levels in short day-induced cotyledons are closely related to flower induction in Pharbitis nil. Plant Cell Physiol 44:35–43

Taki N, Sasaki-Sekimoto Y, Obayashi T, Kikuta A, Kobayashi K, Ainai T, Yagi K, Sakurai N, Suzuki H, Masuda T, Takamiya K, Shibata D, Kobayashi Y, Ohta H (2005) 12-oxo-phytodienoic acid triggers expression of a distinct set of genes and plays a role in wound-induced gene expression in *Arabidopsis*. Plant Physiol 139:1268–1283

Tan X, Calderon-Villalobos LI, Sharon M, Zheng C, Robinson CV, Estelle M, Zheng N (2007) Mechanism of auxin perception by the TIR1 ubiquitin ligase. Nature 446:640–645

Theodoulou FL, Job K, Slocombe SP, Footitt S, Holdsworth M, Baker A, Larson TR, Graham IA (2005) Jasmonic acid levels are reduced in COMATOSE ATP-binding cassette transporter mutants. Implications for transport of jasmonate precursors into peroxisomes. Plant Physiol 137:835–840

Thines B, Katsir L, Melotto M, Niu Y, Mandaokar A, Liu G, Nomura K, He SY, Howe GA, Browse J (2007) JAZ repressor proteins are targets of the SCF(COI1) complex during jasmonate signalling. Nature 448:661–665

Turner JG, Ellis C, Devoto A (2002) The jasmonate signal pathway. Plant Cell 14(Suppl): S153–S164

Van Poecke RM, Posthumus MA, Dicke M (2001) Herbivore-induced volatile production by Arabidopsis thaliana leads to attraction of the parasitoid Cotesia rubecula: chemical, behavioral, and gene-expression analysis. J Chem Ecol 27:1911–1928

van Wees SC, de Swart EA, van Pelt JA, van Loon LC, Pieterse CM (2000) Enhancement of induced disease resistance by simultaneous activation of salicylate- and jasmonate-dependent defense pathways in Arabidopsis thaliana. Proc Natl Acad Sci USA 97:8711–8716

Vancanneyt G, Sanz C, Farmaki T, Paneque M, Ortego F, Castanera P, Sanchez-Serrano JJ (2001) Hydroperoxide lyase depletion in transgenic potato plants leads to an increase in aphid performance. Proc Natl Acad Sci U S A 98:8139–8144

Vellosillo T, Martinez M, Lopez MA, Vicente J, Cascon T, Dolan L, Hamberg M, Castresana C (2007) Oxylipins produced by the 9-lipoxygenase pathway in Arabidopsis regulate lateral root development and defense responses through a specific signalling cascade. Plant Cell 19:831–846

Veronese P, Nakagami H, Bluhm B, Abuqamar S, Chen X, Salmeron J, Dietrich RA, Hirt H, Mengiste T (2006) The membrane-anchored BOTRYTIS-INDUCED KINASE1 plays distinct roles in Arabidopsis resistance to necrotrophic and biotrophic pathogens. Plant Cell 18:257–273

Vollenweider S, Weber H, Stolz S, Chetelat A, Farmer EE (2000) Fatty acid ketodienes and fatty acid ketotrienes: Michael addition acceptors that accumulate in wounded and diseased Arabidopsis leaves. Plant J 24:467–476

Wasternack C (2007) Jasmonates: an update on biosynthesis, signal transduction and action in plant stress response, growth and development. Ann Bot 100:681–697

Wasternack C, Stenzel I, Hause B, Hause G, Kutter C, Maucher H, Neumerkel J, Feussner I, Miersch O (2006) The wound response in tomato–role of jasmonic acid. J Plant Physiol 163:297–306

Weaver LM, Gan S, Quirino B, Amasino RM (1998) A comparison of the expression patterns of several senescence-associated genes in response to stress and hormone treatment. Plant Mol Biol 37:455–469

Weichert H, Stenzel I, Berndt E, Wasternack C, Feussner I (1999) Metabolic profiling of oxylipins upon salicylate treatment in barley leaves–preferential induction of the reductase pathway by salicylate(1). FEBS Lett 464:133–137

Weichert H, Kohlmann M, Wasternack C, Feussner I (2000) Metabolic profiling of oxylipins upon sorbitol treatment in barley leaves. Biochem Soc Trans 28:861–862

Weiler EW (1997) Octadecanoid-mediated signal transduction in higher plants. Naturwissenschaften 84:340–349
Weiler EW, Laudert D, Stelmach BA, Hennig P, Biesgen C, Kubigsteltig I (1999) Octadecanoid and hexadecanoid signalling in plant defence. Novartis Found Symp 223:191–204
Xie DX, Feys BF, James S, Nieto-Rostro M, Turner JG (1998) COI1: an Arabidopsis gene required for jasmonate-regulated defense and fertility. Science 280:1091–1094

Fatty Acid Amide Hydrolase and the Metabolism of *N*-Acylethanolamine Lipid Mediators in Plants

Kent D. Chapman and Elison B. Blancaflor

Abstract *N*-Acylethanolamines (NAEs) are a group of fatty acid derivatives that have been identified in a wide range of multicellular eukaryotes, some unicellular eukaryotes, and in a limited number of prokaryotes. The precise acyl composition of the NAE pool in organisms is variable and the overall levels of NAEs fluctuate with changes in development or in response to cellular stresses, especially where it has been studied in animal and plant systems. In animals, these lipids belong to the endocannabinoid pathway where they regulate diverse behavioral and physiological processes. In plant systems, these NAEs have potent growth-regulating activities, which are terminated by their hydrolysis. The inactivation of NAEs, in part, is accomplished by an enzyme identified as a functional homolog of the fatty acid amide hydrolase (FAAH) that regulates endocannabinoid metabolism in vertebrates. Here, the molecular and biochemical characteristics of this enzyme and its role in NAE metabolism in plants are reviewed.

1 Introduction

N-Acylethanolamines (NAEs) are fatty acid derivatives that occur naturally in plant and animal systems (Chapman 2004; Schmid et al. 1996). Recently, fungi were also reported to accumulate NAEs (Merkel et al. 2005). NAEs vary in their acyl chain length and level of unsaturation, but all have amide linkages between the amino group of ethanolamine and carboxyl group of the corresponding fatty acids, and hence carry no net charge at physiological pH (Fig. 1). The principal NAE types in plants are consistent with the fatty acid compositions of plant membranes and so are mostly C-18 unsaturated species although the composition of NAE varies

---

K.D. Chapman (✉)
Department of Biological Sciences, University of North Texas, Center for Plant Lipid Research, Denton, TX 76203, USA
e-mail: chapman@unt.edu

T. Munnik (ed.), *Lipid Signaling in Plants*, Plant Cell Monographs 16,
DOI 10.1007/978-3-642-03873-0_19,

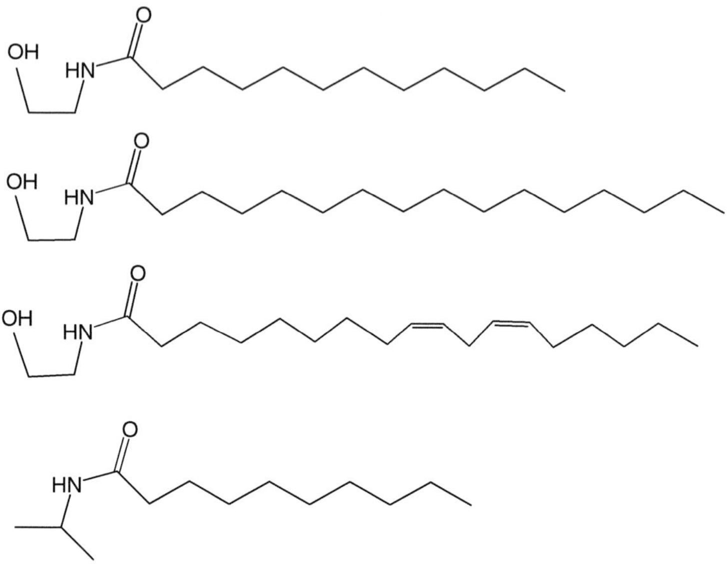

Fig. 1 Structural comparison of *N*-acylethanolamines (*top three*, NAE12:0, NAE16:0, NAE18:2, respectively) identified in plants seeds (Chapman 2004) and one type of alkamide (*bottom*, *N*-isobutyl decanamide) isolated from roots of *Heliopsis longipes* (Lopez-Bucio et al. 2006)

considerably from plant to plant and from tissue to tissue (Chapman 2004; Kilaru et al. 2007; Teaster et al. 2007). The NAEs are structurally similar to alkamides, and together the NAEs and alkamides have been shown to have potent growth-modulating properties when applied to plants exogenously (Blancaflor et al. 2003; Campos-Cuevas et al. 2008; Han et al. 2007; Kilaru et al. 2007; Lopez-Bucio et al. 2006, 2007; Motes et al. 2005; Zhang et al. 2007). While alkamides and NAEs share some similar structural and functional characteristics, NAEs appear to be ubiquitous in eukaryotes and are derived metabolically from a membrane-phospholipid precursor (Blancaflor and Chapman 2006), whereas the alkamides are restricted to a few plant species and their biosynthesis is less clear (Lopez-Bucio et al. 2006). Nevertheless, these two groups of acylamides appear to have important interactions with well-established hormone signaling pathways in plants (e.g., ABA for NAEs and cytokinins for alkamides), suggesting that these lipids are important novel mediators of plant growth and development (Kilaru et al. 2007; Lopez-Bucio et al. 2006).

The hydrolysis of NAEs to form free fatty acids and ethanolamine appears to terminate the signaling activity of NAEs in a variety of organisms. This has, perhaps, been best studied in the mammalian central nervous system where NAE20:4 (numerical designation is number of acyl carbonsto number of double bonds), also referred to as anandamide, functions as a neurotransmitter (Wilson and Nicoll 2002), and its activity is terminated by an NAE hydrolase, designated FAAH, for fatty acid amide hydrolase (McKinney and Cravatt 2005). Consequently, NAE hydrolases have become an important therapeutic target for the treatment of a variety of neurological disorders (Cravatt and Lichtman 2003; Labar and Michaux 2007).

DNA sequences encoding mammalian FAAH enzymes were identified and characterized (Cravatt et al. 1996), and the rat enzyme has been crystallized in the presence of a substrate analog and the three-dimensional structure solved at 2Å resolution (Bracey et al. 2002). FAAH belongs to the amidase superfamily of proteins, and has a Ser–Ser–Lys catalytic mechanism, different from previously characterized serine hydrolases (Labar and Michaux 2007; McKinney and Cravatt 2005).

2 Fatty Acid Amide Hydrolase and the Metabolism of *N*-Acylethanolamine Lipid Mediators in Plants

2.1 *Identification of Plant* N-*Acylethanolamine Hydrolases*

An important advance in the understanding of NAE hydrolysis in plant systems came with the identification of a DNA sequence in plants encoding a homologue of the rat FAAH (Shrestha et al. 2003). Candidate plant FAAHs were identified through bioinformatic approaches using the amidase signature sequence of the rat FAAH to search the *Arabidopsis* genome initially. A DNA sequence with approximately 45% identity in the amidase signature domain was identified in *Arabidopsis* (corresponding to locus At5g64440), although identity over the full length of the predicted protein (compared to rat) was limited to about 18%. However, the Ser–Ser–Lys residues essential for rat FAAH catalysis were absolutely conserved in the *Arabidopsis* sequence. Orthologous sequences in numerous plant species were present in DNA databases, and representative full-length cDNA sequences from the rice, *Medicago*, and *Arabidopsis* genomes were isolated, cloned into heterologous expression vectors, and expressed in *E. coli* (Shrestha et al. 2006). *E. coli* has no endogenous NAE hydrolase activity, so the hydrolysis of ^{14}C-labeled NAEs to ^{14}C-labeled free fatty acids was confirmed in *E. coli* lysates from cells expressing all three of the full-length cDNAs (Fig. 2).

2.2 *Structural and Biochemical Properties of Plant FAAH*

Expression of recombinant plant NAE hydrolases in *E. coli* afforded the opportunity to characterize their biochemical properties and make comparisons to rat FAAH (Shrestha et al. 2003, 2006). The purified *Arabidopsis* recombinant FAAH enzyme hydrolyzed a wide range of NAE types, including those that are generally most prevalent in plants (e.g., NAE 16:0, NAE18:1, NAE18:2) and those that are often less prevalent (e.g., NAE12:0, NAE14:0, NAE18:0). Moreover, the *Arabidopsis*

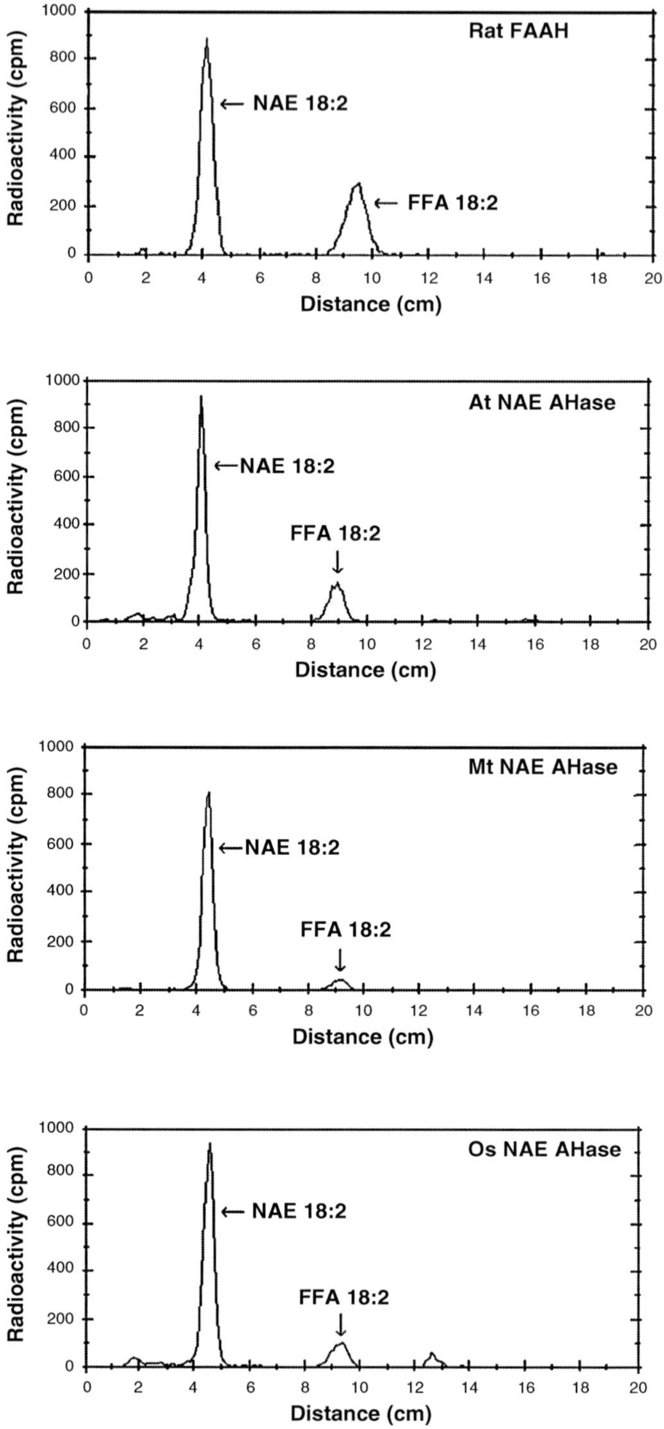
Rat FAAH
← NAE 18:2
← FFA 18:2
At NAE AHase
←NAE 18:2
FFA 18:2
Mt NAE AHase
←NAE 18:2
FFA 18:2
Os NAE AHase
← NAE 18:2
FFA 18:2
Radioactivity (cpm)
Distance (cm)
0
200
400
600
800
1000
0
2
4
6
8
10
12
14
16
18
20

enzyme readily hydrolyzed the mammalian neurotransmitter, NAE20:4, not normally found in plant tissues, and even the mammalian sleep-inducing fatty acid primary amide, oleamide (Neu et al. 2007; Pollmann et al. 2006). This broad substrate specificity is typical of the recombinant FAAH enzymes from rat and other mammals, and is the reason why the enzyme is termed fatty acid amide hydrolase. In addition to acylamides, the *Arabidopsis* FAAH, like the rat FAAH, hydrolyzed a variety of acyl esters, including monoacylglycerols and fatty acid methyl esters. So both enzymes had broad substrate specificities and possessed amidase and esterase activities. It should be emphasized however, that these enzymes are not general esterases or amidases as the substrates must be lipophilic and even lipophilic amides such as ceramide and sphingolipid metabolites were not suitable substrates for either the rat FAAH or the *Arabidopsis* FAAH in vitro.

The kinetic properties of *Arabidopsis* FAAH were measured with respect to different NAE types. Affinities of the *Arabidopsis* FAAH were similar for all NAE types tested, including anandamide, which is not normally found in plants. The apparent K_m of the *Arabidopsis* FAAH toward various NAE types was between 10 μM and 50 μM depending upon NAE type (Shrestha et al. 2003, 2006), and this was similar to the ranges reported for rat FAAH (McKinney and Cravatt 2005) and other recombinant plant FAAH proteins (Shrestha et al. 2006). The apparent V_{max} values estimated for *Arabidopsis* FAAH and rat FAAH varied by a factor of 10, but this is likely due to differences in the relative purity of the proteins. Likewise, *Arabidopsis* FAAH V_{max} values were reported to be about 20 times higher than rice FAAH; however, the rice FAAH protein was unstable in *E. coli*, and purified fractions contained a substantial amount of truncated enzyme (Shrestha et al. 2006). In such cases, the apparent V_{max} would be underestimated, which makes direct comparison of this parameter of FAAH behavior difficult. In any case, affinities for different NAE substrates were generally similar among all FAAH proteins (animal and plant).

In addition to similarities in substrate preference and affinity, plant and animal FAAH were inactivated in a similar fashion by several key inhibitors. Phenylmethylsulphonyl fluoride (PMSF), a common serine hydrolase inhibitor effectively inhibited both *Arabidopsis* and rat FAAH enzyme activity in a concentration-dependent manner. Perhaps, more importantly, both the *Arabidopsis* and rat FAAH enzymes were inhibited at nanomolar concentrations of the substrate analog, methyl arachidonyl fluorophosphonate (MAFP), an irreversible inhibitor of FAAH that has been crystallized in the active site (Bracey et al. 2002). These results strongly argue for a highly conserved active site structure between plant

Fig. 2 Representative radiochromatograms illustrating NAE hydrolysis by recombinant FAAH proteins expressed in *E.coli*. *Top panel*, Rat FAAH; *second panel*, *Arabidopsis* FAAH (At NAE AHase); *third panel*, *Medicago truncatula* FAAH (Mt AHase); *bottom panel*, Oryza sativa FAAH (rice, OS AHase). *E.coli* lyates (10 μg total protein) from cells expressing the respective FAAH proteins were incubated with ^{14}C-NAE18:2 for 20 min. Conversion of ^{14}C-NAE18:2 to ^{14}C-linoleic acid (FFA 18:2) is visualized by radiometric scanning of lipid extracts separated by thin layer chromatography (after Shrestha et al. 2006)

and mammalian FAAH proteins. Indeed, computer modeling of the amidase signature domain of three plant FAAH proteins relative to the three-dimensional structure of rat FAAH superimposed the three candidate catalytic residues, Lys205, Ser281, and Ser305 (numbers in *Arabidopsis* sequence), in an appropriate orientation for catalysis and in a configuration essentially indistinguishable from the rat FAAH structure (Kilaru et al. 2007; Shrestha et al. 2006). While the modeling must inherently remain hypothetical, it certainly points out the high degree of conservation between plant and animal FAAH enzymes. Future experiments to mutate these putative active-site residues in the plant FAAH enzymes will help support the conclusions that despite the limited overall sequence identity between *Arabidopsis* and rat FAAH, the amidase domain is structurally and functionally conserved.

Some differences have been noted between the *Arabidopsis* and rat FAAH enzymes that are noteworthy, and emphasize that, while these enzymes share general catalytic features, there are structural and functional differences between these enzymes as well. One interesting observation is that the rat enzyme activity is reversible, that under conditions of excess ethanolamine and free fatty acids, it can function in the synthesis of NAE (Kurahashi et al. 1997). However, the formation of NAE by plant FAAH enzymes has not been demonstrated in vitro. In addition, there must be some subtle differences in the active site configuration or substrate/product channels because one potent inhibitor of rat FAAH, URB597, does not inhibit plant FAAH activity (Shrestha et al. 2006). URB597 is [3-(3-carbamoylphenyl)phenyl] *N*-cyclohexylcarbamate, and while this compound is reported to be a relatively selective irreversible inhibitor of mammalian FAAH (Kathuria et al. 2003), its absolute lack of activity toward rice or *Arabidopsis* FAAH suggests also its inability to access the active site of the plant enzymes. In summary, while there are numerous structural and functional similarities between the well-characterized mammalian FAAH and the plant *N*-acylethanolamine hydrolases characterized to date, there are some apparent differences as well. Such differences have, no doubt, arisen through the adaptation of this enzyme for different roles in mammalian and plant physiology, and their mechanistic explanation may lead to new insights into the function of this enzyme in lipid signal termination.

2.3 *In Planta Expression of FAAH in Arabidopsis*

FAAH was expressed ubiquitously in *Arabidopsis* tissues, although transcript levels were highest in seedlings and in siliques measured by quantitative RT-PCR. Expression evaluated in plants expressing FAAH promoter::GUS constructs generally agreed with transcript analyses, and indicated that seedling roots were most active in FAAH expression (Wang et al. 2006). Activity measurements in *Arabidopsis* seedling extracts indicated that FAAH enzyme activity increased with seedling development in agreement with expression studies and consistent with the timing of NAE depletion during seed germination and seedling growth

(Chapman 2004; Wang et al. 2006). Overall, the endogenous expression of FAAH in *Arabidopsis* was consistent with a role for NAE metabolism in seedling development, a conclusion that was supported from a variety of approaches and is most dramatically illustrated by the arrest of seedling development and alteration of root cell organization and function in seedlings treated with exogenous, micromolar concentrations of *N*-lauroylethanolamine (NAE12:0; Blancaflor et al. 2003; Motes et al. 2005).

Characterizing the biochemical properties of plant FAAH enzymes has helped to establish the enzymatic behavior of this family of plant proteins, but new understanding of the function of NAE hydrolysis in planta has come from altering FAAH expression in transgenic plants (Wang et al. 2006). In plants, NAE levels normally decline during seed germination and seedling growth (Chapman 2004, 1999) and exogenous application of NAEs during this stage severely retards the growth of *Arabidopsis* seedlings (Blancaflor et al. 2003; Motes et al. 2005). *Arabidopsis* plants with T-DNA disruptions in the FAAH structural gene (At5g64440) were identified in the SALK collection (O'Malley et al. 2007). Also plants overexpressing the *Arabidopsis* FAAH cDNA (behind the CaMV35S promoter) were generated. These FAAH-altered plants exhibited predictable differences in their sensitivity to growth inhibition by NAE (Wang et al. 2006). T-DNA FAAH knockout seedlings were substantially more sensitive to NAE inhibition than the wildtype (Fig. 3a, b). By contrast, growth of *Arabidopsis* seedlings overexpressing FAAH showed substantial resistance to inhibition by NAE, growing much better than wildtype in the presence of NAE12:0 (Fig. 3C). These results supported the

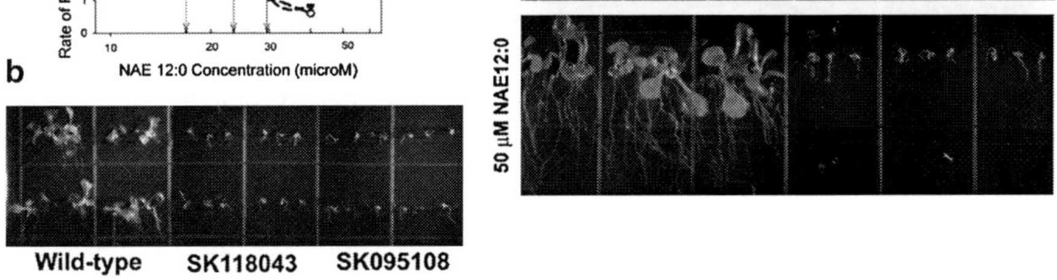

Fig. 3 Differential sensitivity of AtFAAH knockouts and AtFAAH overexpressors to NAE12:0. (**a**) Dose–response curves of root growth inhibition by NAE 12:0 indicates lower EC 50 values for the FAAH knockouts. (**b**) Wild-type seedlings are able to recover from exogenous NAE after 2 weeks of exposure while AtFAAH knockouts (SK118043 and SK095108) remain stunted. (**c**) Fourteen-day-old seedlings of AtFAAH overexpressors are generally larger than wild-type of vector controls. AtFAAH overexpressors grow better on 50 μM NAE12:0 compared with vector only controls, which show severely stunted growth

conclusions from enzyme studies in vitro that the At5g64440 gene encodes a functional FAAH enzyme that is capable of hydrolyzing NAEs in vivo (Wang et al. 2006).

The FAAH overexpressing lines demonstrated the most dramatic phenotypes with respect to growth and development in the absence of exogenous NAE, whereas the T-DNA knockout lines showed only modest differences from wildtype (Wang et al. 2006). All FAAH overexpressing *Arabidopsis* lines examined exhibited an accelerated growth phenotype. Seedling organs developed more rapidly than wildtype and expanded to a larger overall size (Fig. 3C; Wang et al. 2006). The increase in organ size was presumed to be due to enhanced cell expansion since the average cell size was larger in overexpressing lines than in wildtype. This phenotype of larger organ size occurred at later developmental stages as rosette leaves were also larger in overexpressing lines than in wildtype. Overexpressing lines flowered earlier than wildtype under long-day conditions suggesting that flowering responses had been altered by overexpressing FAAH protein. It should be emphasized that multiple lines of FAAH overexpressors exhibited enhanced growth, a somewhat unusual phenotype for transgenic plants. Overall, these results are consistent with the concept that NAEs are negative regulators of plant growth and their enhanced turnover (by overexpressing FAAH) promotes enhanced growth.

In addition to enhanced plant growth, the constitutive overexpression of FAAH in *Arabidopsis* compromised stress responses in these plants (Kilaru et al. 2007). In other words, the cost of enhanced growth in FAAH overexpressing lines came at the expense of reduced capacity to respond to several abiotic stresses, suggesting that NAE metabolism may impinge upon important regulatory mechanisms of plant growth and stress. Indeed, recent results have indicated an interaction of NAE metabolism with ABA signaling, and a combination of physiological, biochemical, molecular, and genetic approaches demonstrated that NAE regulation of plant growth requires an intact ABA signaling pathway (Teaster et al. 2007). *Arabidopsis* seedlings exposed to NAE12:0 exhibited arrested growth, similar to the secondary dormancy induced with ABA (Lopez-Molina et al. 2001, 2002), and the combination of low levels of both compounds showed a synergistic reduction of seedling growth (Teaster et al. 2007). In addition, microarray analyses of 4-day-old seedlings exposed to NAE12:0 alone activated embryo-specific and ABA-regulated gene programs; 32% of transcripts that were higher in NAE-treated seedlings (than untreated seedlings) were annotated as ABA-responsive or had ABA-responsive elements in their 1 Kb upstream promoter sequences (see ArrayExpress: E-MEXP-1310; Teaster et al. 2007). Several ABA-insensitive mutants (*abi3-1*, *abi2-1*, and *abi1-1*) exhibited reduced sensitivity to growth inhibition by NAE12:0 or to synergistic action of NAE and ABA, providing genetic evidence for ABA signaling pathways as a target of NAE. FAAH overexpressing seedlings, although tolerant toward NAE12:0, displayed a marked hypersensitivity toward ABA (Fig. 4), supporting an interaction, albeit complex, between NAE metabolism and ABA regulation of seedling growth. The interaction of NAE and ABA signaling may be via the regulation of *ABI3* gene expression, since NAE/FAAH-mediated growth was closely

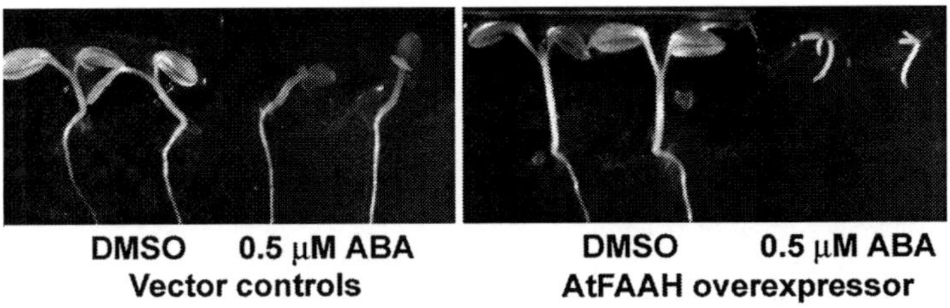

Fig. 4 Differential sensitivity of AtFAAH overexpressors and vector only controls to exogenous ABA. Note that at upon exposure to similar ABA concentrations, expansion of cotyledon, hypocotyls, and roots of AtFAAH overexpressors more significantly impaired compared with vector only controls

associated with the expression of this key regulator of embryo-to-seedling transition. Indeed, FAAH overexpressing lines with enhanced growth showed little detectable expression of *ABI3* in seedlings even when challenged with NAE, whereas the FAAH T-DNA knockout, that was more severely impaired by NAE than wildtype, showed markedly higher levels of *ABI3* transcripts (Teaster et al. 2007). Taken together, these results support an interaction between NAE metabolism and ABA regulation of plant processes and may explain the intriguing combination of growth and stress phenotypes resulting from overexpression of FAAH.

2.4 Regulation of N*-Acylethanolamine Levels in Planta by NAE Hydrolases*

In plants, NAE content is generally highest in desiccated seeds, and NAE profiles have been characterized in seeds of a wide range of plant species (Chapman et al. 1999; Venables et al. 2005). Alteration of FAAH expression in *Arabidopsis* resulted in an alteration of NAE profiles that was particularly evident in seeds. As expected, NAE content was lower in seeds of FAAH overexpressing lines compared with that of wildtype, while NAE content in FAAH-TDNA knockouts seeds was higher than in wildtype. The specific NAE types in *Arabidopsis* seeds were mostly 18C unsaturated species and the differences in NAE content among genotypes also was reflected similarly in these major types, indicating no real endogenous preference of NAE type by the FAAH enzyme. These results indicated that the At5g64440, at least in part, is involved in the modulation of endogenous NAE levels. However, like nearly all important regulatory pathways, there appears to be substantial redundancy in the NAE hydrolytic capacity of plants, since considerable NAE depletion during seedling development occurred in FAAH T-DNA knockouts like in wildtype and FAAH overexpressors. This is not surprising, but points to the likelihood of other NAE hydrolases that can compensate for the absence of At5g64440.

2.5 *Other* N*-Acylethanolamine Amidases*

As noted, recent experimental evidence has clearly demonstrated that *Arabidopsis* FAAH encoded by the gene At5g64440 is a bonafide NAE hydrolase (Shrestha et al. 2003, 2006; Wang et al. 2006). The ability of *Arabidopsis* FAAH knockouts to deplete their endogenous seed NAEs during germination, however, point to the existence of alternative pathways for NAE catabolism in plants (Wang et al. 2006). Although it has yet to be determined whether plants have other enzymes with NAE hydrolytic activity, a FAAH-2 enzyme was recently identified in human cancer cell lines via activity-based protein profiling. The human FAAH-2 protein was shown to share 20% sequence similarity with FAAH-1 and comparison of the enzymatic properties of FAAH-1 and FAAH-2 revealed differences in substrate selectivity. For example, FAAH-1 was more active in hydrolyzing NAE18:1, NAE16:0, and anandamide (NAE20:4) compared to FAAH-2. FAAH-2 also was unable to hydrolyze other fatty acid amides such as the *N*-acyl taurines (NAT), which are substrates of FAAH-1. Thus, it appears that differences in substrate selectivity between the two animal FAAHs are dictated by the amine leaving group and degree of acyl chain saturation (Wei et al. 2006). The precise mechanisms by which FAAH-1 and FAAH-2 coordinate their activities to regulate NAE metabolism and signaling remain to be elucidated.

Another enzyme capable of hydrolyzing NAEs was recently identified in animals. This enzyme designated as NAE acid amidase (NAAA) had no homology to FAAH-1 and was more closely related to the family of acid ceramidase proteins. Not surprisingly, the catalytic properties of NAAA differed significantly from that of FAAH-1 in that NAAA was most active at a pH range of 4.5–5 while FAAH-1 was more active at a broader pH range with optimal activity at more alkaline pH (Tsuboi et al. 2005). Substrate preferences of NAAA and FAAH-1 were also different in that FAAH-1 was highly reactive to anandamide (NAE20:4), NAE18:2, and NAE16:0 while NAAA hydrolyzed NAE14:0 with greater efficiency (Tsuboi et al. 2005). Furthermore, consistent with its preference for acidic environments, NAAA was shown to be localized at lysosomes in animal cells (Tsuboi et al. 2005, 2007). NAAA is widely distributed in various organs of mice and could degrade various NAEs in macrophages in cooperation with FAAH-1 (Sun et al. 2005). Like the recently discovered FAAH-2, additional studies will be needed to determine the precise physiological role of NAAA in NAE metabolism and how it contributes toward defining overall NAE profiles in the cell.

No close homology of the animal NAAA and FAAH-2 enzymes could be identified in plant databases. Other plant genes, however, encode for proteins with amidase signature domains that could be a possible NAE hydrolases. Among the seven amidase signature proteins in the *Arabidopsis* genome, only AtFAAH and amidase 1 (AT-AMI1) have been characterized (Kilaru et al. 2007). Unlike AtFAAH, which efficiently hydrolyzed NAEs, AT-AMI1 was highly specific for indole-3-acetamide and 1-naphthaleneacetamide, and exhibited only minimal hydrolytic activity toward NAEs (Pollmann et al. 2006). Furthermore, while

AtFAAH appears to be a bifunctional enzyme by having both amidase and esterase activity (Shrestha et al. 2006), AMI-1 had no detectable esterase activity (Neu et al. 2007). The in vitro enzymatic properties of the other five plant amidases have yet to be determined but expression data from microarray studies are pointing toward potential coordination of some of these amidases with AtFAAH. The At5g07360 gene, for example, is a particularly intriguing candidate amidase because its in silico expression patterns, as determined from gene chip data from Genevestigator, show high expression in dessicated seeds (Zimmermann et al. 2004). Furthermore, this gene is one of the most highly expressed in seeds of the ABA hypersensitive mutant *ahg1* (Nishimura et al. 2007). Since recent data shows that NAE interacts with ABA signaling during seed germination and early seedling development (Teaster et al. 2007), detailed studies of the amidase encoded by the At5g07360 should provide additional insights into the impact of NAE catabolism on plant stress responses. Similar to AtFAAH, knockouts to the At5g07360 gene display hypersensitivity to the growth inhibitory effects of exogenous NAE12:0, indicating that this gene could contribute to NAE degradation in planta (Blancaflor and Chapman, unpublished observations). Expression of the amidase encoded by the At5g07360 gene in a heterologous system combined with genetic analysis of plants altered in expression of this potential amidase provide exciting avenues for future research.

2.6 Subcellular Location of N-Acylethanolamine Amidases

The mammalian FAAH is distinct from other amidase signature-containing enzymes in that it carries an N-terminal transmembrane domain that predicts it to be localized in cellular membranes. The continued association of FAAH with cellular membranes despite deleting this transmembrane domain, however, suggests that membrane binding could be facilitated by other mechanisms (McKinney and Cravatt 2005). Two helices in the FAAH protein that contain several hydrophobic residues could reinforce membrane association (Bracey et al. 2002). Indeed, a confocal microscopic study of FAAH in human keratinocytes using anti-FAAH polyclonal antibodies revealed its association with the endoplasmic reticulum. Consistent with these immunological results, biochemical fractionation work demonstrated that FAAH enzymatic activity was highest in microsomal membranes (Oddi et al. 2005). Like the mammalian FAAH, AtFAAH contains a transmembrane domain that predicts its localization to membranes (Shrestha et al. 2003) Indeed, in vitro biochemical studies show the highest NAE hydrolase activity in plant microsomal fractions (Shrestha et al. 2002) and GFP fusions to AtFAAH labels the endomembrane system (Kang et al. 2008). Furthermore, proteonomic localization of organelle proteins by isotope tagging (LOPIT) predicts AtFAAH to be associated with the endoplasmic reticulum (Dunkley et al. 2006). It is not known, however, whether membrane binding of plant FAAH is facilitated by similar mechanism as mammalian FAAH.

3 Summary and Conclusions

Recent evidence has revealed the existence of a plant FAAH enzyme that is capable of hydrolyzing NAEs in vitro and in vivo. Further, the potent biological activity of micromolar concentrations of exogenous NAE and the interaction of NAE metabolism with ABA signaling have raised the interest in this class of lipid mediators in the control of plant growth and development. The wide range of physiological and behavioral events that are regulated by NAE metabolism in vertebrates and the interaction of this pathway with phytohormone action in plants continue to suggest that this class of lipid mediators represents an evolutionarily conserved regulatory mechanism that is fundamental to cellular physiology. The machinery for NAE formation and depletion is conserved in diverse eukaryotic organisms, and understanding the regulation of NAE depletion has important implications for general lipid-based regulation of various physiological processes. The key to unraveling this regulation is a better understanding of the FAAH enzyme and the other proteins that influence the profiles of NAEs in cellular systems. Future work should be aimed at identifying additional factors that participate in FAAH function, so a clearer, more precise role of NAE lipid mediators can be identified in plant systems.

Acknowledgments Work in the authors' laboratories on *N*-acylethanolamine metabolism has been supported by grants from the USDA-NRI competitive grants program and the U.S. Department of Energy, Energy Biosciences Program. We thank Dr. Charlene Case-Richardson for assistance with manuscript preparation.

References

Blancaflor EB, Chapman KD (2006) Similarities between endocannabinoid signaling in animal systems and *N*-acylethanolamine metabolism in plants. In: Communication in plants: neuronal aspects of plant life. Springer, Berlin, pp 205–219

Blancaflor EB, Hou G, Chapman KD (2003) Elevated levels of *N*-lauroylethanolamine, an endogenous constituent of desiccated seeds, disrupt normal root development in *Arabidopsis thaliana* seedlings. Planta 217:206–217

Bracey MH, Hanson MA, Masuda KR, Stevens RC, Cravatt BF (2002) Structural adaptations in a membrane enzyme that terminates endocannabinoid signaling. Science 298:1793–1796

Campos-Cuevas JC, Pelagio-Flores R, Raya-González J, Méndez-Bravo A, Ortiz-Castro R, López-Bucio J (2008) Tissue culture of Arabidopsis thaliana explants reveals a stimulatory effect of alkamides on adventitious root formation and nitric oxide accumulation. Plant Sci 174:165

Chapman KD (2004) Occurrence, metabolism, and prospective functions of *N*-acylethanolamines in plants. Prog Lipid Res 43:302–327

Chapman KD, Venables B, Markovic R, Blair RW Jr, Bettinger C (1999) *N*-Acylethanolamines in seeds. Quantification of molecular species and their degradation upon imbibition. Plant Physiol 120:1157–1164

Cravatt BF, Lichtman AH (2003) Fatty acid amide hydrolase: an emerging therapeutic target in the endocannabinoid system. Curr Opin Chem Biol 7:469–475

Cravatt BF, Giang DK, Mayfield SP, Boger DL, Lerner RA, Gilula NB (1996) Molecular characterization of an enzyme that degrades neuromodulatory fatty-acid amides. Nature 384:83–87

Dunkley TP, Hester S, Shadforth IP, Runions J, Weimar T, Hanton SL, Griffin JL, Bessant C, Brandizzi F, Hawes C, Watson RB, Dupree P, Lilley KS (2006) Mapping the Arabidopsis organelle proteome. Proc Natl Acad Sci U S A 103:6518–6523

Han L, Gao JR, Li ZM, Zhang Y, Guo WM (2007) Synthesis of new plant growth regulator: *N*-(fatty acid) *O*-aryloxyacetyl ethanolamine. Bioorg Med Chem Lett 17:3231–3234

Kang L, Wang Y-S, Uppalapati SR, Wang K, Tang Y, Vadapalli V, Venables BJ, Chapman KD, Blancaflor EB, Mysore KS (2008) Overexpression of a fatty acid amide hydrolase compromises innate immunity in *Arabidopsis*. Plant J 56:336–349

Kathuria S, Gaetani S, Fegley D, Valino F, Duranti A, Tontini A, Mor M, Tarzia G, La Rana G, Calignano A, Giustino A, Tattoli M, Palmery M, Cuomo V, Piomelli D (2003) Modulation of anxiety through blockade of anandamide hydrolysis. Nat Med 9:76–81

Kilaru A, Blancaflor EB, Venables B, Tripathy S, Mysore K, Chapman KD (2007) The *N*-Acylethanolamine-mediated regulatory pathway in plants. Chem Biodivers 4(8):1933–1955

Kurahashi Y, Ueda N, Suzuki H, Suzuki M, Yamamoto S (1997) Reversible hydrolysis and synthesis of anandamide demonstrated by recombinant rat fatty-acid amide hydrolase. Biochem Biophys Res Commun 237:512–515

Labar G, Michaux C (2007) Fatty acid amide hydrolase: from characterization to therapeutics. Chem Biodivers 4:1882–1902

Lopez-Bucio J, Acevedo-Hernandez G, Ramirez-Chavez E, Molina-Torres J, Herrera-Estrella L (2006) Novel signals for plant development. Curr Opin Plant Biol 9:523–529

Lopez-Bucio J, Millan-Godinez M, Mendez-Bravo A, Morquecho-Contreras A, Ramirez-Chavez E, Molina-Torres J, Perez-Torres A, Higuchi M, Kakimoto T, Herrera-Estrella L (2007) Cytokinin receptors are involved in alkamide regulation of root and shoot development in Arabidopsis. Plant Physiol 145:1703–1713

Lopez-Molina L, Mongrand S, Chua NH (2001) A postgermination developmental arrest checkpoint is mediated by abscisic acid and requires the ABI5 transcription factor in Arabidopsis. Proc Natl Acad Sci U S A 98:4782–4787

Lopez-Molina L, Mongrand S, McLachlin DT, Chait BT, Chua N-H (2002) ABI5 acts downstream of ABI3 to execute an ABA-dependent growth arrest during germination. Plant J 32:317–328

McKinney MK, Cravatt BF (2005) Structure and function of fatty acid amide hydrolase. Annu Rev Biochem 74:411–432

Merkel O, Schmid PC, Paltauf F, Schmid HH (2005) Presence and potential signaling function of *N*-acylethanolamines and their phospholipid precursors in the yeast *Saccharomyces cerevisiae*. Biochim Biophys Acta 1734:215–219

Motes CM, Pechter P, Yoo CM, Wang YS, Chapman KD, Blancaflor EB (2005) Differential effects of two phospholipase D inhibitors, 1-butanol and *N*-acylethanolamine, on in vivo cytoskeletal organization and Arabidopsis seedling growth. Protoplasma 226:109–123

Neu D, Lehmann T, Elleuche S, Pollmann S (2007) Arabidopsis amidase 1, a member of the amidase signature family. FEBS J 274:3440–3451

Nishimura N, Yoshida T, Kitahata N, Asami T, Shinozaki K, Hirayama T (2007) ABA-Hypersensitive Germination1 encodes a protein phosphatase 2C, an essential component of abscisic acid signaling in Arabidopsis seed. Plant J 50:935–949

Oddi S, Bari M, Battista N, Barsacchi D, Cozzani I, Maccarrone M (2005) Confocal microscopy and biochemical analysis reveal spatial and functional separation between anandamide uptake and hydrolysis in human keratinocytes. Cell Mol Life Sci 62:386–395

O'Malley RC, Alonso JM, Kim CJ, Leisse TJ, Ecker JR (2007) An adapter ligation-mediated PCR method for high-throughput mapping of T-DNA inserts in the Arabidopsis genome. Nat Protoc 2:2910–2917

Pollmann S, Neu D, Lehmann T, Berkowitz O, Schafer T, Weiler EW (2006) Subcellular localization and tissue specific expression of amidase 1 from Arabidopsis thaliana. Planta 224:1241–1253

Schmid HH, Schmid PC, Natarajan V (1996) The *N*-acylation-phosphodiesterase pathway and cell signalling. Chem Phys Lipids 80:133–142

Shrestha R, Noordermeer MA, van der Stelt M, Veldink GA, Chapman KD (2002) *N*-acylethanolamines are metabolized by lipoxygenase and amidohydrolase in competing pathways during cottonseed imbibition. Plant Physiol 130:391–401

Shrestha R, Dixon RA, Chapman KD (2003) Molecular identification of a functional homologue of the mammalian fatty acid amide hydrolase in *Arabidopsis thaliana*. J Biol Chem 278:34990–34997

Shrestha R, Kim SC, Dyer JM, Dixon RA, Chapman KD (2006) Plant fatty acid (ethanol) amide hydrolases. Biochim Biophys Acta 1761:324–334

Sun YX, Tsuboi K, Zhao LY, Okamoto Y, Lambert DM, Ueda N (2005) Involvement of *N*-acylethanolamine-hydrolyzing acid amidase in the degradation of anandamide and other *N*-acylethanolamines in macrophages. Biochim Biophys Acta 1736:211–220

Teaster ND, Motes CM, Tang Y, Wiant WC, Cotter MQ, Wang YS, Kilaru A, Venables BJ, Hasenstein KH, Gonzalez G, Blancaflor EB, Chapman KD (2007) *N*-Acylethanolamine metabolism interacts with abscisic acid signaling in *Arabidopsis thaliana* seedlings. Plant Cell 19:2454–2469

Tsuboi K, Sun YX, Okamoto Y, Araki N, Tonai T, Ueda N (2005) Molecular characterization of *N*-acylethanolamine-hydrolyzing acid amidase, a novel member of the choloylglycine hydrolase family with structural and functional similarity to acid ceramidase. J Biol Chem 280:11082–11092

Tsuboi K, Zhao LY, Okamoto Y, Araki N, Ueno M, Sakamoto H, Ueda N (2007) Predominant expression of lysosomal N-acylethanolamine-hydrolyzing acid amidase in macrophages revealed by immunochemical studies. Biochim Biophys Acta 1771:623–632

Venables BJ, Waggoner CA, Chapman KD (2005) *N*-acylethanolamines in seeds of selected legumes. Phytochemistry 66:1913–1918

Wang YS, Shrestha R, Kilaru A, Wiant W, Venables BJ, Chapman KD, Blancaflor EB (2006) Manipulation of Arabidopsis fatty acid amide hydrolase expression modifies plant growth and sensitivity to *N*-acylethanolamines. Proc Natl Acad Sci U S A 103:12197–12202

Wei BQ, Mikkelsen TS, McKinney MK, Lander ES, Cravatt BF (2006) A second fatty acid amide hydrolase with variable distribution among placental mammals. J Biol Chem 281:36569–36578

Wilson RI, Nicoll RA (2002) Endocannabinoid signaling in the brain. Science 296:678–682

Zhang Y, Guo WM, Chen SM, Han L, Li ZM (2007) The role of *N*-lauroylethanolamine in the regulation of senescence of cut carnations (*Dianthus caryophyllus*). J Plant Physiol 164: 993–1001

Zimmermann P, Hirsch-Hoffmann M, Hennig L, Gruissem W (2004) GENEVESTIGATOR. Arabidopsis microarray database and analysis toolbox. Plant Physiol 136:2621–2632

Sphingolipid Signaling in Plants

Louise V. Michaelson and Johnathan A. Napier

Abstract Until relatively recently, plant sphingolipids were generally considered to be minor cellular components of limited importance. However, research over the last decade has shown that sphingolipids play key roles in many aspects of plant biology, including reproduction, development and also biotic and abiotic stress. Sphingolipids play an important role as structural components of membranes and have been shown to be enriched in membrane microdomains synonomous with so-called lipid rafts. In addition to their structural role, sphingolipid metabolites are, by analogy with other non-plant systems, likely to act as signaling compounds. This emerging area is discussed in this chapter and the evidence for sphingolipid signaling in plants is considered.

1 Introduction

Sphingolipids are ubiquitous membrane lipids that have been shown to be essential in many different eukaryotes (Dunn et al. 2004). They have considerable natural chemical diversity and were generally believed to be the structural components of cell membranes until approximately 20 years ago. The discovery that protein kinase C was inhibited by sphingosine in human platelets generated considerable research interest in sphingolipids as bioactive molecules (Hannun et al. 1986), and since that observation, they have been implicated in many cellular processes in yeast and animal cells, including aspects of signaling, apoptosis, and senescence (Huwiler et al. 2000; Obeid et al. 1993; Venable et al. 1995). Sphingolipids form a significant proportion of the lipids present in higher plants, with some studies suggesting that they constitute up to 10% of plant lipids (reviewed in Dunn et al. 2004). However, their roles in plants have only recently started to be elucidated, inevitably building on

L.V. Michaelson (✉)
Biological Chemistry Department, Rothamsted Research, Harpenden, Herts AL5 2JQ, UK
e-mail: louise.michaelson@bbsrc.ac.uk

T. Munnik (ed.), *Lipid Signaling in Plants*, Plant Cell Monographs 16,
DOI 10.1007/978-3-642-03873-0_20,

observations made in other systems but recently also focussing on plant-specific processes. The aim of this short article is to review our current understanding of sphingolipid synthesis in plants and to consider the evidence for sphingolipids acting as signaling molecules. We will also discuss the suitability of using nonplant paradigms as models for conceptualizing sphingolipid signaling in plants.

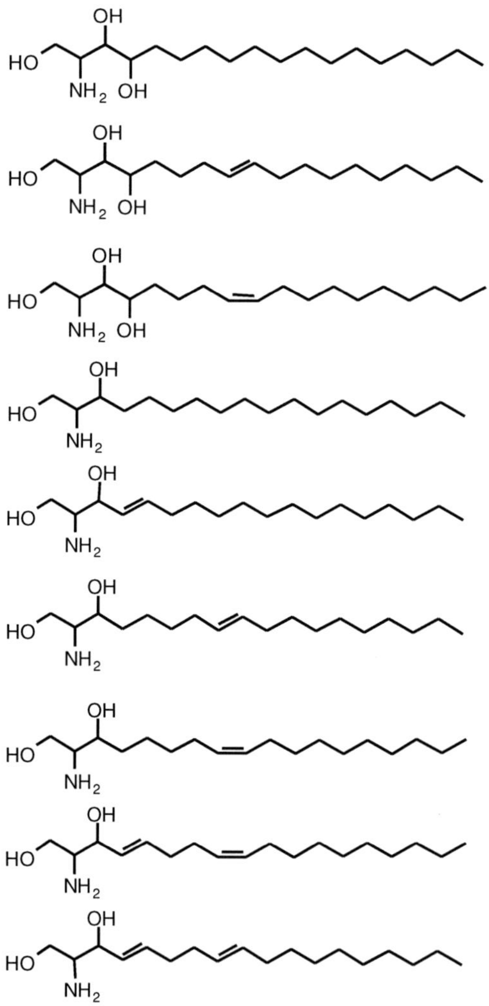

4-hydroxysphinganine (t18:0)

4-hydroxy-Δ^8*trans*-sphingenine (t18:1Δ^{8t})

4-hydroxy-Δ^8*cis*-sphingenine (t18:1Δ^{8c})

sphinganine (d18:0)

Δ^4*trans*-sphingenine(d18:1Δ^{4t})

Δ^8*trans*-sphingenine (d18:1Δ^{8t})

Δ^8*cis*-sphingenine (d18:1Δ^{8c})

Δ^4*trans*,^{8}*cis*-sphingadiene (d18:2$\Delta^{4t,8c}$)

Δ^4*trans*,^{8}*trans*-sphingadiene (18:2$\Delta^{4t,8t}$)

Fig. 1 The diversity of long-chain bases present in higher plant sphingolipids. Note that the double bond at the Δ8-position may be in either the *cis* or the *trans* configuration, resulting from the action of the stereo-unspecific Δ8-desaturase

2 Sphingolipid Structure and Biosynthesis

2.1 Sphingolipid Structure

The basic structure of the sphingolipid molecule consists of a ceramide linked via the terminal hydroxyl group to phosphate or to carbohydrate moieties. The ceramide is made up of a long-chain base (LCB; Fig. 1) linked to a fatty acid via an amide bond. Over 300 different sphingolipids have been structurally characterized (Merrill and Sweeley 1996). The variability in each of these three components required for the assembly of a mature sphingolipid allows for the possibility of development of thousands of different molecular species. In addition, the sphingolipid components, LCBs and ceramides, can have important biological activities in their own right and are often included when discussing the sphingolipid "family" of lipids (as they will be here). These components can be generated through de novo synthesis (i.e., as a part of the biosynthesis of sphingolipids) or via catabolism of the mature lipid.

The cerebrosides (glucosylceramides) and glycosylated inositolphosphorylceramides (GIPCs) are considered to be the predominant sphingolipids in plant tissue (Lynch 1993; Markham et al. 2006). This contrasts to the situation in animals, where the main classes are sphingomyelin (a class of phosphorylceramide not found in higher plants; Tafesse et al. 2006) and the neutral and acidic glycolipids, and to that in yeast where inositolphosphorylceramides (IPCs) and their mannosylated derivatives predominate (Lester and Dickson 1993). In higher plants, the absolute amounts of the sphingolipid classes tend to vary in a species- and a tissue-dependent manner (Huwiler et al. 2000; Markham et al. 2006), though in *Arabidopsis* leaves, GIPCs are the predominant form (~60% of total sphingolipids), with GlcCers present at approximately half this level (30%). The remaining sphingolipids are present mainly as ceramides, with free LCBs and phosphorylated LCBs representing very minor components (Markham and Jaworski 2007).

The LCB and the fatty acid components of sphingolipids are subject to compositional variation depending on the organism. The predominant LCBs in plants are C_{18} amino alcohols. These are largely comprised of 4-hydroxysphinganine (t18.0), commonly known as phytosphingosine, and its desaturated form 4-hydroxy-8-sphingenine (t18:1Δ^8), sphinganine (d18.0, dihydrosphingosine) and its desaturated forms 8-sphingenine (d18.1Δ^8) and 4,8-sphingadienine (d18.1$\Delta^{4,8}$); 4-sphingenine (d18.1Δ^4 sphingosine) is present in moderate levels in some plant species, but the predominant form of Δ^4-unsaturated is as the $\Delta^{4,8}$diene (Markham et al. 2006) (Fig. 1). A few other rarer LCBs have also been reported in plants (Karlsson 1970), but it is these nine LCBs that represent the predominant forms of LCBs so far detected in plants.

Unusually, the configuration of the double bond in the Δ^8 position of plant sphingolipid LCBs may be either *cis* (*Z*) or *trans* (*E*), whereas the double bond at the Δ^4 position invariably has only the *trans* configuration. The ratio of *cis* to *trans* isomers of Δ^8-unsaturated LCBs varies from species to species in plants and can be influenced by the type of sphingolipid (GIPC vs. glucosylceramide) the LCB is a component of, and this, in turn, may influence the subcellular location of the sphingolipid, not least of all due to the difference in charge between GIPC and neutral cerebrosides (Sperling et al. 2005). It has previously been suggested that the ratio of *cis/trans* Δ^8-unsaturated LCBs is correlated with the chilling tolerance in plants (Kawaguchi et al. 2000). It is also worth noting that the Δ^8-unsaturated LCBs are only widely found in the plant kingdom; they are absent from animals and the yeast *Saccharomyces cerevisiae* (which represents the best characterised organism in terms of sphingolipid biosynthesis), though some fungi such as *Pichia pastoris* and *Candida albicans* have been shown to contain Δ^8-unsaturated LCBs (Ternes et al. 2006). It was also recently demonstrated that the marine diatom *Thalisosira pseudonana* has the capacity to synthesis such LCBs (Tonon et al. 2005). In contrast, the predominant LCB of many animal sphingolipids is sphingosine (d18.1^4); as noted above, this is usually only found as a minority component of plant sphingolipids (Markham et al. 2006).

The fatty acyl component of the ceramide, and therefore the sphingolipid, is generally α-hydroxylated and tends to vary in chain length from 14 to 26 carbons, including some chains of odd carbon numbers (Lynch 1993; Imai et al. 2000). Chains may be desaturated at the ω-9 position, and there is evidence to suggest that this modification of the fatty acid predominates in the glucosylceramides of cold-adapted cereal plants (Imai et al. 1997). The significance of this chemical diversity, however, remains to be explored (see Sperling and Heinz 2003; Lynch and Dunn 2004 for further considerations of this topic).

There are two main types of complex plant sphingolipids, as mentioned above. The glucosylceramides (GlcCers) carry between 1 and 4 glycosyl residues attached to C1 of the *N*-acyl hydroxyl group of the LCB of the ceramide. The GIPCs carry inositol-1-phosphate linked as a phosphodiester to the primary carbon of the ceramide. This can be extended by oligosaccharide chains and these are usually linked at position 2 and/or position 6 of the inositol moiety.

Much of our current understanding of the biosynthesis and function of sphingolipids has stemmed from the isolation and study of sphingolipid metabolism mutants in *S. cerevisiae* and the identification of the genes encoding the enzymes responsible for sphingolipid metabolism. It is, however, important to highlight some important differences between this simple model system and higher plants. Notably, in plant species, there is much greater heterogeneity in the composition of sphingolipids in terms of the constituents (LCBs, fatty acids, and the polar head group), which define a sphingolipid, and this extends beyond just variation in LCB unsaturation (though it must be born in mind that *S. cerevisiae* lacks any unsaturated LCBs, and in that respect can be considered atypical in its sphingolipid composition). For example, higher plants and fungi contain significant amounts of glucosylceramides (GlcCers), yet this class of sphingolipids are completely absent from the yeast *S. cerevisiae*. Interestingly, animal sphingolipids such as sphingomyelin have not been detected in fungi or higher plants (Tafesse et al. 2006). Moreover, the LCB composition of fungal GlcCers differs from that observed in higher plants, with the former containing a C9-methyl branch not present in the latter (Ternes et al. 2006).

2.2 Biosynthesis

The biosynthetic pathway for simple sphingolipids (i.e., ceramides) has been elucidated from genetic studies in yeast and this particular topic has been covered in-depth in a number of recent reviews (Dunn et al. 2004; Sperling and Heinz 2003), and will therefore only be covered briefly here. The first committed step in sphingolipid biosynthesis is the synthesis of the LCB; this requires the activities of two distinct enzymes, both believed to reside in the ER. First, an acyl-CoA substrate (usually palmitoyl-CoA) undergoes a condensation reaction with serine to generate 3-ketosphinganine. This is catalyzed by serine palmitoyl transferase (SPT), which, in yeast, is encoded by two genes LCB1 and LCB2 that form a heterodimer (Hanada 2003). *Arabidopsis* orthologs of both LCB1 and LCB2 have been functionally characterized in vivo and confirmed their critical role in LCB synthesis and the essential role of sphingolipids in higher plants (Chen et al. 2006; Dietrich et al. 2008). 3-ketosphingaine is converted to dihydrosphingosine by 3-ketosphinganine reductase (encoded by TSC10 in yeast), and although orthologs have been identified in the *Arabidopsis* genome, no functional characterisation has yet been published (Beeler et al. 1998; Dunn et al. 2004). In yeast and *Arabidopsis*, it is believed that dihydrosphingosine then undergoes C4-hydroxylation to yield phytosphingosine, though it is possible that this hydroxylation reaction (catalysed by SUR2 in yeast) also occurs on *N*-acylated LCBs (i.e., ceramides) as opposed to the free LCB. In *Arabidopsis*, there are two SUR2 functional orthologs (Sperling et al. 2001), and it appears that phytosphingosine is essential for normal plant growth and development as disruption of these two genes is lethal (Chen et al. 2008). There is also evidence of a role for phytosphingosine in stomatal guard cell closure (Coursol et al. 2005), see below.

Sphingolipids (ceramides and their subsequent derivatives) are predominantly formed by the acyl-CoA: sphinganine *N*-acyltransferase activity which generates a ceramide moiety by *N*-linking a fatty acid to an LCB. This ceramide synthase reaction is encoded by the members of the LAG/LAC class of ER-membrane proteins that are essential in yeast (Guillas et al. 2001). In vitro reconstitution of ceramide synthesis using purified LAG1 and LAC1 confirmed the function of these two proteins, though a third subunit (called LIP1) was found to co-purify with this complex (Schorling et al. 2001; Vallee and Riezman 2005). In plants, reduced sensitivity to fungal toxins such as Fb1 and AAL (known to inhibit ceramide synthesis) has been shown to result from mutations in orthologs of LAG1/LAC1 (Spassieva et al. 2002). Similarly, plant orthologs can rescue the *lag1Δ/lac1Δ* yeast mutant. Thus, the basic machinery for ceramide synthesis is conserved between plants and yeast, though it is likely that some degree of substrate-specificity (in terms of the acceptance of acyl chains for ceramide synthesis) is conferred by the plant enzymes – however, this remains to be experimentally demonstrated.

It must also be noted that the biosynthesis of the acyl-component (usually, but certainly not exclusively, a saturated C_{20+} very long-chain fatty acid) results from a conserved pathway based around the ER fatty acid elongase. This biochemical complex is made up of four activities: a condensing enzyme, a ketoreductase, a dehydratase, and an enoyl-reductase. The *Arabidopsis* orthologs for the "core" components of the elongase (ketoreductase, dehydratase, enoyl-reductase) have all been recently identified and shown to play significant roles in plant growth and development. For example, the dehydratase is encoded by PASTICCINO2 (PAS2); disruption of this gene is embryo lethal (Bach et al. 2008), as is the case for the ketoreductase (Beaudoin et al. 2009). Interestingly, the enoyl reductase was identified as one of the well-known eceriferum mutants which have an altered cuticular wax composition (waxes contain very long-chain fatty acids). Mutations in CER10 resulted in reduced synthesis of sphingolipid VLCFAs, and also a reduced growth (Zheng et al. 2005). Surprisingly, *cer10-2* T-DNA insertion mutants are viable, indicating a likely functional redundancy with an as-yet unidentified enoyl-reductase (though as noted by Gable et al. 2004, such an activity must be structurally unrelated to the CER10 gene and its yeast ortholog TSC13). In addition to demonstrating that VLCFAs are required for plant growth and development, these elongase mutants have also revealed some important insights into the hierarchal significance of different sphingolipid classes. Specifically, the *pas2-1* partially viable mutant displays greatly reduced levels of GlcCers, but in contrast, levels of GIPCs remain unchanged. This would indicate that GIPCs are maintained at the expense of GlcCers in the *pas2-1* mutant, implying that the former are more critical for cell functions.

3 Sphingolipid Function

3.1 Membrane Functions

It is fair to say that it is only over the last 10 years that sphingolipids in plants have been thought to have a role beyond the "structural" role envisaged for them as membrane components. It is thought that most complex sphingolipids are located in the membranes but they and their metabolites have also been shown to have a role in intracellular and extra cellular signal transduction. Previous studies have shown that GlcCers are a quantitatively important component of plasma membrane (Lynch 1993) and tonoplast (Yoshida and Uemura 1986; Tavernier et al. 1993). It has also been suggested that the heavily hydroxylated species of GlcCers present in many plant tissues may contribute to the overall integrity of the plasma membrane and tonoplast (Boggs 1987). Biophysically, GlcCers demonstrate unusual behaviour (Lynch et al. 1992; Norberg et al. 1996) and have been implicated to play a role in chilling and freezing tolerance in plants (Steponkus and Lynch 1989; Uemura and Steponkus 1994; Uemura et al. 1995). The plasma membranes of freezing-tolerant plants have been shown to contain lower levels of glucosylceramide, and cold acclimatisation can also reduce the level of glucosylceramide. The desaturation status of the acyl chains was shown to be important in regulating this cold tolerance.

It has been demonstrated recently that the way that sphingolipids are distributed in cell membranes is not uniform and that they are concentrated in specialised microdomains. These domains or the so-called "lipid rafts" are thought to play important roles in protein sorting, signal transduction, and infection by pathogens. The microdomains are clusters of sphingolipids and sterols (sitosterol and stigmasterol) and are characterised by their insolubility at low temperature and in nonionic detergents (Beck et al. 2007). Such properties are likely to be derived from the differences in structure (and hence biophysical properties) between sphingolipids and phosphoglycerolipids. They have been found in tobacco, (Peskan et al. 2000; Mongrand et al. 2004), *Medicargo truncatula* (Lefebvre et al. 2007), and *Arabidopsis* (Borner et al. 2005). Analysis of the proteins of these microdomains has identified the presence of a number of GPI-anchored proteins, receptor kinases, and proteins related to signaling and cellular trafficking (Borner et al. 2005; Lefebvre et al. 2007). It seems that sphingolipids are one of the main components in establishing the microdomain as they are able to associate with each other through interactions between their carbohydrate heads and their long hydrophobic tails (Simons and Ikonen 1997). However, a degree of controversy exists as to whether the detergent-resistant fractions that have been identified as the so-called lipid rafts actually represent membrane microdomains (Munro 2003). There is also little obvious overlap in the protein complements of detergent-resistant membranes (DRMs) isolated from different plant species. Perhaps more surprising, the sphingolipid composition of *Arabidopsis* DRMs appeared to be enriched in GIPCs (as judged by LCB profiles) whereas tobacco DRMs contained GluCers (Borner et al. 2005; Lefebvre et al. 2007; Mongrand et al. 2004). Thus, there is a need for further, more detailed studies, on the role of plant lipid rafts and their relationship to DRMs. It would also be interesting to determine the origins of the GPI moiety present in many PM proteins, since it is possible that this can be derived from acyl-exchange with GIPC sphingolipids (Bosson and Conzelmann 2007).

Another intriguing observation regarding the role of sphingolipids and membrane function was reported by Ryan et al. (2007), demonstrating that changes in the sphingolipid composition of cell membranes can protect plants from aluminium stress. Expression of a stereo-biased Δ^8 LCB desaturase from *Stylosanthes hamata* conferred aluminium tolerance to yeast and plants. Expression of this cDNA leads to the accumulation of 8(*Z*/*E*)-C_{18}-phytosphingenine and 8(*Z*/*E*)-C_{20}-phytopshingenine in yeast and to the accumulation of 8(*Z*/*E*)-C_{18}-phytosphingenine in the leaves and roots of *Arabidopsis* plants. The accumulation of 8(*Z*/*E*)-C_{18}-phytosphingenine in transgenic *Arabidopsis* shifted the ratio of the 8(Z):8(E) isomers from 1:4 in wild-type plants to 1:1 in transgenic plants (Ryan et al. 2007). This could change the permeability of the plasma membrane to aluminium (generally (*Z*)-unsaturated lipids are thought to pack less well); it could change the microdomain structure which favours tolerance to aluminium. In addition to this work, the LCB-1-phosphate lyase that degrades the LCB-1-P to C_{16} fatty aldehydes and phosphoethanolamine has been shown to play a role in dehydration stress in *Arabidopsis* (Nishikawa et al. 2008).

3.2 Signaling and Cell Regulation

There is much evidence from the animal and yeast fields to suggest that sphingolipids and their metabolites are integral to cell regulation and signaling. In animal cells, the phosphorylated sphingolipid metabolite sphingosine-1-phosphate (S1P) acts as a potent messenger, modulating a range of processes such as proliferation and apoptosis (Saba and Hla 2004). A number of roles for phosphorylated LCBs (LCB-1P) have been observed in fungi, Drosophila, and *C. elegans* (Oskouian and Saba 2004). Because of this body of work, there has been a substantial effort to see if sphingolipids in plants perform an analogous role.

The first demonstration of a sphingolipid metabolite acting in plants cells was that of Ng et al. (2001), in which it was reported that S1P is a calcium-mobilizing molecule in plants. Specifically, exogenous application of micromolar concentration of S1P resulted in calcium-mediated guard cell close, whereas similar treatment with dihydrosphingosine-1-P did not elicit such a response. They also presented data to show that after drought treatment, S1P levels increased. These data imply that S1P is involved in the signal-transduction pathway that links the perception of abscisic acid to reductions in guard cell turgor. However, S1P is present at very low levels in plant tissues, dependent on whether the cells accumulated sphingosine. The study of Ng et al. (2001) illustrates that even at low levels, sphingolipids can have a large effect. Coursol et al. (2003) also looked at the signaling potential of S1P. They showed that the enzyme responsible for S1P production, sphingosine kinase, is stimulated by the abscisic acid in guard cells of *Arabidopsis* and that S1P is effective in regulating guard cell turgor. LCBs can also modulate nuclear calcium signaling in tobacco cells (Xiong et al. 2008). Further work published by Coursol et al. 2005 found that phytosphingosine-1-phosphate regulates stomatal aperture and that its action was impaired in guard cells of *Arabidopsis* plants harbouring T-DNA null mutations in the sole prototypical G-protein α-subunit gene. G-protein $\alpha 1$ regulates guard-cell function (Wang et al. 2001; Coursol et al. 2003) and plant cell division (Ullah et al. 2001; Perfus-Barbeoch et al. 2004). Sphingolipid-derived signals could, therefore, play a role in other G-protein-mediated processes in plants, but it is unlikely that it acts in the same way as in animal systems since the components of the signaling pathway for LCB-1Ps are not conserved across eukaryotes. In animals, the receptor for S1P is a specific EDG-class G-protein-coupled receptor, yet there is very limited evidence for non-mammalian systems using such a receptor (Saba and Hla 2004), not least of all in *Arabidopsis* which lacks any ortholog of this class of protein (Assmann 2005; Coursol et al. 2005). Thus, although it is clear that LCBs and LCB-1Ps play important roles in many cellular processes, it is likely that such processes are phyla- or even species-specific and as such, a single unifying paradigm for sphingolipid signaling is unlikely to prove reliable.

In animals, sphingosine (the direct precursor of S1P; Fig. 1) is the most abundant LCB (Pyne and Pyne 2000). However, sphingosine is a very minor component of most higher plant sphingolipids (Imai et al. 1997; Lynch and Dunn 2004;

Dunn et al. 2004; Markham et al. 2006; Sullards et al. 2000) with the predominant LCB modifications being either C4-hydroxylation, Δ^8-desaturation or both, yielding phytosphingosine, sphing-8-enine, or 4-hydroxysphing-8-enine, respectively (Fig. 1). While Δ^4-unsaturated sphingolipids are abundant in some plant species (such as soybean), these predominantly occur in conjunction with Δ^8-unsaturation, in the form of sphinga-4,8-dienine (Markham et al. 2006). In order to better define the role of sphingosine and S1P in *Arabidopsis*, the single ortholog of the sphingolipid Δ^4-desaturase has been functionally characterised (Michaelson et al. 2009). Sphingolipidomic profiles were obtained from leaf and floral tissue of WT and insertion mutant alleles of the sphingolipid Δ^4-desaturase At4g04930. No Δ^4-unsaturated sphingolipids or sphingolipid metabolites were detected in these leaf samples, indicating the absence of sphingosine and sphinga-4,8-dienine in all samples and a Col-0-background insertion mutant line lacked sphinga-4,8-dienine in sphingolipids of floral tissue. Insertional mutagenesis of the *Arabidopsis* sphingolipid Δ4-desaturase did not result in any phenotypic alterations to growth or development, likely indicating a very limited role for S1P and other Δ^4-unsaturated LCBs and their phosphorylated metabolites in this plant.

This data, however, indicated a role for Δ^4-unsaturated LCBs in the channelling of ceramides for the synthesis of GlcCers in certain tissues, though these insertion mutants displayed normal growth and development and did not indicate any perturbation to drought-tolerance, transpiration rate, or pollen viability. Though S1P was shown not to be essential in *Arabidopsis*, it is very likely that other LCBs and LCB-1Ps play the important roles in plant-specific processes. The importance of phytosphingosine-1-P was mentioned earlier (Coursol et al. 2005). Recently published work looking at the involvement of sphingosine kinase has shown that the stomata of sphingosine kinase 1-disrupted *Arabidopsis* plants were less sensitive than wild type to ABA (Worrall et al. 2008). Transgenic *Arabidopsis* plants over-expressing the sphingosine kinase were more sensitive than wild type to ABA. Germination rates in these plants were also affected. This work suggests that phosphorylated long-chain LCB bases can act as messengers in plants, though it remains to be demonstrated what the in vivo substrates of the LCB kinase are.

Work by Chen et al. (2008) has further shown the importance of the trihydroxy LCBs in *Arabidopsis* since these are the precursors to phytosphingosine-1-P. T-DNA mutants and RNA interference suppression lines for the two *Arabidopsis* LCB C4 hydroxylase genes resulted in plants that displayed reductions in growth. The double mutant, completely lacked trihydroxy LCBs, was severely dwarfed. The double mutant was unable to progress from vegetative to reproductive growth. In addition, it showed enhanced expression of programmed cell-death-associated genes. The amount of sphingolipids on a per weight basis increased as the amount of trihydroxy LCBs decreased. This work shows that the hydroxylation of LCBs is important in growth and is tightly regulated.

Mycotoxins such as AAL have been used as experimental tools to disrupt sphingolipid metabolism. They were found to induce a lethal accumulation of LCBs in a variety of plant species (Abbas et al. 1994) These mycotoxins promote apoptosis during cell death in tomato (Wang et al. 1996). AAL toxin prevents ceramide and therefore "mature" sphingolipid biosynthesis, causing the accumulation of LCBs and leading to cell death (Wang et al. 1996). If LCB synthesis via SPT is inhibited, some of these toxic effects are reduced (Spassieva et al. 2002). LCB biosynthesis has a role in programmed cell death in *Arabidopsis* (Brodersen et al. 2002). The *accelerated-cell-death11 Arabidopsis* mutant (*acd11*) constitutively expresses defence-related genes that accompany the hypersensitive response normally triggered by pathogens. ACD11 encodes a protein that may play a role in the transfer of LCBs but not glycosphingolipids, between membranes in vitro. Very recently, the *Arabidopsis* IPC synthase which converts ceramide to IPC has been identified and has been shown to be involved in enhancing the hypersensitive response triggered by the RPW8 gene (Wang et al. 2008). RPW8 triggers the hypersensitive response to restrict powdery mildew infection via the salicylic acid-dependent signaling pathway. Mutation in the IPC synthase causes ceramide accumulation in plants expressing RPW8. This data strengthen the case that sphingolipid metabolism plays a role in modulating programmed cell death, associated with defence, in plants.

Work by Shi et al. (2007) has shown that an *Arabidopsis* mutant disrupted in the LCB1 subunit of serine palmitoyltransferase (SPT) fails to generate reactive oxygen species (ROS) and is incapable of initiating programmed cell death or apoptosis when the mutant is challenged by inhibitors of ceramide synthesis such as fumonisin. They showed that free LCB bases dihydrosphingosine, phytosphingosine, and sphingosine reduced ROS generation followed by cell death. ROS generation and cell death induced by dihydrosphingosine was shown to be blocked by its phosphorylated form dihydrosphingosine-1-phosphate in a dose-dependent manner, suggesting that the maintenance of homeostasis between a free LCB and its phosphorylated derivative is critical to determining the cell fate. They suggest that free LCBs are involved in the control of programmed cell death in *Arabidopsis* through the regulation of ROS levels upon receiving different developmental or environmental cues. In addition, an *Arabidopsis* ceramide kinase mutant (*acd5*; *accelerated-cell-death-5*) shows enhanced disease symptoms during pathogen attack and apoptotic-like cell death dependent on defence signaling late in development. (Liang et al. 2003) It was found that overaccumulation of ceramide induces programmed cell death, whereas its phosphorylated derivative partially blocks this response, supporting a role for ceramide phosphorylation in modulating cell death in plants. These results collectively suggest that sphingolipids and their metabolites play important roles in plant defence and cell death.

The *Arabidopsis* SPT is a heteromeric enzyme comprised of two subunits similar to that which is found in yeast and mammals. *Arabidopsis* plants knocked out for the gene encoding the LCB1 subunit caused the formation of abortive microspores and initiated apoptotic cell death in binucleated microspores (Teng et al. 2008). These results suggest that SPT-modulated programmed cell death plays an important role in the regulation of male gametophyte development (Teng et al. 2008). Dietrich et al. (2008) demonstrated that for the two *Arabidopsis* genes encoding the LCB2 activity, it was not possible to recover homozygous double mutants and that this lethality is associated primarily with the inability to transmit the *lcb2Δ* genotype through the haploid pollen. Similar to the work on LCB1, when pollen was obtained from plants homozygous for a mutation in one gene and heterozygous for a mutation in the second gene, it was shown to have arrested during transition from uni-nucleate microspore to bicellular pollen. This is partially analogous to some of the findings in Drosophila where flies disrupted in sphingolipid biosynthesis were found to be male sterile (Endo et al. 1996; Phan et al. 2007). In a different study when the LCB1 gene was partially suppressed in *Arabidopsis*, there was a reduction in plant size which was as the result of reduced cell expansion (Chen et al. 2006). In these plants, the amount of sphingolipid present by weight was maintained, so it seems that the plants compensated for the lack of sphingolipids with reduced growth, showing the importance of the sphingolipids in metabolism and the tight control that they are under.

4 Conclusions

The understanding of sphingolipid signaling and its regulation in plants is still in its infancy. Work is required to better understand the compartmentation of biosynthetic activities, substrate-channelling, and assembly of metabolic complexes. It is clear that sphingolipids play important role, as bioactive molecules in development and metabolism, and it is expected that new functions and roles will emerge to elucidate the mechanisms that are used for the transmission and integration of developmental and environmental cues. New developments in analytical techniques which allow detailed sphingolipidomic profiling in mutant and metabolically altered plants have recently increased the knowledge base (Markham and Jaworski 2007). As these techniques are combined with available sphingolipid mutants, further precise roles are expected to be discovered for sphingolipids in a range of signaling pathways in plant metabolism. In addition, forward genetic screens continue to indicate the importance of sphingolipids in many processes – such unbiased approaches widen the scope of our understanding of sphingolipid function in plant biology. Finally, we would add that while studies on sphingolipid metabolism in plants have clearly benefited from drawing from studies in other systems, it is equally clear that plant-specific facets of this pathway need to be factored into our considerations of the role of these lipids in non-animal systems.

Reference

Abbas HK, Tanaka T, Duke SO, Porter JK, Wray EM, Hodges L, Sessions AE, Wang E, Merrill AH Jr, Riley RT (1994) Fumonisin- and AAL-toxin-induced disruption of sphingolipid metabolism with accumulation of free sphingoid bases. Plant Physiol 106:1085–1093

Assmann SM (2005) G proteins go green: a plant G protein signalling FAQ sheet. Science 310:71–73

Bach L, Michaelson LV, Haslam R, Bellec Y, Gissot L, Marion J, Da Costa M, Boutin JP, Miquel M, Tellier F, Domergue F, Markham JE, Beaudoin F, Napier JA, Faure JD (2008) The very-long-chain hydroxy fatty acyl-CoA dehydratase PASTICCINO2 is essential and limiting for plant development. Proc Natl Acad Sci 105:14727–14731

Beaudoin F, Wu X, Li F, Haslam RP, Markham JE, Zheng H, Napier JA, Kunst L (2009) Functional characterization of the Arabidopsis beta-ketoacyl-coenzyme A reductase candidates of the fatty acid elongase. Plant Physiol 150:1174–1191

Beck JG, Mathieu D, Loudet C, Buchoux S, Dufourc EJ (2007) Plant sterols in "rafts": a better way to regulate membrane thermal shocks. FASEB J 21:1714–1723

Beeler T, Bacikova D, Gable K, Hopkins L, Johnson C, Slife H, Dunn T (1998) The *Saccharomyces cerevisiae* TSC10/YBR265w gene encoding 3-ketosphinganine reductase is identified in a screen for temperature-sensitive suppressors of the Ca2 + -sensitive csg2Δ mutant. J Biol Chem 273:30688–30694

Boggs JM (1987) Lipid intermolecular hydrogen bonding: influence on structural organization and membrane function. Biochim Biophys Acta 906:353–404

Borner GH, Sherrier DJ, Weimar T, Michaelson LV, Hawkins ND, Macaskill A, Napier JA, Beale MH, Lilley KS, Dupree P (2005) Analysis of detergent-resistant membranes in Arabidopsis. Evidence for plasma membrane lipid rafts. Plant Physiol 137:104–116

Bosson R, Conzelmann A (2007) Multiple functions of inositolphosphorylceramides in the formation and intracellular transport of glycosylphosphatidylinositol-anchored proteins in yeast. Biochem Soc Symp 74:199–209

Brodersen P, Petersen M, Pike HM, Olszak B, Skov S, Odum N, Jørgensen LB, Brown RE, Mundy J (2002) Knockout of Arabidopsis accelerated-cell-death11 encoding a sphingosine transfer protein causes activation of programmed cell death and defense. Genes Dev 16:490–502

Chen M, Han G, Dietrich CR, Dunn TM, Cahoon EB (2006) The essential nature of sphingolipids in plants as revealed by the functional identification and characterization of the Arabidopsis LCB1 subunit of serine palmitoyltransferase. Plant Cell 18:3576–3593

Chen M, Markham JE, Dietrich CR, Jaworski JG, Cahoon EB (2008) Sphingolipid long-chain base hydroxylation is important for growth and regulation of sphingolipid content and composition in Arabidopsis. Plant Cell 20:1862–1878

Coursol S, Fan LM, Le Stunff H, Spiegel S, Gilroy S, Assmann SM (2003) Sphingolipid signalling in Arabidopsis guard cells involves heterotrimeric G proteins. Nature 423:651–654

Coursol S, Le Stunff H, Lynch DV, Gilroy S, Assmann SM, Spiegel S (2005) Arabidopsis sphingosine kinase and the effects of phytosphingosine-1-phosphate on stomatal aperture. Plant Physiol 137:724–737

Dietrich CR, Han G, Chen M, Berg RH, Dunn TM, Cahoon EB (2008) Loss-of-function mutations and inducible RNAi suppression of Arabidopsis LCB2 genes reveal the critical role of sphingolipids in gametophytic and sporophytic cell viability. Plant J. 54:284–298

Dunn TM, Lynch DV, Michaelson LV, Napier JA (2004) A post-genomic approach to understanding sphingolipid metabolism in Arabidopsis thaliana. Ann Bot 93:483–497

Endo K, Akiyama T, Kobayashi S, Okada M (1996) Degenerative spermatocyte, a novel gene encoding a transmembrane protein required for the initiation of meiosis in Drosophila spermatogenesis. Mol Gen Genet 253:157–165

Gable K, Garton S, Napier JA, Dunn TM (2004) Functional characterization of the Arabidopsis thaliana orthologue of Tsc13p, the enoyl reductase of the yeast microsomal fatty acid elongating system. J Exp Bot 55:543–545

Guillas I, Kirchman PA, Chuard R, Pfefferli M, Jiang JC, Jazwinski SM, Conzelmann A (2001) C26-CoA-dependent ceramide synthesis of *Saccharomyces cerevisiae* is operated by Lag1p and Lac1p. EMBO J 20:2655–2665

Hanada K (2003) Serine palmitoyltransferase, a key enzyme of sphingolipid metabolism. Biochim Biophys Acta 1632:16–30

Hannun YA, Loomis CR, Merrill AH Jr, Bell RM (1986) Sphingosine inhibition of protein kinase C activity and of phorbol dibutyrate binding in vitro and in human platelets. J Biol Chem 261:12604–12609

Huwiler A, Kolter T, Pfeilschifter J, Sandhoff K (2000) Physiology and pathophysiology of sphingolipid metabolism and signalling. Biochim Biophys Acta 1485:63–99

Imai H, Ohnishi M, Hotsubo K, Kojima M, Ito S (1997) Sphingoid base composition of cerebrosides from plant leaves. Biosci Biotechnol Biochem 61:351–353

Imai H, Yamamoto K, Shibahara A, Miyatani S, Nakayama T (2000) Determining double-bond positions in monoenoic 2-hydroxy fatty acids of glucosylceramides by gas chromatography-mass spectrometry. Lipids 35:233–236

Karlsson KA (1970) Sphingolipid long chain bases. Lipids 5:878–891

Kawaguchi M, Imai H, Naoe M, Yasui Y, Ohnishi M (2000) Cerebrosides in grapevine leaves: distinct composition of sphingoid bases among the grapevine species having different tolerances to freezing temperature. Biosci Biotechnol Biochem 64:1271–1273

Lefebvre B, Furt F, Hartmann MA, Michaelson LV, Carde JP, Sargueil-Boiron F, Rossignol M, Napier JA, Cullimore J, Bessoule JJ, Mongrand S (2007) Characterization of lipid rafts from *Medicago truncatula* root plasma membranes: a proteomic study reveals the presence of a raft-associated redox system. Plant Physiol 144:402–418

Lester RL, Dickson RC (1993) Sphingolipids with inositolphosphate-containing head groups. Adv Lipid Res 26:253–274

Liang H, Yao N, Song JT, Luo S, Lu H, Greenberg JT (2003) Ceramides modulate programmed cell death in plants. Genes Dev 17:2636–2641

Lynch D (1993) Sphingolipids. In: Moore TS Jr (ed) Lipid metabolism in plants. CRC, Boca Raton, FL, pp 285–308

Lynch DV, Dunn TM (2004) An introduction to plant sphingolipids. New Phytol 161:677–702

Lynch DV, Caffrey M, Hogan JL, Steponkus PL (1992) Calorimetric and x-ray diffraction studies of rye glucocerebroside mesomorphism. Biophys J 61:1289–1300

Markham JE, Jaworski JG (2007) Rapid measurement of sphingolipids from *Arabidopsis thaliana* by reversed-phase high-performance liquid chromatography coupled to electrospray ionization tandem mass spectrometry. Rapid Commun Mass Spectrom 21:1304–1314

Markham JE, Li J, Cahoon EB, Jaworski JG (2006) Separation and identification of major plant sphingolipid classes from leaves. J Biol Chem 281:22684–22694

Merrill AH Jr, Sweeley CC (1996) Sphingolipid metabolism and cell signalling. In: Vance DE, Vance JE (eds) New comprehensive biochemistry: biochemistry of lipids, lipoproteins, and membranes. Elsevier Science, Amsterdam, The Netherlands, pp 309–338

Michaelson LV, Zauner S, Markham JE, Haslam R, Desikan R, Mugford S, Albrecht S, Warnecke D, Sperling P, Heinz E, Napier JA (2009) Functional characterisation of a higher plant sphingolipid Δ^4-desaturase: defining the role of sphingosine and sphingosine-1-phosphate in *Arabidopsis* thaliana. Plant Physiol 149:487–498

Mongrand S, Morel J, Laroche J, Claverol S, Carde JP, Hartmann MA, Bonneu M, Simon-Plas F, Lessire R, Bessoule JJ (2004) Lipid rafts in higher plant cells: purification and characterization of Triton X-100-insoluble microdomains from tobacco plasma membrane. J Biol Chem 279:36277–36286

Munro S (2003) Lipid rafts: elusive or illusive? Cell 115:377–388

Ng CK, Carr K, McAinsh MR, Powell B, Hetherington AM (2001) Drought-induced guard cell signal transduction involves sphingosine-1-phosphate. Nature 410:596–599

Nishikawa M, Hosokawa K, Ishiguro M, Minamioka H, Tamura K, Hara-Nishimura I, Takahashi Y, Shimazaki K, Imai H (2008) Degradation of sphingoid long-chain base 1-phosphates (LCB-1Ps): functional characterization and expression of AtDPL1 encoding LCB-1P lyase involved in the dehydration stress response in Arabidopsis. Plant Cell Physiol 49:1758–1763

Norberg P, Nilsson R, Nyiredy S, Liljenberg C (1996) Glucosylceramides of oat root plasma membranes-physicochemical behaviour in natural and in model systems. Biochim Biophys Acta 1299:80–86

Obeid LM, Linardic CM, Karolak LA, Hannun YA (1993) Programmed cell death induced by ceramide. Science 259:1769–1771

Oskouian B, Saba JD (2004) Death and taxis: what non-mammalian models tell us about sphingosine-1-phosphate. Semin Cell Dev Biol 15:529–540

Perfus-Barbeoch L, Jones AM, Assmann SM (2004) Plant heterotrimeric G protein function: insights from Arabidopsis and rice mutants. Curr Opin Plant Biol 7:719–731

Peskan T, Westermann M, Oelmüller R (2000) Identification of low-density Triton X-100-insoluble plasma membrane microdomains in higher plants. Eur J Biochem 267:6989–6995

Phan VH, Herr DR, Panton D, Fyrst H, Saba JD, Harris GL (2007) Disruption of sphingolipid metabolism elicits apoptosis-associated reproductive defects in Drosophila. Dev Biol 309:329–341

Pyne S, Pyne NJ (2000) Sphingosine 1-phosphate signalling in mammalian cells. Biochem J 349:385–402

Ryan PR, Liu Q, Sperling P, Dong B, Franke S, Delhaize E (2007) A higher plant Δ8 sphingolipid desaturase with a preference for (Z)-isomer formation confers aluminum tolerance to yeast and plants. Plant Physiol 144:1968–1977

Saba JD, Hla T (2004) Point-counterpoint of sphingosine 1-phosphate metabolism. Circ Res 94:724–734

Schorling S, Vallee B, Barz WP, Riezman H, Oesterhelt D (2001) Lag1p and Lac1p are essential for the acyl-CoA-dependent ceramide synthase reaction in Saccharomyces cerevisiae. Mol Biol Cell 12:3417–3427

Shi L, Bielawski J, Mu J, Dong H, Teng C, Zhang J, Yang X, Tomishige N, Hanada K, Hannun YA, Zuo J (2007) Involvement of sphingoid bases in mediating reactive oxygen intermediate production and programmed cell death in Arabidopsis. Cell Res 17:1030–1040

Simons K, Ikonen E (1997) Functional rafts in cell membranes. Nature 387:569–572

Spassieva SD, Markham JE, Hille J (2002) The plant disease resistance gene Asc-1 prevents disruption of sphingolipid metabolism during AAL-toxin-induced programmed cell death. Plant J 32:561–572

Sperling P, Heinz E (2003) Plant sphingolipids: structural diversity, biosynthesis, first genes and functions. Biochim Biophys Acta 1632:1–15

Sperling P, Ternes P, Moll H, Franke S, Zahringer U, Heinz E (2001) Functional characterization of sphingolipid C4-hydroxylase genes from *Arabidopsis thaliana*. FEBS Lett 494:90–94

Sperling P, Franke S, Lüthje S, Heinz E (2005) Are glucocerebrosides the predominant sphingolipids in plant plasma membranes? Plant Physiol Biochem 43:1031–1038

Steponkus PL, Lynch DV (1989) Freeze/thaw-induced destabilization of the plasma membrane and the effects of cold acclimation. J Bioenerg Biomembr 21:21–41

Sullards MC, Lynch DV, Merrill AH, Adams J (2000) Structure determination of soybean and wheat glucosylceramides by tandem mass spectrometry. J Mass Spectrom 35:347–353

Tafesse FG, Ternes P, Holthuis JC (2006) The multigenic sphingomyelin synthase family. J Biol Chem 281:29421–29425

Tavernier E, Lê Quôc D, Lê Quôc K (1993) Lipid composition of the vacuolar membrane of *Acer pseudoplatanus* cultured cells. Biochim Biophys Acta 1167:242–247

Teng C, Dong H, Shi L, Deng Y, Mu J, Zhang J, Yang X, Zuo J (2008) Serine palmitoyltransferase, a key enzyme for de novo synthesis of sphingolipids, is essential for male gametophyte development in Arabidopsis. Plant Physiol 146:1322–1332

Ternes P, Sperling P, Albrecht S, Franke S, Cregg JM, Warnecke D, Heinz E (2006) Identification of fungal sphingolipid C9-methyltransferases by phylogenetic profiling. J Biol Chem 281: 5582–5592

Tonon T, Sayanova O, Michaelson LV, Qing R, Harvey D, Larson TR, Li Y, Napier JA, Graham IA (2005) Fatty acid desaturases from the microalga *Thalassiosira pseudonana*. FEBS J 272:3401–3412

Uemura M, Steponkus PL (1994) A contrast of the plasma membrane lipid composition of oat and rye leaves in relation to freezing tolerance. Plant Physiol 104:479–496

Uemura M, Joseph RA, Steponkus PL (1995) Cold acclimation of Arabidopsis thaliana (effect on plasma membrane lipid composition and freeze-induced lesions). Plant Physiol 109:15–30

Ullah H, Chen JG, Young JC, Im KH, Sussman MR, Jones AM (2001) Modulation of cell proliferation by heterotrimeric G protein in Arabidopsis. Science 292:2066–2069

Vallee B, Riezman H (2005) Lip1p: a novel subunit of acyl-CoA ceramide synthase. EMBO J 24:730–741

Venable ME, Lee JY, Smyth MJ, Bielawska A, Obeid LM (1995) Role of ceramide in cellular senescence. J Biol Chem 270:30701–30708

Wang H, Li J, Bostock RM, Gilchrist DG (1996) Apoptosis: a functional paradigm for programmed plant cell death induced by a host-selective phytotoxin and invoked during development. Plant Cell 8:375–391

Wang XQ, Ullah H, Jones AM, Assmann SM (2001) G protein regulation of ion channels and abscisic acid signaling in Arabidopsis guard cells. Science 292:2070–2072

Wang W, Yang X, Tangchaiburana S, Ndeh R, Markham JE, Tsegaye Y, Dunn TM, Wang GL, Bellizzi M, Parsons JF, Morrissey D, Bravo JE, Lynch DV, Xiao S. (2008) An inositolphosphorylceramide synthase is involved in regulation of plant programmed cell death associated with defense in Arabidopsis. Plant Cell 20:3163–3179

Worrall D, Liang YK, Alvarez S, Holroyd GH, Spiegel S, Panagopulos M, Gray JE, Hetherington AM (2008) Involvement of sphingosine kinase in plant cell signalling. Plant J 56(1):64–72

Xiong TC, Coursol S, Grat S, Ranjeva R, Mazars C (2008) Sphingolipid metabolites selectively elicit increases in nuclear calcium concentration in cell suspension cultures and in isolated nuclei of tobacco. Cell Calc 43:29–37

Yoshida S, Uemura M (1986) Lipid composition of plasma membranes and tonoplasts isolated from etiolated seedlings of Mung Bean (Vigna radiata L.). Plant Physiol 82:807–812

Zheng H, Rowland O, Kunst L (2005) Disruptions of the Arabidopsis Enoyl-CoA reductase gene reveal an essential role for very-long-chain fatty acid synthesis in cell expansion during plant morphogenesis. Plant Cell 17:1467–1481

Index

D

E

F

G

H

I

J

K

L

R

S